NEW DEVELOPMENTS IN MEDICAL RESEARCH

ANTIMICROBIAL RESISTANCE

OPPORTUNITIES AND CHALLENGES

New Developments in Medical Research

Additional books and e-books in this series can be found
on Nova's website under the Series tab.

NEW DEVELOPMENTS IN MEDICAL RESEARCH

ANTIMICROBIAL RESISTANCE

OPPORTUNITIES AND CHALLENGES

PRANAV KUMAR PRABHAKAR, PHD

AND

VIJAY MISHRA, PHD

EDITORS

nova
Medicine & Health
New York

Library of Congress Cataloging-in-Publication Data

Names: Kumar Prabhakar, Pranav, editor. | Mishra, Vijay, Dr. editor.
Title: Antimicrobial resistance : opportunities and challenges / Pranav Kumar Prabhakar, Vijay Mishra, editors.
Description: New York : Nova Science Publishers, [2020] | Series: New developments in medical research | Includes bibliographical references and index. | Summary: "This book contains comprehensive and up-to-date reviews of multidrug resistance mechanisms. The book intends to provide a state-of-the-art collection of reviews and methods for both basic and clinician investigators who are interested in multidrug resistance mechanisms and reversal strategies. We believe that this information will be of value to clinicians, epidemiologists, microbiologists, virologists, parasitologists, public health authorities, medical students, and fellows in training. Each chapter begins with a summary of the concepts, so that those not actively working in the field can readily gain an overall picture of what follows. The book contains 13 chapters which deal with the antibiotic resistance mechanism in bacteria, fungus, virus and also methicillin resistance S.aureus. The book also explains the futuristic strategy to deal with the antibiotic resistance. We have endeavoured to provide this information in a style that is accessible to the broad community of persons who are concerned with the impact of drug resistance in our clinics and across broader global communities"-- Provided by publisher.
Identifiers: LCCN 2020018264 (print) | LCCN 2020018265 (ebook) | ISBN 9781536179439 (hardcover) | ISBN 9781536179446 (adobe pdf)
Subjects: LCSH: Drug resistance in microorganisms.
Classification: LCC QR177 .A587 2020 (print) | LCC QR177 (ebook) | DDC 616.9/041--dc23
LC record available at https://lccn.loc.gov/2020018264
LC ebook record available at https://lccn.loc.gov/2020018265

Published by Nova Science Publishers, Inc. † New York

CONTENTS

PREFACE

Pranav Kumar Prabhakar, PhD and Vijay Mishra, PhD
Lovely Professional University, Phagwara, Punjab, India

Microbes enlist a variety of means to defend themselves, but perhaps most significant is the fact that mechanisms evolved by one species can, through genetic exchange, be shared throughout the bacterial community. Antimicrobial drug resistance is a global health problem that continues to expand as microorganisms adapt to the antibiotics used to treat them. The discovery of antimicrobial agents has created a new era of medicine. The control of infectious diseases is now based on the choice and prudent use of a large group of low molecular weight inhibitors with diverse mechanisms of action and varying spectra of antibacterial and antifungal activity. To withstand this ever-increasing menace to their existence and survival, microorganisms have developed a variety of defence mechanisms. New and better antibiotics are sought continually because of the capacity of microorganisms to survive this threat. However, new classes of antimicrobial agents have been harder to discover and advance into the clinic. In addition, the development of agents or strategies to overcome resistance has been the subject of intense study.

This book contains comprehensive and up-to-date reviews of multidrug resistance mechanisms. The book intends to provide a state-of-the-art collection of reviews and methods for both basic and clinician investigators who are interested in multidrug resistance mechanisms and reversal strategies. We believe that this information will be of value to clinicians, epidemiologists, microbiologists, virologists, parasitologists, public health authorities, medical students, and fellows in training. Each chapter begins with a summary of the concepts, so that those not actively working in the field can readily gain an overall picture of what follows. We have endeavored to provide this information in a style that is accessible to the broad community of persons who are concerned with the impact of drug resistance in our clinics and across broader global communities.

In: Antimicrobial Resistance: Opportunities and Challenges ISBN: 978-1-53617-943-9
Editors: P. Kumar Prabhakar and V. Mishra

Chapter 1

HISTORY OF DRUG-RESISTANT MICROBES

Anuj Modi[1,*], Monica Chopra[2], Vimal Kumar[3], Laxmikant Gautam[4], Anshul Das[4] and Shiv Kumar Prajapati[5]

[1]Department of Pharmacy, Integrated Institute of Technology, New Delhi, India

[2]Department of Pharmacy, Meera Bai Institute of Technology, New Delhi, India

[3]ITM School of Pharmacy, ITM Universe, Vadodara, India

[4]Department of Pharmaceutical Sciences, Dr. Hari Singh Gour University, Sagar, India

[5]Bhagyoday Tirth Pharmacy College, Sagar, India

ABSTRACT

Antibiotic resistance is one of the biggest threats to global health, food security, and development today. It can affect humans of any age worldwide. As per the World Health Organization (WHO), antibiotic resistance is rising to dangerously high levels in all parts of the world. WHO has declared a list of drug-resistant bacteria that shows utmost danger to human health and for which new drugs/antibiotics are urgently required. Drug resistance generally begins due to the misapplication and over use of antibiotics. Substitutions to antibiotics should be used to succeed the problem of drug resistance. New resistance mechanisms are emerging and spreading globally, threatening our ability to treat common infectious diseases. A growing list of infections, such as pneumonia, tuberculosis, blood poisoning, gonorrhoea, and food borne diseases-are becoming harder, and sometimes impossible, to treat as antibiotics are becoming less effective. This unexpected danger demands the development of new drugs and strategies for combating drug resistance exposed by many microbial species. Recent developments in nanotechnology based drug

* Corresponding Author's E-mail: pharmaanuj@gmail.com.

carrier system with anticipated physicochemical properties have been expected as a new line of defense against drug resistant microorganisms. This chapter includes detail study about history, causes, mechanism, diagnostic and potential threats due to development of drug resistance as well as various plans to control drug resistance.

Keywords: drug resistance, microorganism, antibiotics

1. Introduction

The advantageous utilization of any medicinal specialist is undermined by the potential advancement of resilience or protection from that compound from the time it is first used. This is valid for the drugs which are utilized in the treatment of everlasting illnesses, for example, cancer and diabetes. It also applies to disease caused by any living being, including humans, animals, fish, plants, creepy crawlies, etc. In general, thinking back on the historical backdrop of human diseases, infectious diseases represent an enormous extent of illnesses; brought about by the microbes, fungi, parasites, viruses and viral infection and so forth (Davies and Davies, 2010). The drug which is used for the treatment of microbial/bacterial infection is called antimicrobial/antibacterial. Antimicrobials are either cytotoxic or cytostatic in nature and are widely used to treat various infectious diseases caused by microorganisms. They stimulate the body's natural defense mechanism, such as the immune system, to eliminate them. Sometimes, antimicrobials inhibit the synthesis of the cell wall, deoxyribonucleic acid (DNA) or ribonucleic acid (RNA) and act as membrane destabilizing agents or other specific mechanisms to produce the therapeutic effects. Furthermore, they often bind with the cell wall and enter inside the cell using the energy-dependent transport mechanism. Thereafter, they bind with the ribosome, which leads to the inhibition of the protein synthesis (Zaman et al., 2017). Multiple varieties of the antimicrobials have been used for therapeutic purposes since time immemorial. Repeated use, or use over a longer period of time, leads to the development of resistance known as antimicrobial resistance (AMR). AMR is arising after a wide range of complex biochemical and physiological mechanisms involved in the development of resistance. However, one of the three mechanisms are mainly involved in the development of AMR namely natural resistance, genetic mutation, or one species acquiring resistance from another (Jacoby, 2017).

Resistance to antibiotic agents has been perceived since the beginning of the era. However in the past 20 years, dangerous resistant strains have developed with disturbing regularity. In the specific instance of antibiotics, the multifaceted nature of the procedures that add to the rise and scattering of obstruction cannot be overemphasized (Fair and Tor, 2014). They are not only more severe and require longer and progressively complex medications, however, they are likewise more expensive to diagnose and to treat. Antibiotics resistance, at first an issue of the emergency clinic settings has now expanded to a number of medical clinics in fundamentally sick and immunosuppressed patients, has now stretched out into the network causing serious contaminations hard to analyze and treat (Alanis, 2005). Recently, antibiotic discovery, its mode of action, and mechanism of resistance have become productive research topics in academia as well as in the pharmaceutical industry. Many resolutions and recommendations have been propounded, and numerous reports have been written, but the development of antibiotic resistance is persistent (Davies and Davies, 2010).

Table 1. Development of antibiotics resistance (Ventola, 2015)

Antibiotics approved	Year	Identified Resistant Germ	Identified in Year
1943	Penicillin	Penicillin resistant-*Streptococcus pneumoniae*	1967
		Penicillinase producing- *Neisseria gonorrhoeae*	1976
1950	Tetracycline	Tetracycline resistant-*Shigella*	1959
1953	Erythromycin	Erythromycin resistant-*Streptococcus*	1968
1958	Vancomycin	Vancomycin resistant-*Enterococcus faecium*	1988
		Vancomycin resistant-*Staphylococcus aureus*	2002
1959	Amphotericin B	Amphotericin B -resistant-*Candida auris*	2016
1960	Methicillin	Methicillin-resistant-*Staphylococcus aureus*	1960
1967	Gentamicin	Gentamicin resistance-*Entrococcus*	
1980	Broad-spectrum cephalosporins (Cefotaxime)	Broad spectrum beta lactamase producing - *Escherichia coli*	1983
1980	Azithromycin	Azithromycin resistant-*Neisseria gonorrhoeae*	2011
1985	Imipenem	Klebsiella pneumonia carbapenemase (KPC) producing -*Klebsiella pneumoniae*	1996
1985	Ceftazidime	Ceftazidime resistant -*Enterobacteria*	1987
1987	Ciprofloxacin	Ciprofloxacin resistant -*Neisseria gonorrhoeae*	2007
1990	Fluconazole	Fluconazole resistant -*Candida*	1988
1996	Levofloxacin	Levofloxacin resistant-*Pneumococcus*	1996
2000	Linezolid	Linezolid resistant-*Staphylococcus*	2001
2001	Caspofungin	Caspofungin resistant-*Candida*	2004
2003	Daptomycin	Daptomycin resistant methicillin resistant - *Staphylococcus aureus*	2004
2010	Ceftaroline	Ceftaroline resistant -*Staphylococcus*	2011
2015	Ceftazidime- avibactam	Ceftazidime-avibactam resistant KPC producing -*Klebsiella pneumoniae*	2015

2. History of Bacterial Drug Resistance

In the later half of the 19th century, it was very common to die from minor infectious diseases that had been tormenting mankind from the old days. Thus, antibiotics were developed with an aim to kill the causative infectious bacteria as the main therapeutic target (Saga and Yamaguchi, 2009). In 1910, Salvarsan was the first antimicrobial drug, synthesized by Ehrlich from cloths dye and used in syphilis treatment. Further, in 1928,

Alexander Fleming discovered penicillin. In 1941, Ernst Chain and Howard Flory conducted first clinical trials of penicillin. During World War II, penicillin led the era of antimicrobial therapy by saving thousands of soldier's lives by controlling bacterial infections in wounds. Once penicillin was used widely, soon after that resistant strains capable of inactivating penicillin turned out to be a significant clinical concern. However, to maintain the use of antibiotic treatment strategies, the pharmaceutical industry soon discovered, developed and introduced many novel antibiotics to resolve the resistance issue. With the passage of time, the pace of antibiotic development staggered, so very few new antibiotics were introduced antibiotics resistance also developed (Table 1) (Aslam et al., 2018; Piddock, 2012; Sengupta et al., 2013; Rossolini et al., 2014).

Antibiotics had a parallel helpful effect over the globe, for example, creating general wellbeing of human beings as well as diminish the morbidity and mortality rate brought about by bacterial contaminations in humans and domesticated animals. These enchantment projectiles have additionally changed the result of contaminations which broadened the normal life expectancy of an individual past hope (Spellberg and Gilbert, 2014).

3. Causes for Development of Resistance

Resistance is a means whereby a naturally susceptible microorganism acquires ways of not being affected by the drug. Understanding the causes of resistance in bacteria is very important in order to define better ways to keep existing agents useful for a little longer time, and also to help in the design of better antimicrobial agents that are not affected by the currently known, predicted, or unknown causes (Michael et al., 2014). A variety of factors play an important role in the generation of AMR are:

3.1. Microbial Causes

3.1.1. Environmental Factor

Unlike multicellular organisms like humans, microbes have a very large number of genomes, which contain very few varieties of genes required for surviving. But the ability of microbes, to adapt rapidly and effectively to the environmental change, makes them survive under unfavourable environmental conditions. Presence of enormous number of microbes, in small volumes and their high speed generation rate are solely responsible for their adaptive capacity. This in turn, ensures that advantageous changes will become prevalent rapidly in continually growing and evolving microbes (Gogarten et al., 2002).

3.1.2. Mutation

Mutation is nothing but the random changes in the genetic sequence of an organism. It arises during replication and generally has either no effect or produces new responses to adaptive challenges on the organism. Mutations arising in the microbe under antimicrobial treatment may help an individual microbe to survive exposure to an antimicrobial agent (Michael et al., 2004).

3.1.3. Gene Transfer

The gene of any microorganism acts as a reservoir of its adaptive capacity which can be transferred within microbial species through horizontal gene transfer (HGT) mechanism. Such transferred adaptive genes have shown to provide "resistance" to many different adaptive challenges, such as resistance to heat, oxidative stress, U.V. light, and antibiotics. Thus, the adaptive gene containing microorganism multiplies and thrives in the presence of antimicrobial drugs (Baharoglu et al., 2013).

3.1.4. Selective Pressure of Antimicrobial Therapy

Antimicrobials are employed to kill or diminish the growth of microbes. Thus, they produce a "selective pressure" upon microbes. Under such conditions, microbes which are able to survive and reproduce become prevailed. The widespread, persistent and intense use of antimicrobials provides a strong selective pressure that will continue to generate more adaptive solutions provoking strong adaptive responses in the microbes (Michael et al., 2014).

3.2. Human Causes

3.2.1 Human Population

Increase in the human population in a short period of time, increase urbanization and ability to transport rapidly from one place to another, increases the transfer of their attendant microbes and pathogens without any impediment globally. Apart from this large numbers of people living in close proximity provide a significant opportunity for the rapid proliferation of infectious diseases (WHO, 2019).

3.2.2. Over Use of Antibiotics

Generally, antimicrobial is considered as "silver bullet" which acts as an effective antidote against majority of illnesses. However, the efficacy of antimicrobials depends upon the drugs given, its dosing frequency and duration of treatment prescribed by the physician. Widespread advertisement or non-prescription purchase and use by the people lead to a strong and polarized selective pressure on the microbial world. This has resulted in increasing rates of AMR (Lipsitch et al., 2002; Nesme et al., 2014).

3.3. Clinical Uses

3.3.1. Over-Prescription

Nowadays many physicians prescribe antibiotic therapy either for a longer duration of time or give several antibiotics simultaneously for the treatment of majority of illnesses in human beings. Over prescription also occurs either due to inaccurate diagnosis or lack of laboratory based test reports or in emergency situations. The time of treatment may also be reduced based on the past experience of the patient or if the patient demands an immediate solution. This process of treatment might provide a prompt cure but leads to an intense and extreme pressure on non-pathogenic organisms which promote the development of AMR. This resistance, then, may be transferred via HGT crosswise over other microbial communities and importantly, to previously sensitive pathogenic organisms (Tellier et al., 2004; Zhu et al., 2014; Kuehn, 2013).

3.3.2. Entrenched Methodologies

Antimicrobials are generally prescribed in a fixed regimen, dictating the dose, rate, and period. Prescription of higher dosages of extended regimes by the physician for a long period of time not only eradicates the infecting pathogen from the body but also increases the chance of AMR (McNulty et al., 2007).

3.4. Vaccination Reluctance

Decreased vaccination or unvaccinated individuals prompt to the re-emergence of illness, such as small pox and polio. Such re-emergence prompts an expansion in the numbers of pathogenic microorganisms within the pre-defined population. This inturn, may then more promptly contaminate the people who are not yet vaccinated such as children, or the individuals who are immunocompromised, such as older people (Pless et al., 2000).

3.5. Agricultural Use

Scientists also believe that the practice of adding antibiotics to agricultural feed promotes drug resistance. However, there is still much debate about whether drug-resistant microbes in animals pose a significant public health burden (WHO, 2019).

3.6. Miscellaneous

Although antimicrobial resistance is a natural biological phenomenon, it is often enhanced as a consequence of infectious agents' adaptation to exposure to antimicrobials used in humans or agriculture, and the widespread use of disinfectants at the farm and the household levels. Some other important factors which play an important role in the development of resistance are:

1. Poor infection control in healthcare settings
2. Poor hygiene and sanitation
3. Absence of new antibiotics being discovered.

4. Mechanism of Bacterial Resistance Development

Bacteria are living organisms that develop over a prolonged period of time. Their primary function is to duplicate, increase, and spread rapidly and efficiently. Antibiotics fight with bacteria to eliminate them. Hence, bacteria tend to develop a natural process that empowers resistance. Bacteria build up resistance through numerous mechanisms that empower them to dodge the effect of antibiotics. Developed resistance either can be intrinsic or acquired. Intrinsic resistance is a basic characteristic that naturally occurs in the bacteria, and is essentially always there. However, acquired resistance means that the bacterium is not naturally resistant but becomes resistant somehow overtime. A random mutation in the genetic sequence of normal cellular genes by acquirement of foreign resistance genes or transferring plasmids between the bacterium, or a combination of these two mechanisms, is commonly employed in the development of resistance in the bacterium. Common mechanisms involved in the development of antibiotic resistance in bacteria are (Figure 1):

1. Alterations in drug target or activating enzyme
2. Inactivation by enzymes
3. Changes in cellular permeability towards the drug
4. Active efflux
5. Overproduction of target enzymes
6. Bypass of drug action
7. Intercellular cooperation (Mckeegan, 2002)

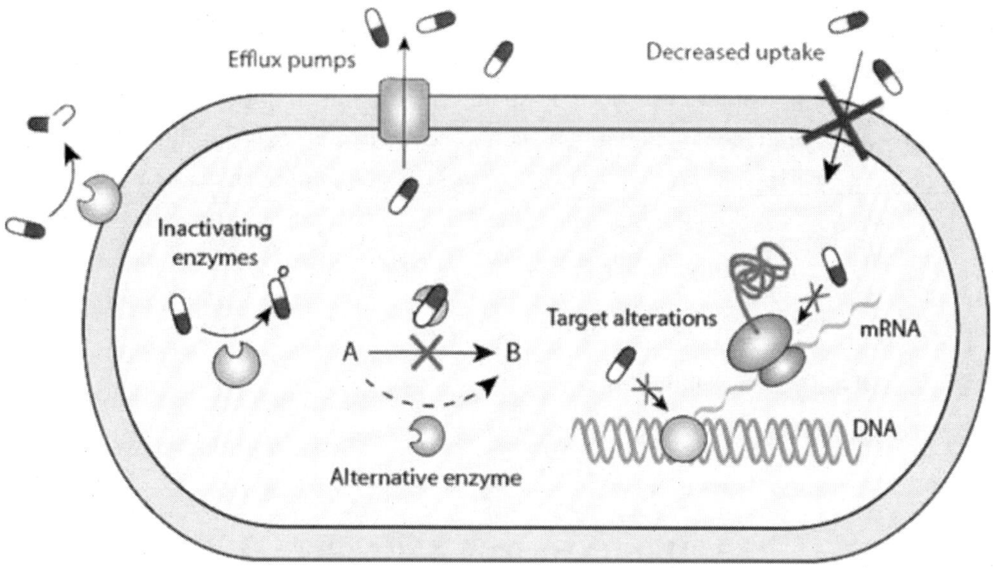

Figure 1. Common Mechanisms involved in the Development of Drug Resistance.

Table 2. Mechanism of drug resistance of common antibiotics (Zaman et al., 2017)

Antibiotic class	Examples	Mechanism of resistance
Aminoglycosides	Gentamicin, Streptomycin, Spectinomycin	Phosphorylation, acetylation, nucleotidylation, efflux, altered target
Cationic peptides	Colistin	Altered target, efflux
Glycopeptides	Vancomycin, Teicoplanin	Reprogramming peptidoglycan biosynthesis
Lincosamides	Clindamycin	Nucleotidylation, efflux, altered target
Lipopeptides	Daptomycin	Altered target
Macrolides	Erythromycin, Azithromycin	Hydrolysis, glycosylation, phosphorylation, efflux, altered target
Oxazolidinones	Linezolid	Efflux, altered target
Phenicols	Chloramphenicol	Acetylation, efflux, altered target
β-Lactams	Penicillins, Cephalosporins, Penems, Monobactams	Hydrolysis, efflux, altered target
Pyrimidines	Trimethoprim	Efflux, altered target
Quinolones	Ciprofloxacin	Acetylation, efflux, altered target
Rifamycins	Rifampin	ADP-ribosylation, efflux, altered target
Streptogramins	Synercid	Carbon-oxygen lyase, acetylation, efflux, altered target
Sulfonamides	Sulfamethoxazole	Efflux, altered target
Tetracyclines	Minocycline, Tigecycline	Monooxygenation, efflux, altered target

Antibiotic-resistant organisms are known as superbugs. Inspite of all facts, this issue is not new or not just laboratory concern. It is getting progressively intense and is becoming a worldwide danger as it is liable for high losses of life. The different mechanisms involved in the generation of drug resistance of commonly used antibiotics are given in Table 2. AMR is carried by antibiotic-resistant bacteria (ARBs) and expressed through antibiotic-resistant genes (ARGs) as given in Table 3 (Klein, 2018; Hiller, 2019).

Antibiotic resistance is driven by antibiotic consumption which is expressed by "defined daily doses" (DDD). Between 2000-2015, consumption of antibiotics in 76 low- and middle-income countries increased by 65% (21.1-34.8 billion DDDs), and the antibiotic consumption rate increased by 39% (11.3-15.7 DDDs per 1,000 inhabitants per day). Still mainly countries are continued to be burdened with high rates of infectious disease related mortality and low rates of antibiotic consumption (Klein et al., 2018). It is presently evident that a systematic comprehension of the mechanisms by which microbes unite in the development of resistance is essential, if we are to increase new bits of knowledge into approaches to solve this issue (Pillay, 1998).

Table 3. Antibiotic-resistant genes (Hiller, 2019)

S. No.	Antibiotic groups	Antibiotics	Resistance gene
1	Broad spectrum antibiotics	Penicillin	ampC
		Tetracycline	tet
		Sulphonamide	sul
		Phenicole	cml
		Aminoglycoside	aac
		1st & 2nd generation Cephalosporins	blaROB-1
		Macrolide	erm
		Quinolone and fluoroquinolone	qnr
2	Antibiotics of emergency	3rd & 4th generation Cephalosporins	-
		Polymyxin	-
		Antipseudomonal penicillin	-
		Beta-lactamase inhibitor	-
		Oxazolidinones	-
		Tigecycline	-
		Daptomycin	-
		Imipenem	blaVIM
		Vancomycin	van
3	Narrow spectrum antibiotics	Nitrofurantoin	nfsA
		Trimethoprim	dfr
		Clindamycin	lnu
		Tiamulin	rplC
		Oxacillin	oxa1
		Metronidazole	rdxA

World Health Organization (WHO), estimated that more than 700,000 individuals die on every year due to drug-resistant diseases, including 230,000 individuals who die on from multidrug-resistant tuberculosis. WHO has also cautioned that a post-antibiotic era will bring about the regular disease and little wounds may bring about death in the event of negligence against antibiotic resistance. Drug-resistant diseases could cause mortality of 10 million by 2050 while antimicrobial resistance could force up to 24 million individuals from extreme poverty by 2030 (Anonymous, 2019).

5. Diagnostic Tests of Microbes and Diseases

Many infectious diseases are characteristic enough to be identified clinically and biologically. Most pathogens, however, can cause a wide spectrum of diseases and syndromes in human beings. Therefore, it is necessary to use methods of the microbiological laboratory to recognise a specific etiologic agent. Diagnostic medical microbiology is the regulation that identifies etiologic agents of infectious disease. The job of the clinical microbiologist is to test specimens from different infected patients for microorganisms that are, or maybe, a cause of the illness and to provide information (when appropriate) about the *in vitro* activity of antimicrobial drugs against the microbes identified (Washington, 1996; Koneman et al., 1997). The most frequently used methods for immunological detection of microbial antigens are latex particle agglutination, co-agglutination, and enzyme-linked immunosorbent assay (ELISA) in the microbiological laboratory. Antibody to a specific antigen is bound to latex particles or to a heat-killed microbe and treated with protein A-rich strain of *Staphylococcus aureus* to produce agglutination. Genetic probes are based on the detection of unique nucleotide or base pair sequences with the DNA or RNA of a microorganism (Washington, 1996). PCR and real-time PCR technology have grown from monoplex to multiplex assays with master mixes comprising several primer/probes sets for multiple pathogen detection. Array technology has revisited monoplex assays but utilising them as ‘simultaneous single plex’ to detect more pathogens from the full spectrum of microorganisms, thus introducing a syndromic approach to diagnosis rather than pathogen focused approach. The TaqMan® Array Card (TAC) formally known as TaqMan® Low-density Array (TLDA) is a microfluidic card which utilises primers and probes pre-loaded and lyophilized onto wells. The TAC rostrum provides a 384-well system, single-plate real-time PCR in 8 samples by 48-well format along with a single target tested in each of 48 wells/pods (including control groups).

Table 4. Commonly used tests to detect microbes in a specimen (Washington, 1996; Koneman et al., 1997)

S. No.	Specimen	Test	Application
1	Blood	Giemsa EIA	Plasmodia, microfilaria, hepatitis A and B virus, human immunodeficiency virus
	Cerebrospinal fluid	Gram stain LA; COA India ink wet mount or LA	Bacteria *Haemophilus influenzae, Neisseria meningitides, Streptococcus pneumoniae, Cryptococcus neoformans*
2	Wound exudates, pus	Gram stain	Bacteria
3	Respiratory secretions	Gram stain Acid-fast stain IFA or Genetic probe KOH wet mount Gormeri silver Methenamine stain FA, EIA	Bacteria Mycobacteria, Nocardiae *Legionella* species, *Streptococcus pyogens* Fungi Fungi, *Pneumocystis canni* Respiratory synovial virus
4	Urine	Gram stain	Bacteria
5	Urethral or cervical scrapings or exudates	Gram stain IFA, EIA or genetic probe	*Neisseria gonorrhoeae* *Chlamydia trachomatis*, papilloma viruses
6	Genital ulcer	FA, EIA or Genetic probe	Herpes simplex virus
7	Faeces	Methylene blue stain Eosin wet mount, trichome stain EM, LA, EIA	Leukocytes Parasites Rotavirus Adenovirus

Abbreviations: COA- coagulation, EIA- enzyme immunoassay, IFA- immuno-fluorescent antibody, LA- latex agglutination.

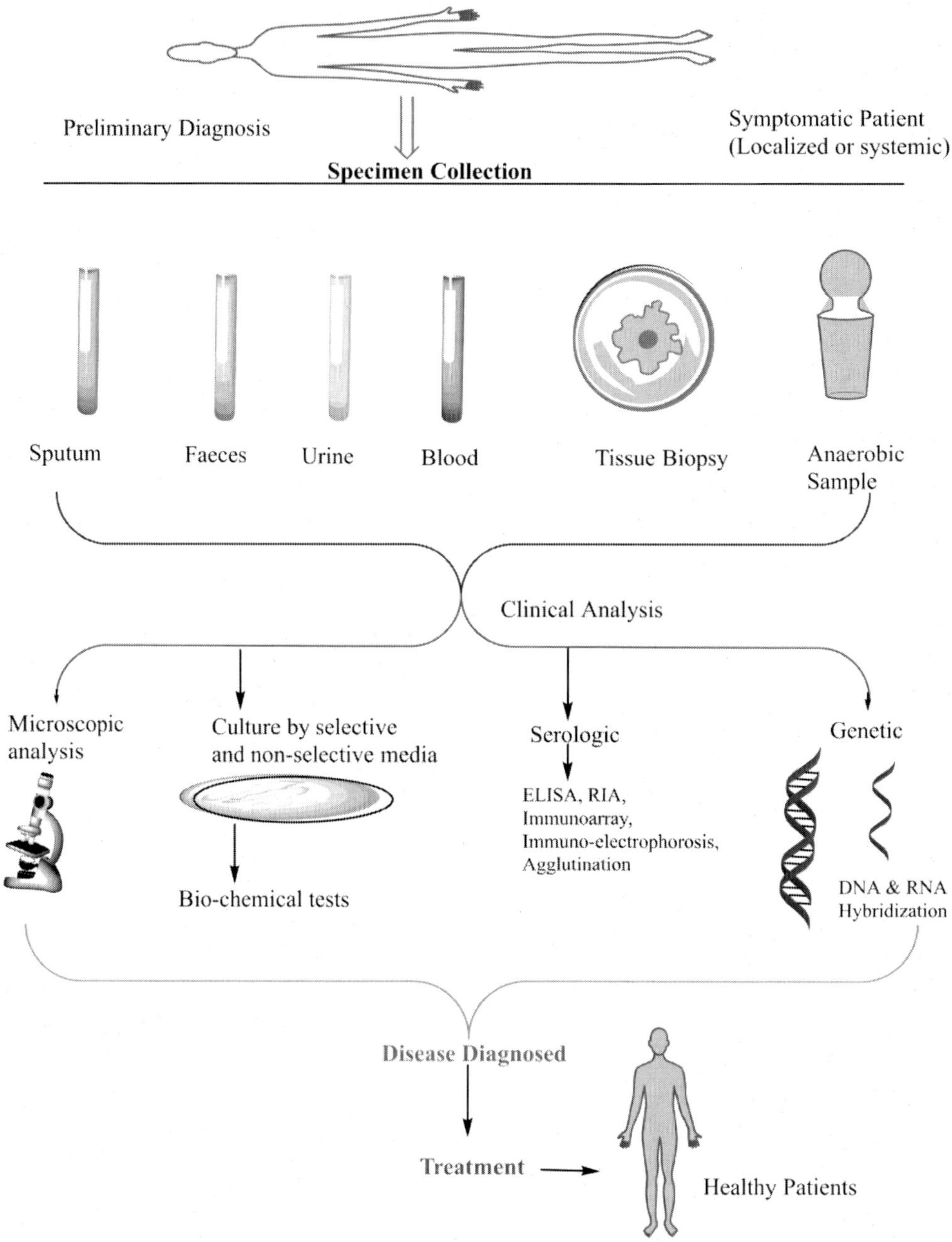

Figure 2. Procedure adopted in the clinical diagnosis of infectious disease with bacterial etiology in laboratory.

The sample is distributed by micro fluidic technology into individual PCRs and real-time PCR and detection takes place with the help of an analyzer, for example, ViiA7 or QuantStudio7. The results are generated within 1 hr (Heaney et al., 2015). Commonly used procedures and tests in the clinical diagnosis of infectious disease with bacterial etiology in the laboratory are given in Table 4 and Figure 2.

6. Examples of Drug Resistance and Diagnosis in Bacterial Diseases

6.1. Tuberculosis

According to the WHO report, tuberculosis (TB) being a reparative disease, continues to be a serious public health problem, with an infection rate of 10.4 million people (including 0.4 million people suffering from HIV) and a death toll of 1.8 million. (WHO, 2016; Unissa et al., 2017). For the treatment of TB, the intensive regimen includes isoniazid (INH), pyrazinamide (PZA), rifampicin (RIF), and ethambutol (EMB). The regimen of some of the drugs given continuously includes INH and RIF for four months, or INH and EMB or thiacetazone for six months (Anonymous, 2016). The success of TB treatment and control programs are threatened by the evolution and transmission of drug resistant (DR) *Mycobacterium tuberculosis* (MTB). The phenomenon of drug resistance may be multi-factorial and can be indicated by recent studies of whole-genome sequencing (WGS) of MTB. This may explain the controversy between the phenotype and genotype analyzed in selective cases (Zhang et al., 2015; Casali et al., 2014). It is believed that drug resistance in TB is mediated exclusively by chromosomal mutations. These affect the bacterial enzymes, or activate prodrugs, or the drug target itself. Most significant amongst drug resistance is the phenomenon of multidrug resistance (MDR) and is defined as resistance to the two most potent anti-TB drugs such as RIF and INH. Extensive drug resistance (XDR) is defined as MDR plus additional resistance to second line anti-TB drugs, such as fluoroquinolones, and at least, one of the parenteral drugs, such as capreomycin, kanamycin or amikacin. As per the report analyzed globally in 2015, 480,000 people were found to be multi drug-resistant of TB and 170,000 deaths occurred.

These results concluded that 9.5% of MDR-TB converts to XDR-TB (WHO; 2016). Unissa and Hanna reported that rifampicin resistance is considered as a proxy marker for MDR-TB (Unissa and Hanna, 2017; Drobniewski et al., 1998). Kanji and their colleagues reported bioinformatics based evidence for the prokaryotic and eukaryotic cells (genomic size). They stated that for innate and acquired drug resistance, the ubiquitous mechanism is responsible for their efflux pumps (Kanji et al., 2019).

6.1.1. Drug Resistance in MTB

Following rapid chromosomal mutations, the drug resistant mutants can be selected by physician's poor prescription, poor patient compliance, and poor supply or quality of drugs, as well as other factors such as the difference in metabolism and nutrition. Thus, drug resistant TB is a man**m**ade phenomenon (Islam et al., 2017).

6.1.2. First Line Anti-TB drugs

Drug-susceptible TB is effectively curable by the first-line drugs: H, R, Z, E and S. Resistance to the first line drugs are found to be associated with several gene mutations, as such as:

- *katG, inhA, ahpC, kasA* and *ndh* for H-resistance,
- *rpoB* for R-resistance,
- *c* for Z-resistance,
- *embB* and *embC* for E-resistance, and
- *rpsL* and *rrs* for S-resistance

First-line drugs, gene names, minimum inhibitory concentrations (MICs) and resistance mechanisms are summarized in Table 5.

6.1.3. Second Line Anti-TB drugs (Islam et al., 2017; Lohrasbi et al., 2018)

Some of the drugs used for the second-line treatment for *M. tuberculosis* are listed in Table 6, along with their minimum inhibitory concentration (MIC), resistant genes, role of gene product, resistance mechanism of action and frequency of mutation.

Table 5. Mechanism of resistance in first-line drugs of *M. tuberculosis* (Unissa et al., 2017; Islam et al., 2017; Lohrasbi et al., 2018)

S. No.	Drug	MIC (μg/ml)	Gene involved in resistance	Role of gene product	Resistance mechanism of action	Frequency of mutation (%)
1	Isoniazid	0.02-0.2	katG	Catalase peroxidase	Inhibition of synthesis of mycolic acid	40-96%
			inhA	Enoyl ACP reductase		
			ahpC	Alkyl hydroperoxide reductase		
			iniA	Efflux pump associated		
			kasA	Ketoacyl acyl carrier protein synthase		
			Ndh	NADH dehydrogenase		
2	Rifampicin	0.05-0.50	rpoB rpoA rpoC	β-subunit of RNA polymerase α- subunit of RNA polymerase β- subunit of RNA polymerase	Preventing synthesis of RNA	40-95%
3	Pyrazinamide	10.0-100.0 (lower pH)	pncA rpsA panD	PZase Ribosomal S1 protein Aspartate decarboxylase	PZase is responsible for activation RpsA involves in trans- translation panD is responsible for pantothenate and CoA synthesis	60-90%
4	Ethambutol	1.0-5.0	embB embA embC	Arabinosyl transferase	Inhibit the synthesis of arabinogalactan	40- 80%
			embR	Regulator of embCAB operon expression		
			rmlD	dTDP-4-dehydro-rhamnose		
5	Streptomycin	1.0-8.0	Rpsl	S12 ribosomal protein	Inhibition of protein synthesis	30-60%
			Rrs	16S rRNA		
			gidB	7-methylguanosine methyltransferase		

Table 6. Mechanisms of resistances in second-line drugs of *M. tuberculosis* (Islam et al., 2017; Lohrasbi et al., 2018)

S. No.	Drug	MIC (μg/ml)	Resistant gene	Role of the gene product	Resistance mechanism of action	Frequency of mutation (%)
1	Fluoroquinolones	0.50-2.0	ubiA gyrA gyrB	DPPR synthase DNA gyrase subunit A and B	Inhibition of DNA synthesis	42-95
2	Kanamycin/Amikacin	1.0-4.0	Rrs eis whiB7	16S rRNA Aminoglycoside acetyltransferase Transcriptional regulator	Inhibition of protein synthesis	30-75
3	Para-aminosalicylic acid	0.50-8.0	thyA folC dfrA ribD	Thymidylate synthase A Dihydrofolate synthase Dihydrofolate reductase Riboflavin biosynhtesis	Inhibition of folic acid and thymine nucleotide metabolism	<40
4	Clofazimine	0.01-0.25	Rv0678 Rv1979c Rv2535c	Transcription repressor for efflux pump MmpL5 Unknown Unknown	Production of reactive oxygen species	<40
5	Ethionamide	2.5-10.5	inhA ethA ethR	Fatty acid enoyl acyl carrier protein reductase A Flavin monooxygenase Transcriptional repressor	Disrupts cell wall biosynthesis	37-100

6.1.4. Diagnostic Tests

Generally, for the diagnosis of tuberculosis, conventional methodologies take two to three weeks. The delay in diagnosis and increase in detection complexity is due to the prominent risks of XDR-TB (extensively drug resistant TB) and MDR-TB (multi-drug resistant TB). At present, the researchers have developed modern detection technologies for TB, such as markers/biosensing and other point-of-care (POC) techniques (Table 7). The use of nanotechnology along with biosensing approaches have enormous potential to boost the MTB detection and for overall management in clinical diagnosis (Gupta and Kakkar, 2018).

Table 7. Conventional detection tests for tuberculosis (Gupta and Kakkar, 2018)

S. No.	Name of Test	Sensitivity
1	Adenosine deaminase enzyme level and nucleic acid-based amplification tests	35%
2	Drug susceptibility test for mutants	89-98%
3	ELISA (Enzyme-linked immunosorbent assay)	68%
4	Flow cytometry	3.5 × 103 cells/mL
5	Latex agglutination	73.6%
6	PCR technique	95.5%
7	The AMTDT (Gene probe amplified M. tuberculosis direct test)	94.3%
8	Tuberculin skin test	42-60%
9	Ziehl-Neelsen method	22-42%

6.1.5. Bio-sensing Techniques for Detection of Tuberculosis (Gupta and Kakkar, 2018)

Table 8. Classification of biosensors for TB diagnosis

Based on Analyte used for detection		**Based on Transducer**	
Sample	Biomarker	Transducer	Principle
Blood	Host antibodies against TB	Mass dependent	• Surface acoustic wave • Piezoelectric • Magnetoelastic
Sputum	• Secreted and cell internal antigen • Use of whole cells • DNA • Other TB biomarkers	Optical	• Fluorometric • Wavelength • Optical Fibre • Chemiluminescence
Urine	• DNA • Volatile form MTB cells • Secreted and cell internal antigen	Electro-chemical	• Amperometric • Conductometric • Potentiometric • Impedimetric
		Magnetic	• Nucleic magnetic resonance

Biosensing technique has high sensitivity and specificity in detecting analytes (biomarkers) in complex samples (urine, serum, saliva, blood, etc.) (Gupta and Kakkar, 2018; Martinkova et al., 2017; Rogers et al., 1996). Zhou and their researchers showed that biosensor (a compact analytical device) is composed of a bio-recognition component and a bio-transducer (physio-chemical transducer) component (Zhou et al., 2011). The bio-recognition element used in framing a biosensor can be an enzyme complex, antigen-antibody, nucleic acid, whole-cell etc. This element is bound tightly onto the bio-transducer by means of different physical or chemical immobilization processes, which can then measure the arising signals precisely and accurately (Table 8) (Martinkova et al., 2017; Bueno, 2014).

6.2. HIV

In 1981, a new disease was introduced in the society known as AIDS (Acquired immunodeficiency syndrome) infected by Human immunodeficiency virus and presently one of the most significant pathogens affecting the life of human beings (Sharp and Hahn, 2018). As per the WHO report, it affects >35 million lives globally, and 9.4 million population were died due to this virus in 2017 (WHO; 2018). Currently, approved HIV drugs, such as target HIV protease, reverse transcriptase (RT), integrase, the C-C motif chemokine receptor type 5 (CCR) are found over the host cells along with the process of membrane fusion (Kinch and Patridge, 2014). The antiretroviral treatment is used for the treatment of HIV in which 2-4 drugs are taken regularly i.e., atroiza, tenofovir, lamivudine emtricitabine etc. However, side effects of these drugs were also noticed in the last decade and now it is treated as a chronic disease (Tan and McArthur, 2012).

Table 9. Various virologic responses in HIV infection

S. No.	Virologic Response	Features
1	Viral suppression	HIV-1 RNA <200 copies/ml
2	Virologic failure	HIV-1 RNA >=200 copies/ml
3	Incomplete virologic response	Failure to achieve viral suppression to <200 copies /ml on 2 measurements after 24 wk on antiretroviral therapy
4	Virologic rebound	Sustained HIV-1 RNA >= 200 copies /ml on at least 2 HIV-1 RNA measurements after a previous period of viral suppression
5	Viral blips	Brief, isolated episode of detectable HIV-1 RNA, between 2 suppressed HIV-1RNA measurements

Optimal management of treatment failure plays a critical role in the ability to improve viral suppression rates and to achieve epidemic control (McCluskey et al., 2019). Terminology regarding levels of the human immunodeficiency virus (HIV)-1 viremia and virologic suppression are presented in Table 9 (Gergiso and Erango, 2019).

6.2.1. Diagnostic Tests

6.2.1.1. First Generation

The first HIV antibody test was introduced in 1985, in which the assay detected IgG antibody and the patient was positive 6 to 12 weeks post-infection. The description related to the first-generation diagnostic test is shown below:

- Antigens used to bind anti-HIV antibodies: Lysate of whole HIV-1 viruses grown in cell culture
- Format: Indirect immunoassay using labelled anti-human IgG
- Detects: IgG anti-HIV antibodies

The earliest HIV EIAs used lysate of the whole HIV-1 purified from cell cultures as the source of antigens to bind antibodies to HIV-1. Anti-human IgG conjugated to an enzyme binds to HIV-1 antibodies, if present, and produces a colour change when the enzyme's substrate is added. The viral lysate EIAs has high empirical sensitivity for HIV IgG antibodies but can remain negative for 12 weeks after infection, before antibodies develop. False-positive test also occur, associated with other infections such as pregnancy, autoimmune disease, and unspecified conditions (Petersen et al., 1994; Meyer, 1987).

6.2.1.2. Second Generation

In this generation, the detection of IgG antibody is confirmed. The details related to the analysis are shown below:

- Antigens for anti-HIV antibodies: recombinant viral proteins or synthetic peptide
- Format: indirect immunoassay using labelled anti-human IgG or, in rapid tests, protein A, which binds to human IgG with high affinity
- Detects: IgG anti-HIV antibodies
- Advantages: removing cellular contaminants from viral lysates expands specificity by removing cross-reactivity with cellular proteins; improves sensitivity for HIV-1, group O and HIV-2 by designing specific epitopes (Branson, 2019; Hopkins et al., 2019). Various second generation tests are shown in Table 10.

Table 10. Second generation tests

Test	Specimen Types	Markers Used for Detection	Principle	Generation
Oraquick advance rapid HIV- ½ antibody test	Oral fluid, whole blood plasma	Gp41,gp36	Lateral flow	Second
Reveal G4 Rapid HIV-1 antibody test	Whole blood, serum, plasma	Gp41,p120	Flow-through	Second
Multisport HIV-1/ HIV-2 rapid test	Serum, plasma	Gp41, gp36	Flow-through	Second
Genius HIV ½ supplemental assay	Whole blood, serum, plasma	P24, p31, gp41, gp160, gp36, gp140	Dual-path platform	Second
Chembio DPP HIV ½ assay	Oral fluid, whole blood, serum, plasma	gp41, gp120, gp36	Dual-path platform	Second

6.2.1.3. Third Generation

In this generation the detection of both IgG and IgM are confirmed. The other details related to this generation are shown below:

- Antigens used to bind anti-HIV antibodies: recombinant viral proteins or synthetic peptides
- Format: On the substrate, HIV antibodies of sample binds to HIV antigen termed as "antigen sandwich".
- Detects: IgG and IgM anti-HIV antibodies, which grow more radially.
- Advantage: becomes reactive earlier during seroconversion.

Table 12. Third generation tests

Test	Antigenic Markers Used for Detection	Analyte Detected	Generation
Bio-rad GS HIV -1/2 PLUS O	HIV-2 gp36, gp160, Recombinant p24, synthetic group O peptide	IgG and IgM antibodies	Third
Ortho vitros anti-HIV 1+2	HIV-2 gp36, gp41/120, gp41, Recombinant p24,	IgG and IgM antibodies	Third
Siemens Advia Centaur HIV 1/O/2	HIV-2 gp36, p24, Recombinant p41/120 synthetic group O peptide	IgG and IgM antibodies	Third
Uni-Gold recombigen HIV- ½	Gp41, gp120, gp36	-	Third

6.2.1.4. Fourth Generation

In this generation, detection of both HIV antibody and the p24 antigen is confirmed.

- Antigens used to bind anti-HIV antibodies: Synthetic peptide or recombinant viral proteins; monoclonal anti-p24 antibodies
- Format: Antigen sandwich
- Detects: IgG and IgM anti-HIV antibodies, HIV-1 p24 antigen
- Advantage: becomes reactive shortly before HIV-1 seroconversion

Table 13. Fourth generation tests

Test	Antigenic markers used for detection	Analytes detected	Generation
Abbott Architect HIV Ag/Ab combo	Synthetic and recombinant gp41 and HIV-2 gp36, group O peptide, anti-p24 monoclonal antibodies	IgG and IgM antibodies P24 antigen	Fourth
Siemens Advia Centaur HIV Ag/Ab combo	Recombinant p41/120, HIV-2 gp36, group O peptide, anti-p24 monoclonal antibodies	IgG and IgM antibodies P24 antigen	Fourth
Roche Elecys combi PT	Recombinant gp41, HIV-2 gp36, HIV-1 RT, HIV-2 RT	IgG and IgM antibodies P24 antigen	Fourth
Ortho vitros combi	Recombinant p41/120, HIV-2 gp36, group O peptide, anti-p24 monoclonal antibodies	-	Fourth
BioRad Bioplex 2200 HIV Ag-Ab	Recombinant gp160, HIV-2 gp36, group O peptide, anti-p24 monoclonal antibodies	IgG and IgM antibodies P24 antigen (detection and differentiation)	Fourth

6.3. Malaria

Generally, resistance occurs through an instinctive mutation that endues diminished sensitivity to a given drug or class of drugs. For a few drugs, only a single point mutation results in resistance, while for other drugs, resistance can be achieved only by multiple mutations. Mutations are not harmful to the existence or reproduction of the parasite. The drug given eradicates susceptible parasites while resistant parasites survive (Bloland, 2001).

6.3.1. Chloroquine Resistance

Excessive quantity of toxic by-product is formed as a result of digestion of hemoglobin by the malaria parasite. This by-product gets polymerized by the parasite in the food vacuole, and produces a malaria pigment i.e., "hemozoin." It is believed that, the resistance of *P. falciparum* to chloroquine is associated with augmented efficiency for the parasite to depose chloroquine at a rate that does not permit chloroquine to reach the levels required to prevent haem polymerization (Foley and Tilley, 1997). The efflux of chloroquine occurs 40 to 50 times faster among resistant parasites than sensitive ones (Krogstad et al., 1987). Chloroquine resistance can be upturned by drugs that interfere with this efflux system (Martin et al., 1987). It is uncertain, whether parasite resistance to other quinoline antimalarials (amodiaquine, mefloquine, halofantrine, and quinine) occurs via similar mechanisms (Foley and Tilley, 1997).

Table 14. Clinical diagnosis techniques (Tangpukdee et al., 2009)

	Clinical diagnosis	**PBS**	**QBC**	**RDTs**	**Serological tests**	**PCR**	**Microarrays**	**FCM**	**ACC**
Principle of the method	Based on presenting malarial signs and symptoms	Visualization of morphologically distinguishable stages of parasites under a light microscope by thick and thin blood smear and staining	Blood staining with acridine orange and detection by epifluorescent microscope	Detection of parasite antigens or enzyme	Detection of antibodies against the parasite	Specific amplification of malaria DNA	Hybridization of DNA isolate and quantified by fluorescence-based detection	Detection of hemozoin by flow cytometer	Detection of malarial pigment in activated monocyte
Sensitivity and specificity	Depends on malaria endemicity	Depends on good technique, good reagent and microscopist's skill	Sensitivity and specificity higher than PBS	Moderate, if more than 100parasite/ µl	Relatively high but do not correlate to clinical symptoms of patients	Excellent	Relatively high	Variable sensitivity, high specificity	Variable in both sensitivity and specificity
Time consumed (min)	Depends on 30-60 physician's skill	-	<15	10-15	30-60	45-360 depends 0n the methods	<60	Automated, <1/sample	automated, <1/sample
Detection limit (parasites/µl)	Undetermined	Expert- 5-10	>5	50-100	Undetermined	>=1	Undetermined	Poor correlation with parasitemia	5-20
Expertise required	High in non-endemic areas	High in non - endemic areas	Moderate	Low	Moderate	High	High	High	High

Abbreviations: PBS: peripheral blood smears; QBC: quantitative buffy coat; RDTs: rapid diagnostic tests; PCR: polymerase chain reaction; FCM: flow cytometry; ACC: automated blood cell counter.

6.3.1.1 Antifolate Combination Drugs

Antifolate combination drugs, for instance sulfadoxine with pyrimethamine, act through consecutive and synergistic blockade of two key enzymes involved with folate synthesis. Pyrimethamine and related compounds obstruct the step facilitated by dihydrofolate reductase (DHFR) whereas sulfones and sulfonamides impede the step mediated by dihydropteroate synthase (DHPS) (Bruce-Chwatt et al., 1986). Specific gene mutations encoding for resistance to both DHPS and DHFR have been identified. Various combinations of these mutations have been associated with varying degrees of resistance to antifolate combination drugs (Plowe et al., 1998).

6.3.1.2. Atovaquone

Atovaquone inhibits electron transport at the cytochrome *bc1* complex (Ittarat et al., 1994). Development of resistance against atovaquone is very rapid when used individually. However, it was observed that, resistance develops gradually when atovaquone is combined with other drugs, e.g., proguanil (for Malarone TM) or tetracycline (Looareesuwan et al., 1996). Resistance is conferred by single-point mutations in the cytochrome-b gene.

Table 15. Sulfonamide (sul) and Quinolone (qnr) resistance genes in *B. pertussis* strains (Syed et al., 2016)

S. No.	Sulfamethoxazole Resistance Gene (s)	Qunilone Resistance Gene (s)
1.	Sul 1	qnrA
2.	Sul 2	qnrB
3.	Sul 3	qnrS
4.	Sul 1 + sul2	qnrA + qnrB
5.	Sul 2 + su3	qnrA + qnrS
6.	Sul 1 + sul3	qnrB + qnrS
7.	Sul 1 + sul2+sul3	qnrA + qnrB + qnrS

6.4. Whooping Cough

It is also known as pertussis (100 days' cough). It is a highly contagious disease of the respiratory tract (especially trachea and bronchi) caused by the bacterium *Bordetella pertussis*. This bacterium spreads by airborne respiratory droplets (cough or sneezes), by saliva, or by skin-to-skin contact. Symptoms are blocked nose, dry and irritating cough, malaise, mild fever, runny nose, sore throat, watery eyes, etc. Details of bacterium resistance and the diagnostic tests related to whooping cough are given below (Table 16).

6.4.1. Macrolide-Resistant Bordetella pertussis Infection

B. pertussis resistance to macrolides is unlikely to be the cause of severity because the infecting strain was found twice to be sensitive to all macrolides, and acquired resistance subsequently. Nevertheless, in spite of proper treatments, we could not eliminate *B. pertussis*, and it was observed that cultures were positive after two courses of macrolides and one course of cotrimoxazole given for an associated staphylococcal infection (Guillot et al., 2012).

6.4.2. Diagnostic Tests (Wendelboe and Van, 2006)

Table 16. Diagnostic tests for whooping cough

Diagnostic Method	Sensitivity	Specificity	Recommendations for clinical use	Limitations
Culture	12-60%	Approximately 100%	< 2 weeks post symptom onset	Low sensitivity, 5-day delay between sample collection and results
PCR	70-99%	86-100%	< 4 weeks post symptom onset	No method available that meets international standards, laboratory contamination with subsequent false-positive results
Paired sera	90-92%	72-100%	Not useful in the clinical setting	4-6 week delay in sample collection and results, difficulty in collection of samples at the right time
Single sera	63-76%	99%	>= weeks post symptom onset	Lack of inter laboratory standardization of antibody titre cut-off
Clinical diagnosis symptoms and signs	67-89%	52-94%	Used in conjunction with laboratory test	Low sensitivity in the vaccine era due to the high proportion of atypical and mild pertussis cases

7. Genetics Involved in Development of Antimicrobial Drug Resistance

Antimicrobials are being used for a very long time and are frequently prescribed (Harbottle et al., 2006). More microbial resistance can be seen with the frequent use of antibiotics. Stronger microbes resist the treatment, grow and multiply while the sensitive microbes are killed. Antibiotic resistance is a common health problem stated by the FDA because almost all types of bacteria have become stronger and less responsive to antibiotic treatment (Boerlin et al., 2018).

In previous years, various novel concepts have arisen for a better understanding of the development and enhancement of antimicrobial resistance (AMR) including, the finding of the mutator state and the idea of mutant selection window for resistances evolving mainly via mutations in existing genes. The mechanism for antimicrobial resistance include, (i) mutations in pre-existing genes, and (ii) horizontal gene transfer (HGT).

Resistance begins by the attainment of new resistance genes rather than by instinctive mutations. The source of these genes is unknown, but most of them are produced from ecological organisms. Some AMR genes are produced from natural antibiotics, where they shield the bacterium against its own weapons. Various elements are involved in the movement of AMR genes and it happens at two different levels. Intracellularly, the AMR genes can transfer within the genome, including between chromosome and replicon, for instance, phages and plasmids. In the movement of AMR gene, the elements, transposons and integrons are chiefly involved and depend on both homologous and non-homologous recombination. The intercellular movement, i.e., horizontal spread involves three mechanisms (i) uptake of naked DNA (transformation), (ii) transfer by bacteriophages (transduction) and (iii) transfer by plasmids and other conjugative elements (conjugation).

Due to their ability to capture gene and association with transposons (Tn21 and Tn7), they are very crucial in the development and spread of multi-resistance. Therefore, the recognized integrons are continually increasing to deliver bacteria with the tools to counterattack antimicrobials. Some integrons of class 1 have grownup from their host to develop vectors of resistance; and to feast in commensal and pathogenic bacteria from animals and humans by their associations with transposons (Boerlin et al., 2008). Insertion sequences transposons are crucial in the AMR gene mobilization of AMR genes. Generally, termination ensues outside the limit marked by the terIS site permitting it for the mobilization of sequences adjacent to the ISCR elements, including AMR genes. ISCRs are important in the development and spread of identified AMR genes. For AMR gene, conjugative or mobilizable plasmids are common transmission vectors. Many of them carry multiple resistance genes leading to what can be termed 'infectious multiresistance.' The exponential growths of DNA sequencing aptitudes, informatics for analysis of sequence analysis and clarification have carried inclusive consideration of the molecular epidemiology of AMR plasmids (Alba et al., 2018). For numerous reasons, genetic methods compared to conventional susceptibility methods are consistent for assessment of AMR (Cockerill and Franklin, 1999).

1. Genetic susceptibility testing can be achieved with clinical samples avoiding the requirement for isolation of the organism by culture.
2. These approaches estimate the genotype of the organism, while conventional methods assess the phenotype or expression of the genotype under laboratory circumstances.
3. In some cases, genotypes may be determined long before phenotypes can be determined due to the slow growth of the organism.

4. For some organisms, it is not easy to culture them and so only genotypes can be determined in such cases.
5. Genetic methods may reduce the biohazard which may ascend with the proliferation by culture of a microorganism.

8. Roles of Ecosystem in the Development of Drug Resistance

In previous years, the role of environment as a significant source and distribution route of resistance has been progressively identified. The deficiency of facts regarding how, and under which conditions, the environment enables resistance makes mitigation of the development and distribution of mobile resistance factors challenging. Augmented information of the environmental factors that drive resistance may eventually permit to build models for how resistance arises and spreads (Hiltunen et al., 2017; Bengtsson-Palme et al., 2017). Development of novel resistance factors, mobilization, transfer to human pathogens, and dissemination are four main steps to clinically significant antibiotic resistance. Evidently, these steps need not occur in this specific order; transmission to human pathogens occurs before or after spread to the human microbiome, and few stages in the procedure may repeat. A critical aspect of a resistance gene to reach human pathogens is that it is preserved during all these stages. Recently, mobilized genes that do not provide noticeable fitness advantage are undeniably sorted out from mobile genetic elements such as plasmids (Petersen, 2011). This highlights the weightage of environment in which resistance genes provide a strong selective advantage, for example milieus subjected to industrial pollution with antibiotics. Since these environments present bacteria with conditions that favor augmented mutation rate, one consequence may be that resistance genes could be present in numerous different variants, all selected for detoxification efficiency, and of which only those with a low fitness cost are maintained when the selection pressure is removed. This proposes that once a resistance gene is extensively spread among human pathogens then we are conditioned to be able to manage its spread. Extenuation in the spread of resistance aspects to human pathogens takes place before they get in the human microbiome. Eventually, the most crucial aim to study antibiotic resistance is to gain further insights into health risks for humans and domestic animals that often are dependent on effective antibiotics. To recognize appropriate extenuation tactics, firstly we need to describe what environments and circumstances establish the most severe risks (Bengtsson-Palme et al., 2017). Some have claimed that the severe risk conditions encompass known resistance genes that are previously reported to be located in mobile genetic elements hosted by human bacterial pathogens (Martinez et al., 2015). This is a valid statement when such genes are encountered in the human microbiome. However there are important findings that environmental bacterial communities do not essentially reveal of a high-risk condition. Well-known resistance genes easily spread with human faeces, and their detection in the environment may be a sign of human faecal contamination (Martinez, 2009; Bengtsson-Palme et al., 2017).

9. Threats of Drug Resistance and CDC Action Plans to Control Drug Resistance

The risk of drug resistance encounters progress in healthcare, and life expectancy. Knowing these threats, necessitate preventing infections in the first place, reducing the expansion of resistance through antibiotics, and discontinuing the spread of resistance when it develops. Resistant germs can rapidly expand across settings-communities, food supply, healthcare facilities, environment (e.g., soil, water). This spread enhances the problem of resistance and infections of antibiotic-resistant and can be stopped by preventing the expansion of germs and improving infection prevention and antibiotic use in all settings. Investments from Congress to support actions-such as improving infection, prevention and control and antibiotic use-are working, proving the value of continued action against this threat. Coordinated, aggressive action and a health approaches have been critical to this success (Soltani et al., 2019). To avoid hopeful antibiotic-resistant strains of infections to develop, take precautions, such as (i) Do not use an antibiotic when a health care worker says it is not needed, and (ii) use of antibiotics should be avoided for a viral infection (e.g., common cold). If you are taking antibiotics do not skip doses. It is required to complete the prescribed course of antibiotic treatment. If you are a hospital patient or have a loved one in the hospital, make sure that you and the doctors, nurses, support staff, and all visitors wash their hands or use a hand sanitizer prior to touching the patient (Fleming-Dutra et al., 2018).

References

8th World Health Organization (WHO). *Antimicrobial Resistance: Global Report on Surveillance* (2019). Available from: http://www.who.int/drugresistance/ documents/ surveillancereport/en/.

Alanis AJ. Resistance to antibiotics: are we in the post-antibiotic era? *Archives of medical research*. 2005 Nov 1;36(6):697-705.

Alba P, Leekitcharoenphon P, Franco A, Feltrin F, Ianzano A, Caprioli A, Stravino F, Hendriksen RS, Bortolaia V, Battisti A. Molecular epidemiology of MCR-encoded colistin resistance in *Enterobacteriaceae* from food-producing animals in Italy revealed through the EU harmonized antimicrobial resistance monitoring. *Frontiers in microbiology*. 2018 Jun 12;9:1217.

Aminov RI, Mackie RI. Evolution and ecology of antibiotic resistance genes. *FEMS microbiology letters*. 2007 Jun 1;271(2):147-61.

Aslam B, Wang W, Arshad MI, Khurshid M, Muzammil S, Rasool MH, Nisar MA, Alvi RF, Aslam MA, Qamar MU, Salamat MK. Antibiotic resistance: a rundown of a global crisis. *Infection and drug resistance*. 2018;11:1645.

Baharoglu Z, Garriss G, Mazel D. Multiple pathways of genome plasticity leading to development of antibiotic resistance. *Antibiotics*. 2013 Jun;2(2):288-315.

Bengtsson-Palme J, Kristiansson E, Larsson DJ. Environmental factors influencing the development and spread of antibiotic resistance. *FEMS microbiology reviews*. 2017 Oct 24;42(1):fux053.

Bloland PB, World Health Organization. *Drug resistance in malaria*. Geneva: World Health Organization; 2001.

Boerlin P, Reid-Smith RJ. Antimicrobial resistance: its emergence and transmission. *Animal Health Research Reviews*. 2008 Dec;9(2):115-26.

Branson, B. M. (2019). *Human Immunodeficiency Virus Diagnostics*. Infectious Disease Clinics of North America.

Bruce-Chwatt LJ, Black RH, Canfield CJ, Clyde DF, Peters W, Wernsdorfer WH, World Health Organization. *Chemotherapy of malaria*.

Bueno J. Biosensors in antimicrobial drug discovery: since biology until screening platforms. *J. Microb. Biochem. Technol. S*. 2014;10.

Casali N, Nikolayevskyy V, Balabanova Y, Harris SR, Ignatyeva O, Kontsevaya I, Corander J, Bryant J, Parkhill J, Nejentsev S, Horstmann RD. Evolution and transmission of drug-resistant tuberculosis in a Russian population. *Nature genetics*. 2014 Mar;46(3):279.

Centers for Disease Control and Prevention, *Treatment of tuberculosis*, http://www.cdc.gov/tb/topic/treatment/default.htm accessed on 2.03.16 (Anonymous 2016).

Cockerill FR. Genetic methods for assessing antimicrobial resistance. *Antimicrobial agents and chemotherapy*. 1999 Feb 1;43(2):199-212.

Davies J, Davies D. Origins and evolution of antibiotic resistance. *Microbiol. Mol. Biol. Rev*. 2010 Sep 1;74(3):417-33.

Drobniewski FA, Wilson SM. The rapid diagnosis of isoniazid and rifampicin resistance in Mycobacterium tuberculosis—a molecular story. *Journal of medical microbiology*. 1998 Mar 1;47(3):189-96.

Fair RJ, Tor Y. Antibiotics and bacterial resistance in the 21st century. *Perspectives in medicinal chemistry*. 2014 Jan;6: PMC-S14459.

Fleming-Dutra KE, Linder JA, Hyun D, Iskander JK, Thorpe P, Laird S. *Be antibiotics aware: smart use, best care*, 2018.

Foley M, Tilley L. Quinoline antimalarials: mechanisms of action and resistance. *International journal for parasitology*. 1997 Feb 1;27(2):231-40.

Freire-Moran L, Aronsson B, Manz C, Gyssens IC, So AD, Monnet DL, Cars O, ECDC-EMA Working Group. Critical shortage of new antibiotics in development against multidrug-resistant bacteria—Time to react is now. *Drug resistance updates*. 2011 Apr 1;14(2):118-24.

Gergiso KT, Erango MA. Evaluation of the Trend of CD4 Cell Count Over Time in Case of HIV/AIDS Patients under ART Follow-up. *Journal of Scientific Research and Reports*. 2019 Oct 1:1-8.

Gogarten JP, Doolittle WF, Lawrence JG. Prokaryotic evolution in light of gene transfer. *Molecular biology and evolution*. 2002 Dec 1;19(12):2226-38.

Guillot S, Descours G, Gillet Y, Etienne J, Floret D, Guiso N. Macrolide-resistant Bordetella pertussis infection in newborn girl, France. *Emerging infectious diseases*. 2012 Jun;18(6):966.

Gupta S, Kakkar V. Recent technological advancements in tuberculosis diagnostics–A review. *Biosensors and Bioelectronics*. 2018 Sep 15;115:14-29.

Harbottle H, Thakur S, Zhao S, White DG. Genetics of antimicrobial resistance. *Animal biotechnology*. 2006 Dec 1;17(2):111-24.

Heaney J, Rolfe K, Gleadall NS, Greatorex JS, Curran MD. Low-Density TaqMan® Array Cards for the Detection of Pathogens. In *Methods in Microbiology* 2015 Jan 1 (Vol. 42, pp. 199-218). Academic Press.

Hiller CX, Hübner U, Fajnorova S, Schwartz T, Drewes JE. Antibiotic microbial resistance (AMR) removal efficiencies by conventional and advanced wastewater treatment processes: A review. *Science of the Total Environment*. 2019 May 25.

Hiltunen T, Virta M, Laine AL. Antibiotic resistance in the wild: an eco-evolutionary perspective. *Philosophical Transactions of the Royal Society B: Biological Sciences*. 2017 Jan 19;372(1712):20160039.

Hopkins C, Reid M, Gilmour J, Werder S, Briggs S. Missed opportunities for earlier diagnosis of human immunodeficiency virus infection among adults presenting to Auckland District Health Board hospital services. *Internal medicine journal*. 2019 Apr;49(4):495-501.

Islam MM, Hameed HA, Mugweru J, Chhotaray C, Wang C, Tan Y, Liu J, Li X, Tan S, Ojima I, Yew WW. Drug resistance mechanisms and novel drug targets for tuberculosis therapy. *Journal of Genetics and Genomics*. 2017 Jan 20;44(1):21-37.

Ittarat I, Asawamahasakda W, Meshnick SR. *The effects of antimalarials on the Plasmodium falciparum dihydroorotate dehydrogenase*.

Jacoby GA. History of drug-resistant microbes. In *Antimicrobial drug resistance* 2017 (pp. 3-8). Springer, Cham.

Kanji A, Hasan R, Hasan Z. Efflux pump as alternate mechanism for drug resistance in Mycobacterium tuberculosis. *Indian Journal of Tuberculosis*. 2019 Jan 1;66(1):20-5.

Kinch MS, Patridge E. An analysis of FDA-approved drugs for infectious disease: HIV/AIDS drugs. *Drug discovery today*. 2014 Oct 1;19(10):1510-3.

Klein EY, Van Boeckel TP, Martinez EM, Pant S, Gandra S, Levin SA, Goossens H, Laxminarayan R. Global increase and geographic convergence in antibiotic consumption between 2000 and 2015. *Proceedings of the National Academy of Sciences*. 2018 Apr 10;115(15): E3463-70.

Koneman EW, Allen SD, Janda WM, Schreckenberger PC, Winn WC. Diagnostic microbiology. *The nonfermentative gram-negative bacilli*. Philedelphia: Lippincott-Raven Publishers. 1997:253-320.

Krogstad DJ, Gluzman IY, Kyle DE, Oduola AM, Martin SK, Milhous WK, Schlesinger PH. Efflux of chloroquine from *Plasmodium falciparum*: mechanism of chloroquine resistance. *Science*. 1987 Nov 27;238(4831):1283-5.

Kuehn BM. IDSA: better, faster diagnostics for infectious diseases needed to curb overtreatment, antibiotic resistance. *JAMA*. 2013 Dec 11;310(22):2385-6.

Lipsitch M, Samore MH. Antimicrobial use and antimicrobial resistance: a population perspective. *Emerging infectious diseases*. 2002 Apr;8(4):347.

Lohrasbi V, Talebi M, Bialvaei AZ, Fattorini L, Drancourt M, Heidary M, Darban-Sarokhalil D. Trends in the discovery of new drugs for Mycobacterium tuberculosis therapy with a glance at resistance. *Tuberculosis*. 2018 Mar 1;109:17-27.

Lohrasbi V, Talebi M, Bialvaei AZ, Fattorini L, Drancourt M, Heidary M, Darban-Sarokhalil D. Trends in the discovery of new drugs for Mycobacterium tuberculosis therapy with a glance at resistance. *Tuberculosis*. 2018 Mar 1;109:17-27.

Martin SK, Oduola AM, Milhous WK. Reversal of chloroquine resistance in *Plasmodium falciparum* by verapamil. *Science*. 1987 Feb 20;235(4791):899-901.

Martinez JL, Coque TM, Baquero F. What is a resistance gene? Ranking risk in resistomes. *Nature Reviews Microbiology*. 2015 Feb;13(2):116-23.

Martinez JL. The role of natural environments in the evolution of resistance traits in pathogenic bacteria. *Proceedings of the Royal Society B: Biological Sciences*. 2009 Apr 8;276(1667):2521-30.

Martinkova P, Kostelnik A, Válek T, Pohanka M. Main streams in the construction of biosensors and their applications. *International Journal of Electrochemical Science*, volume 12, issue: 8. 2017.

McCluskey SM, Siedner MJ, Marconi VC. Management of Virologic Failure and Human Immunodeficiency Virus Drug Resistance. *Infectious Disease Clinics*. 2019 Jun 27.

McKeegan KS, Borges-Walmsley MI, Walmsley AR. Microbial and viral drug resistance mechanisms. *Trends in microbiology*. 2002 Oct 1;10(10):S8-S14.

Mcnulty CA, Boyle P, Nichols T, Clappison P, Davey P. The public's attitudes to and compliance with antibiotics. *Journal of Antimicrobial Chemotherapy*. 2007 Aug 1;60(suppl_1):i63-8.

Meyer KB, Pauker SG. Screening for HIV: can we afford the false positive rate? *N Engl J Med* 1987;317:238–41

Michael CA, Gillings MR, Holmes AJ, Hughes L, Andrew NR, Holley MP, Stokes HW. Mobile gene cassettes: a fundamental resource for bacterial evolution. *The American naturalist*. 2004 May 3;164(1):1-2.

National institute of allergy and infectious diseases, antimicrobial (drug) resistance, understanding, https://www.niaid.nih.gov/research/antimicrobial-resistance-causes, (accessed date 20/11/2019) (Ananymous 2019).

Nesme J, Cécillon S, Delmont TO, Monier JM, Vogel TM, Simonet P. Large-scale metagenomic-based study of antibiotic resistance in the environment. *Current biology*. 2014 May 19;24(10):1096-100.

Petersen J. Phylogeny and compatibility: plasmid classification in the genomics era. *Archives of microbiology*. 2011 May 1;193(5):313-21.

Petersen LR, Satten GA, Dodd R, Busch M, Kleinman S, Grindon A, Lenes B, HIV Seroconversion Study Group. Duration of time from onset of human immunodeficiency virus type 1 infectiousness to development of detectable antibody. *Transfusion*. 1994 Apr;34(4):283-9.

Piddock LJ. The crisis of no new antibiotics-what is the way forward? *The Lancet infectious diseases*. 2012 Mar 1;12(3):249-53.

Pillay, D. and Zambon, M. (1998) Antiviral drug resistance. *Br. Med. J.* 317, 660–662.

Pless R. Vaccination benefits, risks and safety: the need for a complete picture. *Bulletin of the World Health Organization*. 2000;78(2):219.

Plowe CV, Kublin JG, Doumbo OK. *P. falciparum* dihydrofolate reductase and dihydropteroate synthase mutations: epidemiology and role in clinical resistance to antifolates. *Drug Resistance Updates*. 1998 Jan 1;1(6):389-96.

Rogers KR, Mulchandani A, Zhou W, editors. Biosensor and Chemical Sensor Technology: process monitoring and control. *American Chemical Society;* 1996 May 5.

Rossolini GM, Arena F, Pecile P, Pollini S. Update on the antibiotic resistance crisis. *Current opinion in pharmacology*. 2014 Oct 1;18:56-60.

Saga T, Yamaguchi K. History of antimicrobial agents and resistant bacteria. *JMAJ*. 2009 Mar;52(2):103-8.

Sengupta S, Chattopadhyay MK, Grossart HP. The multifaceted roles of antibiotics and antibiotic resistance in nature. *Frontiers in microbiology*. 2013 Mar 12;4:47.

Sharp PM, Hahn BH. Origins of HIV and the AIDS pandemic. Cold Spring *Harbor perspectives in medicine*. 2011 Sep 1;1(1):006841.

Soltani J, Versporten A, Goossens H. *Antibiotic Resistance: A Global Concern; Current Situation and Action Plans*.

Spellberg B, Gilbert DN. The future of antibiotics and resistance: a tribute to a career of leadership by John Bartlett. *Clinical infectious diseases*. 2014 Sep 15;59(suppl-2):71-5.

Syed MA, Jamil B, Bokhari H. Shattering a myth: Whooping cough susceptible to antibiotics. *Pakistan journal of pharmaceutical sciences*. 2016 May 1;29(3).

Tan IL, McArthur JC. HIV-Associated neurological disorders. *CNS drugs*. 2012 Feb 1;26(2):123-34.

Tangpukdee N, Duangdee C, Wilairatana P, Krudsood S. Malaria diagnosis: a brief review. *The Korean journal of parasitology*. 2009 Jun;47(2):93.

Tellier G, Niederman MS, Nusrat R, Patel M, Lavin B. Clinical and bacteriological efficacy and safety of 5- and 7-day regimens of telithromycin once daily compared with a 10-day regimen of clarithromycin twice daily in patients with mild to moderate community-acquired pneumonia. *Journal of Antimicrobial Chemotherapy*. 2004 Aug 1;54(2):515-23.

Unissa AN, Hanna LE. Molecular mechanisms of action, resistance, detection to the first-line anti tuberculosis drugs: Rifampicin and pyrazinamide in the post whole genome sequencing era. *Tuberculosis*. 2017 Jul 1; 105:96-107.

Ventola CL. The antibiotic resistance crisis: part 1: causes and threats. *Pharmacy and therapeutics*. 2015 Apr;40(4):277.

Washington John A. Chapter 10- principles of diagnosis, medical microbiology, 4th edition, Galveston (TX): university of Texas medical branch at Galveston; 1996.

Wendelboe AM, Van Rie A. Diagnosis of pertussis: a historical review and recent developments. *Expert review of molecular diagnostics*. 2006 Nov 1;6(6):857-64.

World Health Organization. *Media centre, Fact sheet N'104*. http://www.who.int/mediacentre/factsheets/fs104/en/2016; [accessed 19.01.17].

Zaman SB, Hussain MA, Nye R, Mehta V, Mamun KT, Hossain N. A review on antibiotic resistance: alarm bells are ringing. *Cureus*. 2017 Jun;9(6). e1403.

Zhang H, Li D, Zhao L, Fleming J, Lin N, Wang T, Liu Z, Li C, Galwey N, Deng J, Zhou Y. Genome sequencing of 161 Mycobacterium tuberculosis isolates from China identifies genes and intergenic regions associated with drug resistance. *Nature genetics*. 2013 Oct;45(10):1255.

Zhou L, He X, He D, Wang K, Qin D. Biosensing technologies for Mycobacterium tuberculosis detection: status and new developments. *Clinical and developmental immunology*. 2011 Mar 16;2011.

Zhu YG, Johnson TA, Su JQ, Qiao M, Guo GX, Stedtfeld RD, Hashsham SA, Tiedje JM. Diverse and abundant antibiotic resistance genes in Chinese swine farms. *Proceedings of the National Academy of Sciences*. 2013 Feb 26;110(9):3435-40.

In: Antimicrobial Resistance: Opportunities and Challenges ISBN: 978-1-53617-943-9
Editors: P. Kumar Prabhakar and V. Mishra

Chapter 2

MECHANISM OF ANTIBIOTIC RESISTANCE IN BACTERIA

Manju[1], Atul Kumar Tiwari[2], Dhananjay Shukla[3] and Saurabh Saxena[1,*]

[1]Department of Medical Laboratory Sciences, Lovely Professional University, Jalandhar – Delhi G.T. Road, Phagwara, Punjab, India
[2]Department of Zoology, Dr. Bhanvar Singh Porte Government College, Bilaspur, Chhattisgarh, India
[3]Department of Biotechnology, Guru Ghasidas Vishwavidyalaya, Bilaspur, Chhattisgarh, India

ABSTRACT

Development of antibiotic resistance among the pathogenic bacteria has emerged as a serious problem in the treatment of infectious disorders. Although it is a natural phenomenon developed by bacterium for survival in an environment containing natural antibiotics; the overuse of natural/synthetic antibiotics in clinical practices has enhanced the rate of emergence of resistant microorganisms at an alarming pace. Bacteria are continuously developing new ways to adapt to the hostile environment created by the antibiotic drive against it. Even multi-drug resistant bacteria are appearing that are tolerant to a range of antibiotics. Bacteria can be intrinsically resistant or may acquire it by various mechanisms. These strategies include reducing cell membrane permeability, increased efflux, target/drug modification, target protection, etc. The genes responsible for antibiotic resistance can be transferred from one bacterium to another either horizontally or vertically. Thus, awareness for a controlled use of antibiotics in the society is required. The understanding of the molecular and biochemical mechanism of antibiotic resistance in bacteria is important for the development of strategies to curb the emergence and spread of

* Corresponding Author's E-mail: drsaurabh.bio@gmail.com.

the responsible genes and design a plan of action for the treatment of multi-drug resistant bacteria. This chapter gives details on the various mechanisms of antibiotic resistance in bacteria with specific example.

Keywords: antibiotic resistance, efflux pump, mutation, molecular mechanism

1. Antibiotic Resistance

Antibiotics are the modern medicines that interfere with the structures or processes essential for bacterial growth or survival without harming the eukaryotic host suffering from the bacterial infection (Walsh 2000). The use of these antibiotics has greatly increased the life expectancy and reduced childhood mortality (Blair et al. 2015). Therefore, antibiotics have become the most important medical interventions required for developing complex medical approaches *viz.* invasive surgeries, organ transplantation, chemotherapy, etc. Unfortunately, due to extensive antibiotic usage, antimicrobial resistance is now threatening the therapeutic accomplishments (Munita and Arias 2016).

Antibiotic resistance (ABR) is the resistance developed by bacteria against bactericidal or bacteriostatic drugs, allowing them to withstand against such drugs (Zaman et al. 2017). Several ABR pathogens have been emerged such as multidrug resistant and extensively drug resistant *Mycobacterium tuberculosis* (MDR-MTB and XDR-MTB), extended spectrum beta lactamase (ESBL) producing pathogens, metallo-beta lactamase (MBL) producing bacteria, methicillin resistance *Staphylococcus aureus* (MRSA) and the advent of New Delhi metallo-beta-lactamase (NDM) producing bacteria (Ventola 2015; Dookie et al. 2018; Hrabák 2007; Mojica, Bonomo, and Fast 2016; Lakhundi and Zhang 2018; Khan, Maryam, and Zarrilli 2017). Since 1940, the emergence and wide-spread of ABR has been reported (Alison J Baylay, Piddock, and Webber 2019). Due to the irrational, over usage as well as misuse of antibiotics, ABR is emerging as a worldwide threat leading to extra financial burden and even fatal consequences (Viswanathan et al. 2012).

ABR pathogens may possess either intrinsic resistance to certain antibiotics or can acquire resistance via chromosomal mutations or horizontal gene transfer (Peterson and Kaur 2018). The intrinsic resistance is the ability of bacteria to resist the action of particular antibiotic because of inherent functional or structural characteristics.

Examples of intrinsic resistance include:

1. Absence of a susceptible target of a specific antibiotic – Both Gram positive and Gram negative bacteria can be targeted and treated with biocide triclosan but this drug is not active against *Pseudomonas*, a Gram-negative genus (Chuanchuen, Karkhoff-Schweizer, and Schweizer 2003). The reason for the same is the presence of insensitive allele, *fabl* that is responsible for the production of an additional

enoyl-ACP reductase enzyme — the target for triclosan in sensitive species (Zhu et al. 2010).

2. Difference in the composition of cell membrane – Due to intrinsic differences in the composition of cell membrane of Gram-positive and Gram-negative bacteria; daptomycin, a lipopeptide is active against Gram-positive bacteria but ineffective against Gram-negative ones. As Gram negative bacteria have a lower number of anionic phospholipids in cell membrane, the efficiency of Ca^{2+}-mediated daptomycin insertion into the cell membrane is greatly reduced that decreases antibacterial activity (Randall et al. 2013; Mishra et al. 2013).
3. Inability of drug to cross the outer membrane – Most of the additional benefit provided to the Gram-negative bacteria is due to the presence of outer membrane that is absent in case of Gram-positive bacteria (Delcour 2009; Exner et al. 2017). Vancomycin, a glycopeptide antibiotic inhibits the growth of Gram-positive bacteria by binding to d-Ala-d-Ala peptide target inhibiting peptidoglycan crosslinking. However, due to the presence of outer membrane in Gram negative bacteria, it is not able to cross and bind to its target. Therefore, it is ineffective against Gram-negative bacteria (Tsuchido and Takano 1988).

In addition to intrinsic resistance, bacteria can develop acquired resistance to antibiotics through various mechanisms, which can be sorted into three main categories as given below (Fernández and Hancock 2012; Nikaido 2009; Wright 2011). Figure 1 illustrates the various mechanisms of acquired resistance to antibiotics.

A. Minimizing the intracellular concentrations of the antibiotic as a result of poor permeability or antibiotic efflux.
B. Modification of the antibiotic target either by genetic mutation or post-translational modifications.
C. Inactivation of antibiotic either by hydrolysis or modification

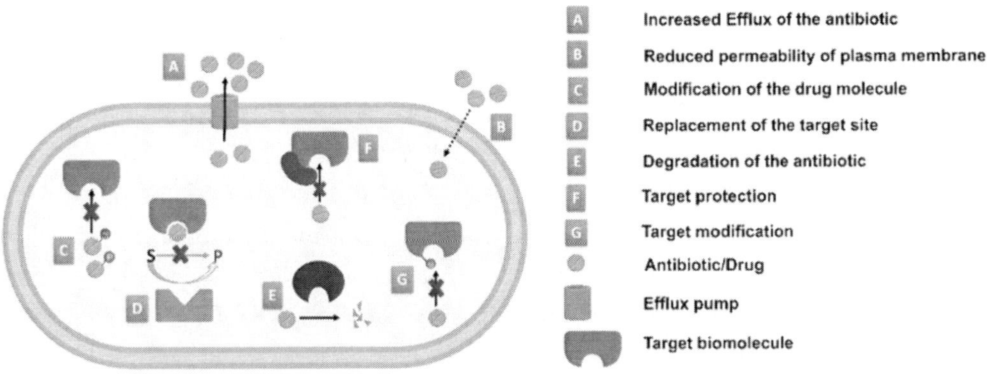

Figure 1. Various mechanisms of acquired antibiotic resistance among bacteria.

A. Minimizing Intracellular Antibiotic Concentrations

1. Reduced permeability – The outer membrane in Gram-negative bacteria serves as a permeability barrier that makes them intrinsically resistant to various antibiotics as compared to Gram-positive species (Kojima and Nikaido 2013; Vargiu and Nikaido 2012). Hydrophilic antibiotics transport through the outer membrane porin proteins by diffusion (Ghai and Ghai 2018; Delcour 2009). In several bacteria (mostly members of Enterobacteriaceae), the major porin proteins *viz.* the outer-membrane proteins OmpF and OmpC of *Escherichia coli* function as non-specific channels (Kojima and Nikaido 2013; Tran et al. 2013; Ghai and Ghai 2018). Therefore, reduction of the outer membrane permeability limits antibiotic entry into the bacterial cell either by replacing porins with more selective channels or by downregulating the expression of porin proteins.

   One of the best example is *Pseudomonas aeruginosa*, which expresses OprF protein, a major non-specific porin embedded in outer membrane that is dedicated to non-specific uptake of ions and saccharides, but having low efficiency for antibiotic permeability (Tran et al. 2013; Nikaido 2003). It also shows high expression level of OprF protein, a homolog of OmpF, in a closed conformation as compared to open confirmation (Sugawara et al. 2006). Reduction in the expression of outer membrane protein OprD either due to the presence of mutant allele or due to mutation causes decreased porin production, leading to carbapenem resistance even in the absence of carbapenemase (Margaret, Drusano, and Standiford 1989; Tamber and Hancock 2003). In *Klebsiella pneumoniae*, porins OmpK35 and OmpK36 are replaced with an alternative porin OmpK37 with a narrower channel, which results in a similar effect to porin loss (Doménech-Sánchez et al. 1999).

   Gram-positive bacteria are intrinsically less tolerant to antibiotics than Gram negative bacteria because of absence of outer membrane (Alison J Baylay, Piddock, and Webber 2019). However, it may show reduced permeability to some drugs. For example, vancomycin intermediate *Staphylococcus aureus* (VISA) possess a thickened cell wall that makes vancomycin unable to prevent peptidoglycan crosslinking by binding to its precursors, resulting increased minimum inhibitory concentration (MIC) (Cui et al. 2006).

2. Increased efflux – Efflux pumps are major contributors to the intrinsic resistance in bacteria as they actively extrude antibiotics outside the cell (Blair et al. 2015). These pumps are more prevalent in Gram-negative bacteria as compared to Gram-positive ones. Efflux pumps also play a significant role in other important cellular processes including virulence and biofilm formation (Alvarez-Ortega, Olivares, and Martínez 2013; Baugh et al. 2012; L.J. Piddock 2006b; Webber et al. 2009). Overexpression of efflux pumps either due to

mutations in global/local regulators (Baranova, Danchin, and Neyfakh 1999; Webber, Talukder, and Piddock 2005; Alekshun, Kim, and Levy 2000) or acquisition of insertion sequence (IS) elements that act as strong promoters of efflux pump genes (Boutoille et al. 2004; Rouquette-Loughlin et al. 2004), results in high resistance against clinically useful antibiotics. These efflux pumps can have narrow substrate specificity (*viz.* Tet pumps) or broad substrate specificity [*viz.* multidrug resistance (MDR) efflux pumps]. The multiple genes for these efflux pumps are carried on chromosomes but some have been mobilized on extrachromosomal DNA and hence undergo horizontal gene transfer (Munita and Arias 2016). For example, genes coding novel tripartite resistance nodulation division (RND) pump were recently found to be carried on an IncH1 plasmid in *Citrobacter freundii* strain that also carried the gene for the antibiotic-targeting enzyme New Delhi metallo-β-lactamase 1 (NDM1) (Blair et al. 2015; Dolejska et al. 2012). This transmissible resistance gene is a worrying development as it can be rapidly disseminated to other clinically relevant pathogens. There are five known families of multidrug efflux pump as shown in Table 1.

Table 1. Classification of efflux pumps

S. No.	Efflux pump	Structure	Mechanism	Example
1.	Resistance nodulation family	Tripartite transporters consisting of an inner membrane pump, an outer membrane pump and a connecting adaptor protein (Alvarez-Ortega, Olivares, and Martínez 2013; Blair et al. 2015)	Proton motive force (I. T. Paulsen, Brown, and Skurray 1996)	AcrAB–TolC pump in *Escherichia coli* (Pos 2009)
2.	Major facilitator superfamily	Soluble transporters made up of single polypeptide chain with 12 or 14 membrane spanning domains (Zhang et al. 2015)	Proton motive force (Zhang et al. 2015)	NorA gives resistance to fluoroquinolone (Ubukata, Itoh-Yamashita, and Konno 1989) QacA exports cationic lipophilic drugs (Brown and Skurray 2001) LmrS exports lncomycin, chloroamphenicol, etc. (Floyd et al. 2010)
3.	Small multidrug resistance transporters	Small functional transporters having 110 – 120 amino acid, proteins with four membrane spanning domains (I. T. Paulsen, Brown, and Skurray 1996)	Proton motive force (I. T. Paulsen, Brown, and Skurray 1996)	QacC from *Staphylococcus aureus* exchanges toxic monovalent cations, quaternary ammonium compounds, etc. (Ian T Paulsen et al. 1995) EmrE from *Escherichia coli* exchanges toxic cations (Yerushalmi, Lebendiker, and Schuldiner 1995)
4.	Multidrug and toxic compound extrusion (MATE) efflux pumps	Consists of single polypeptide chain with 12 membrane spanning domains (Lu 2016)	Proton motive force or sodium antiport mechanism (L.J.V. Piddock 2006a)	VcrM from *Vibrio cholerae* (Huda et al. 2003) MepA from *Staphyloccocus aureus (McAleese* et al. *2005)* PmpM from *Pseudomonas aeruginosa* (He et al. 2004)
5.	ATP-binding cassette (ABC) transporters	Primary transporters possess 4 subunits: 2 transmembrane domains and 2 ATP binding domains (Zolnerciks et al. 2011)	Conserved mechanism of energy coupling (Locher 2008; Zolnerciks et al. 2011)	PatAB in *Streptococcus pneumoniae* gives resistance against some fluoroquinolones (Alison J. Baylay, Ivens, and Piddock 2015) MacAB in *E. coli* is a macrolide transporter (Moussatova et al. 2008)

B. Modifications of the Antibiotic Molecule: Modification of antibiotic is the most commonly used strategy rendering bacterial cell resistant to particular antibiotic. This can be achieved either by biochemical modification or degradation of antibiotics.

1. Biochemical modifications – Introduction of chemical changes to the antibiotic molecule by modifying enzyme is a well-known acquired antibiotic resistance mechanism in both Gram-positive and Gram-negative bacteria. Various modifying enzymes can be produced by the bacterial cell. The most frequent biochemical reactions include the following reactions:
   (i) Acetylation (aminoglycosides, chloramphenicol, streptogramins),
   (ii) Phosphorylation (aminoglycosides, chloramphenicol), and
   (iii) Adenylation (aminoglycosides, lincosamides)

The overall resulting effect is a steric hindrance that decreases the drug avidity for its target and hence higher bacterial MICs. Few examples of biochemical modifications are given in Table 2.

Table 2. Biochemical modifications of antibiotic/drug with enzymes

S. No.	Antibiotic	Mode of action	Resistant gene	Gene location	Resistant mechanism
1.	Chloramphenicol	Inhibition of protein synthesis by interacting with the peptidyl-transfer center of the 50s ribosomal subunit (Krause et al. 2016)	*Cat* genes	Plasmid/ chromosome	Chemical modification of chloroamphenicol by production of chloroamphenicol Acetyltransferase (Shaw et al. 1985)
2.	Aminoglycosides	Inhibition of protein synthesis by binding to the A-site on the 16s ribosomal RNA of the 30s ribosome	*Aac, Ant, Aph*	Plasmid/ chromosome	The enzymes aminoglycoside acetyletransferae (AAC), aminoglycoside nucleotidyl-transferase (ANT) and aminoglycoside phosphotransferase (APH) covalently modify the hydroxyl or amino groups of the aminoglycoside molecule. (C.A. Smith and Baker 2002; Azucena and Mobashery 2001)

2. Destruction of the antibiotic molecule- One of the most important mechanisms of β-lactam resistance includes destruction by the action of antibiotic hydrolyzing enzymes, β- lactamases (Kaur and Peterson 2018). The mechanism of action of these enzymes include breakdown of the amide bond present in β-lactam ring leading to structure destruction and ineffectiveness of antibiotics (Davies, White, and Nicholas 2001; Cag et al. 2016). β-lactam antibiotics bind to Penicillin-binding proteins (PBP) and interfere with cell wall synthesis. β- lactamases mimic PBP in structure (Davies, White, and Nicholas 2001). Thus, they bind to β-lactam and neutralize it. The genes for these enzymes are present on chromosome as well as plasmids. β-lactamases were first described in the early 1940s, however, their existence for millions of years is evident (Abraham and Chain 1940; D'Costa et al. 2011). After the

wide distribution of penicillin, clinically important penicillin resistant strain of *Staphylococcus aureus* emerged (Rammelkamp and Maxon 1942). Its resistance mechanism was found to be the production of penicillinase enzyme encoded on plasmid, which are readily transmitted between *S. aureus* as a result of horizontal gene transfer (Bush 2013). To overcome this problem, new β-lactam compounds were then manufactured that had a broad spectrum and less sensitivity to penicillinase degradation (Kong, Schneper, and Mathee 2010). In 1960s, new plasmid encoded with ampicillin hydrolyzing β-lactamase (TEM-1, named after the person, Temoneira from whom it was first isolated) among Gram-negative bacteria was found (Paterson and Bonomo 2005). Discovery of β-lactamases was followed by extended-spectrum beta lactamases (ESBL) and carbapenemases, including imipenemases, metallo beta lactameses, oxacillinase, *K. pneumoniae* carbapenemase and New Delhi metallo beta lactamase active against β-lactams (Johnson and Woodford 2013; Voulgari et al. 2013; Lynch III, Clark, and Zhanel 2013). This underpins the emergence of β-lactam resistant bacteria and hence, serious complications in the treatment of severe infections.

C. Changes in target site: Bacteria can avoid action of antibiotics by altering the target site of antibiotics by various processes *viz.* protecting it with certain proteins, mutation in the encoding gene or post-translational modification of the target protein (Alison J Baylay, Piddock, and Webber 2019).

    1. Target protection — Genetic determinants for the proteins that are produced by the bacteria to protect the target are present on the chromosomes as well as on Mobile Genetic Elements (MGEs). The examples of target protection are presented in Table 3.

Table 3. Protection of the antibiotic target molecule

S. No.	Antibiotics	Mode of action	Resistant determinants	Mechanism	Examples
1.	Tetracycline	Inhibition of protein synthesis (Chopra and Roberts 2001)	Tet[M] and Tet[O]	These proteins act as homologues of elongation factor (EF-G and EF-Tu). They bind with ribosome and dislodge tetracycline from its binding site in an FTP-dependent manner. (Connell et al. 2003; Dönhöfer et al. 2012; Li et al. 2013)	*Streptococcus* spp. (Ayer et al. 2007) and *Campylobacter jejuni* (Wieczorek and Osek 2013)
2.	Fluroquinoles	Inhibiting DNA synthesis (Aldred, Kerns, and Osheroff 2014)	Qnr	This protein acts as DNA homologue, competing for DNA binding site of DNA gyrase and topoisomerase IV. Thus, the interaction of DNA gyrase with DNA reduces, inhibiting the quinolone complex formation (Martínez-Martínez, Pascual, and Jacoby 1998; Rodríguez-Martínez et al. 2011)	*Klebsiella pneumoniae*, *Escherichia coli* (Wang et al. 2008)

2. Modification of the target site- The other way of bacteria to protect themselves from the action of antibiotic is to modify the target site of the same. These target changes include:
   (i) Point mutations in the genes encoding the target site,
   (ii) Enzymatic alterations of the binding site (e.g., addition of methyl groups), and/or
   (ii) Replacement or bypass of the original target.

   Whichever mechanism is followed by bacteria, final result is reduction in the affinity of antibiotic for the target site.

   (i) Mutations of the target site: Target site of antibiotics undergo mutational changes. One of the most classical examples is Rifampin (RIF) resistance. Rifamycin inhibits the DNA-dependent RNA polymerase, a complex enzyme made up of five subunits (i.e., $\alpha 2\beta\beta'\sigma$). Highly conserved RIF binding pocket is present in its β-subunit (encoded by *rpoB*); to which antibiotic binds, directly interfering with the process of transcription (Campbell et al. 2001; Floss and Yu 2005). A single point mutation in *rpoB* gene results in an amino acid substitution in this RIF-binding pocket, resulting in a decreased affinity of antibiotic for its target (Floss and Yu 2005; Xu et al. 2005). However, this point mutation does not interfere with the catalytic activity of RNA polymerase. Another example involves the Fluoroquinolone (FQ) resistance. FQs inhibit DNA replication by inhibiting the activity of DNA gyrase and topoisomerase IV (Redgrave et al. 2014). Chromosomal mutation in the genes encoding DNA gyrase (*gyrA-gyrB*) and topoisomerase IV (*parC-parE*) inhibits the interaction of drug and the enzymes, leading to an acquired resistance against fluroquinolones (Hooper 2002; Redgrave et al. 2014).

   (ii) Enzymatic alteration of the target site: If the target site of antibiotic is altered by the action of an enzyme, it results in a decreased affinity of the antibiotic. One of the classic examples is macrolide resistance. Macrolides work by inhibiting bacterial protein synthesis either by premature dissociation of peptidyl-tRNA from the ribosome or by preventing peptidyl-transferase from adding the growing peptide attached to tRNA to the next amino acid (Marne and Alexander 2003). Enzymes (erythromycin resistance methylases) encoded by erm genes (erythromycin ribosomal methylation) modify the target site of macrolides by methylation of ribosome resulting in macrolide resistance. These enzymes are capable of mono- or di-methylating an adenine residue in position A2058 of the domain V of the 23s rRNA of the 50S ribosomal subunit. (Leclercq 2002; Weisblum 1995; Alison J Baylay, Piddock, and Webber 2019). The first erythromycin-resistant Streptococci strains were reported in the United Kingdom in 1959 (Lowbury and Hurst 1959).

(iii) Complete replacement or bypass of the target site: Some bacteria employ an alternative strategy to carry out the metabolic activities that are inhibited by the introduction of drug even if the drug has been bound to the target. This can be achieved by the production of an alternative enzyme that is not inhibited by the antibiotic and can carry out the otherwise blocked metabolic processes (Alison J Baylay, Piddock, and Webber 2019). Further, the overproduction of the antibiotic target enzyme can increase the MIC value of given antibiotic. (Alison J Baylay, Piddock, and Webber 2019) Few examples are demonstrated in Table 4.

Table 4. Replacement or bypass of the target molecule by another molecule or pathway

S. No.	Antibiotic	Mode of action	Resistant gene	Resistant mechanism	Example
1.	β-lactam antibiotics	Inhibition of penicillin binding proteins (PBPs) catalysed transpeptidase reaction (Davies, White, and Nicholas 2001)	*mecA*	Production of an alternative PBPs catalyzing transpeptidase reaction, itself not getting inhibited.	Methicillin-Resistant *S. aureus* (MRSA) (Ubukata et al. 1989)
2.	Sulfonamide	Inhibition of dihydropteroate synthase (DHPS) and hence folate synthesis (C.L. Smith and Powell 2000)	*dhps* genes	Acquire a drug resistant version of the relevant biosynthesis enzyme that allows production of folate despite the presence of the drug. (Sköld 2000)	*Streptococcus pyogens* (Swedberg, Ringertz, and Sköld 1998)
3.	Trimethoprim	Competitive inhibition of dihydrofolate reductase (DHFR) and hence folate synthesis (Brogden et al. 1982)	*dhfr*	Chromosomal mutations in promoter of *dhfr* leads to overproduction of antibiotic target. (D.R. Smith and Calvo 1982)	*Escherichia coli* (D.R. Smith and Calvo 1982)

CONCLUSION

The increasing cases of drug resistance in bacteria are a major concern. Although, it is a part of evolution and resistance against naturally occurring antimicrobials have been emerging among bacteria since millions of years. However, uncontrolled use of antibiotics in clinical practices, beekeeping, horticulture and genetic manipulations in the laboratories has aggravated the rise of multi-drug resistant bacteria. There are multiple genes and intricate molecular pathways involved in acquiring the resistance against any drug. Further, resistance genes can be easily transferred, horizontally and vertically, from one bacterium to another. Therefore, a detailed study of the mechanism of antibiotics and its resistance is required to overcome this problem. Further, continuous research is required to develop new antibiotics keeping in mind that resistance will develop against those also. A highly controlled use of these drugs is required to curb the spread of antibiotics among pathogens. Thus, awareness programs should be organized by the governments, healthcare facilities, etc. to spread knowledge about the harmful effects of overuse and 'without prescription intake' of antibiotics.

References

Abraham, Edward P; Ernst Chain. An enzyme from bacteria able to destroy penicillin. *Nature*., 1940, 146(3713), 837-837.

Aldred, Katie J; Robert J Kerns, Neil Osheroff. *Mechanism of quinolone action and resistance. Biochemistry*., 2014, 53(10), 1565-1574.

Alekshun, Michael N; Yang Soo Kim, Stuart B Levy. Mutational analysis of MarR, the negative regulator of marRAB expression in *Escherichia coli*, suggests the presence of two regions required for DNA binding. *Molecular microbiology*., 2000, 35(6), 1394-1404.

Alvarez-Ortega, Carolina, Jorge Olivares, José L Martínez. RND multidrug efflux pumps: what are they good for? *Frontiers in microbiology*., 2013, 4, 7.

Ayer, Vanessa, Wezenet Tewodros, Anand Manoharan, Sini Skariah, Feng Luo, Debra E. Bessen. Tetracycline Resistance in Group A Streptococci: Emergence on a Global Scale and Influence on Multiple-Drug Resistance. *Antimicrobial Agents and Chemotherapy*., 2007, 51(5), 1865-1868.

Azucena, Eduardo; Shahriar, Mobashery. Aminoglycoside-modifying enzymes: mechanisms of catalytic processes and inhibition. *Drug Resistance Updates*., 2001, 4 (2), 106-117.

Baranova, Natalya N; Antoine Danchin, Alexander A Neyfakh. Mta, a global MerR-type regulator of the *Bacillus subtilis* multidrug-efflux transporters. *Molecular microbiology*., 1999, 31(5), 1549-1559.

Baugh, Stephanie, Aruna S Ekanayaka, Laura, JV Piddock; Mark, A Webber. Loss of or inhibition of all multidrug resistance efflux pumps of *Salmonella enterica* serovar *Typhimurium* results in impaired ability to form a biofilm. *Journal of Antimicrobial Chemotherapy*., 2012, 67(10), 2409-2417.

Baylay, Alison J; Laura, JV Piddock; Mark, A Webber. Molecular Mechanisms of Antibiotic Resistance–Part I. *Bacterial Resistance to Antibiotics: From Molecules to Man* (eds. B.B. Bonev and N.M. Brown), 2019, 1-26.

Baylay, Alison J; Alasdair Ivens, Laura, JV. Piddock. A novel gene amplification causes upregulation of the PatAB ABC transporter and fluoroquinolone resistance in *Streptococcus pneumoniae. Antimicrobial agents and chemotherapy.*, 2015, 59(6), 3098-3108.

Blair, Jessica MA; Mark A Webber, Alison J Baylay, David O Ogbolu, Laura JV Piddock. Molecular mechanisms of antibiotic resistance. *Nature reviews microbiology.*, 2015, 13(1), 42-51.

Boutoille, David, Stéphane Corvec, Nathalie Caroff, Cécile Giraudeau, Eric Espaze, Jocelyne Caillon et al. Detection of an IS21 insertion sequence in the mexR gene of *Pseudomonas aeruginosa* increasing β-lactam resistance. *FEMS microbiology letters.*, 2004, 230(1), 143-146.

Brogden, RN; Carmine, AA; Heel, RC; Speight, TM; Avery, GS. Trimethoprim: a review of its antibacterial activity, pharmacokinetics and therapeutic use in urinary tract infections. *Drugs.*, 1982, 23(6), 405-430.

Brown, Melissa H; Ronald, A Skurray. Staphylococcal multidrug efflux protein QacA. *Journal of molecular microbiology and biotechnology.*, 2001, 3(2), 163-170.

Bush, Karen. Proliferation and significance of clinically relevant β-lactamases. *Annals of the New York Academy of Sciences.*, 2013, 1277(1), 84-90.

Cag, Yasemin, Hulya Caskurlu, Yanyan Fan, Bin Cao, Haluk Vahaboglu. Resistance mechanisms. *Annals of translational medicine.*, 2016, 4(17), 326-326.

Campbell, Elizabeth A, Nataliya Korzheva, Arkady Mustaev, Katsuhiko Murakami, Satish Nair, Alex Goldfarb et al. Structural mechanism for rifampicin inhibition of bacterial RNA polymerase. *Cell.*, 2001, 104(6), 901-912.

Chopra, I; Roberts, M. Tetracycline antibiotics: mode of action, applications, molecular biology, and epidemiology of bacterial resistance. *Microbiology and molecular biology reviews: MMBR.*, 2001, 65(2), 232-260.

Chuanchuen, Rungtip, RoxAnn R Karkhoff-Schweizer, Herbert P Schweizer. High-level triclosan resistance in *Pseudomonas aeruginosa* is solely a result of efflux. *American journal of infection control.*, 2003, 31(2), 124-127.

Connell, Sean R; Dobryan M Tracz, Knud H Nierhaus, Diane E Taylor. Ribosomal protection proteins and their mechanism of tetracycline resistance. *Antimicrobial agents and chemotherapy.*, 2003, 47(12), 3675-3681.

Cui, Longzhu, Akira Iwamoto, Jian-Qi Lian, Hui-min Neoh, Toshiki Maruyama, Yataro Horikawa et al. Novel mechanism of antibiotic resistance originating in vancomycin-intermediate *Staphylococcus aureus. Antimicrobial agents and chemotherapy.*, 2006, 50(2), 428-438.

D'Costa, Vanessa M, Christine E King, Lindsay Kalan, Mariya Morar, Wilson WL Sung et al. Antibiotic resistance is ancient. *Nature.*, 2011, 477 (7365), 457-461.

Davies, Christopher, Stephen W White, Robert A Nicholas. Crystal structure of a deacylation-defective mutant of penicillin-binding protein 5 at 2.3-Å resolution. *Journal of Biological Chemistry.*, 2001, 276(1), 616-623.

Delcour, Anne H. Outer membrane permeability and antibiotic resistance. *Biochimica et biophysica acta.*, 2009, 1794(5), 808-816.

Dolejska, Monika, Laura Villa, Laurent Poirel, Patrice Nordmann, Alessandra Carattoli. Complete sequencing of an IncHI1 plasmid encoding the carbapenemase NDM-1, the ArmA 16S RNA methylase and a resistance–nodulation–cell division/multidrug efflux pump. *Journal of Antimicrobial Chemotherapy.*, 2012, 68(1), 34-39.

Doménech-Sánchez, Antonio, Santiago Hernández-Allés, Luis Martínez-Martínez, Vicente J Benedí, Sebastián Albertí. Identification and characterization of a new porin gene of *Klebsiella pneumoniae*, its role in β-lactam antibiotic resistance." *Journal of bacteriology.*, 1999, 181(9), 2726-2732.

Dönhöfer, Alexandra, Sibylle Franckenberg, Stephan Wickles, Otto Berninghausen, Roland Beckmann, and Daniel N Wilson. Structural basis for TetM-mediated tetracycline resistance. *Proceedings of the National Academy of Sciences.*, 2012, 109 (42), 16900-16905.

Dookie, Navisha, Santhuri Rambaran, Nesri Padayatchi, Sharana Mahomed, Kogieleum Naidoo. Evolution of drug resistance in Mycobacterium tuberculosis: a review on the molecular determinants of resistance and implications for personalized care. *Journal of Antimicrobial Chemotherapy.*, 2018, 73(5), 1138-1151.

Exner, Martin, Sanjay Bhattacharya, Bärbel Christiansen, Jürgen Gebel, Peter Goroncy-Bermes, Philippe Hartemann, et al. Antibiotic resistance: What is so special about multidrug-resistant Gram-negative bacteria? *GMS hygiene and infection control.*, 2017, 12, Doc05.

Fernández, Lucía, Robert, EW Hancock. Adaptive and mutational resistance: role of porins and efflux pumps in drug resistance. *Clinical microbiology reviews.*, 2012, 25(4), 661-681.

Floss, Heinz G; Tin-Wein Yu. Rifamycin mode of action, resistance, and biosynthesis. *Chemical reviews.*, 2005, 105(2), 621-632.

Floyd, Jody L; Kenneth P Smith, Sanath H Kumar, Jared T Floyd, Manuel F Varela. LmrS is a multidrug efflux pump of the major facilitator superfamily from *Staphylococcus aureus. Antimicrobial agents and chemotherapy.*, 2010, 54(12), 5406-5412.

Ghai, Ishan, Shashank Ghai. Understanding antibiotic resistance via outer membrane permeability. *Infection and drug resistance.*, 2018, 11, 523-530.

He, Gui-Xin, Teruo Kuroda, Takehiko Mima, Yuji Morita, Tohru Mizushima, Tomofusa Tsuchiya. An H^+-coupled multidrug efflux pump, PmpM, a member of the MATE family of transporters, from *Pseudomonas aeruginosa. Journal of bacteriology.*, 2004, 186(1), 262-265.

Hooper, David C. Fluoroquinolone resistance among Gram-positive cocci. *The Lancet infectious diseases.*, 2002, 2(9), 530-538.

Hrabák, J. [Clinically important beta-lactamases of gram-negative bacteria: extended-spectrum beta-lactamases (ESBL)]. *Epidemiologie, mikrobiologie, imunologie: casopis Spolecnosti pro epidemiologii a mikrobiologii Ceske lekarske spolecnosti J.E. Purkyne.*, 2007, 56(3), 103-111.

Huda, Md Nazmul, Jing Chen, Yuji Morita, Teruo Kuroda, Tohru Mizushima, Tomofusa Tsuchiya. Gene cloning and characterization of VcrM, a Na+-coupled multidrug efflux pump, from Vibrio cholerae non-O1. *Microbiology and immunology.*, 2003, 47(6), 419-427.

Johnson, Alan P; Neil Woodford. Global spread of antibiotic resistance: the example of New Delhi metallo-β-lactamase (NDM)-mediated carbapenem resistance. *Journal of medical microbiology.*, 2013, 62(4), 499-513.

Kaur, Parjit, Elizabeth Peterson. Antibiotic resistance mechanisms in bacteria: relationships between resistance determinants of antibiotic producers, environmental bacteria, and clinical pathogens. *Frontiers in Microbiology.*, 2018, 9, 2928.

Khan, Asad U; Lubna Maryam, Raffaele Zarrilli. Structure, Genetics and Worldwide Spread of New Delhi Metallo-β-lactamase (NDM): a threat to public health. *BMC microbiology.*, 2017, 17(1), 101-101.

Kojima, Seiji, Hiroshi Nikaido. Permeation rates of penicillins indicate that *Escherichia coli* porins function principally as nonspecific channels. *Proceedings of the National Academy of Sciences*, 2013, 110(28), E2629-E2634.

Kong, Kok-Fai, Lisa Schneper, Kalai Mathee. Beta-lactam antibiotics: from antibiosis to resistance and bacteriology. *APMIS: acta pathologica, microbiologica, et immunologica Scandinavica.*, 2010, 118(1), 1-36.

Krause, Kevin M; Alisa W Serio, Timothy R Kane, Lynn E Connolly. Aminoglycosides: an overview. *Cold Spring Harbor perspectives in medicine.*, 2016, 6(6), a027029.

Lakhundi, Sahreena, Kunyan Zhang. Methicillin-resistant *Staphylococcus aureus*: molecular characterization, evolution, and epidemiology. *Clinical Microbiology Reviews.*, 2018, 31(4), e00020-18.

Leclercq, Roland. Mechanisms of resistance to macrolides and lincosamides: nature of the resistance elements and their clinical implications. *Clinical Infectious Diseases*, 2002, 34(4), 482-492.

Li, Wen, Gemma C Atkinson, Nehal S Thakor, Ülar Allas, Chuao-chao Lu, Kwok-Yan Chan et al. Mechanism of tetracycline resistance by ribosomal protection protein Tet (O). *Nature communications.*, 2013, 4, 1477.

Locher, Kaspar P. Structure and mechanism of ATP-binding cassette transporters. *Philosophical Transactions of the Royal Society B: Biological Sciences.*, 2008, 364 (1514), 239-245.

Lowbury, EJL; Hurst, L. The sensitivity of staphylococci and other wound bacteria to erythromycin, oleandomycin, and spiramycin. *Journal of clinical pathology.*, 1959, 12(2), 163-169.

Lu, Min. Structures of multidrug and toxic compound extrusion transporters and their mechanistic implications. *Channels (Austin, Tex.).*, 2016, 10(2), 88-100.

Lynch, III; Joseph, P; Nina M Clark, George G Zhanel. Evolution of antimicrobial resistance among Enterobacteriaceae (focus on extended spectrum β-lactamases and carbapenemases). *Expert opinion on pharmacotherapy.*, 2013, 14(2), 199-210.

Margaret, Barbara S; Drusano, GL; Harold, C Standiford. Emergence of resistance to carbapenem antibiotics in *Pseudomonas aeruginosa. Journal of Antimicrobial Chemotherapy.*, 1989, 24 (suppl_A), 161-167.

Marne, Gaynor, Mankin Alexander, S. Macrolide Antibiotics: Binding Site, Mechanism of Action, Resistance. *Current Topics in Medicinal Chemistry.*, 2003, 3(9), 949-960.

Martínez-Martínez, Luis, Alvaro Pascual, George A Jacoby. Quinolone resistance from a transferable plasmid. *The Lancet.*, 1998, 351(9105), 797-799.

McAleese, Fionnuala, Peter Petersen, Alexey Ruzin, Paul M. Dunman, Ellen Murphy, Steven J. Projan, Patricia A. Bradford. A novel MATE family efflux pump contributes to the reduced susceptibility of laboratory-derived *Staphylococcus aureus* mutants to tigecycline. *Antimicrobial Agents and Chemotherapy.*, 2005, 49(5), 1865-1871.

Mishra, Nagendra N; Soo-Jin Yang, Liang Chen, Claudette Muller, Azzam Saleh-Mghir, Sebastian Kuhn, et al. Emergence of daptomycin resistance in daptomycin-naive rabbits with methicillin-resistant *Staphylococcus aureus* prosthetic joint infection is associated with resistance to host defense cationic peptides and mprF polymorphisms. *PLoS One.*, 2013, 8(8), e71151.

Mojica, Maria F; Robert A. Bonomo, Walter Fast. B1-Metallo-β-Lactamases: Where Do We Stand? *Current drug targets.*, 2016, 17(9), 1029-1050.

Moussatova, Anastassiia, Christian Kandt, Megan L. O'Mara, Peter Tieleman, D. ATP-binding cassette transporters in *Escherichia coli. Biochimica et Biophysica Acta (BBA) – Biomembranes.*, 2008, 1778(9), 1757-1771.

Munita, Jose M; Cesar, A Arias. Mechanisms of antibiotic resistance." *Microbiology spectrum.*, 2016, 4(2), 481-511.

Nikaido, Hiroshi. Molecular basis of bacterial outer membrane permeability revisited. *Microbiol. Mol. Biol. Rev.*, 2003, 67(4), 593-656.

Nikaido, Hiroshi. Multidrug resistance in bacteria. *Annual review of biochemistry.*, 2009, 78, 119-146.

Paterson, David L; Robert, A Bonomo. Extended-spectrum β-lactamases: a clinical update. *Clinical microbiology reviews.*, 2005, 18(4), 657-686.

Paulsen, IT; Brown, MH; Skurray, RA. Proton-dependent multidrug efflux systems. *Microbiological reviews.*, 1996, 60(4), 575-608.

Paulsen, Ian T, Melissa H Brown, Sarah J Dunstan, Ronald A Skurray. Molecular characterization of the staphylococcal multidrug resistance export protein QacC. *Journal of Bacteriology.*, 1995, 177(10), 2827-2833.

Peterson, Elizabeth, Parjit Kaur. Antibiotic Resistance Mechanisms in Bacteria: Relationships Between Resistance Determinants of Antibiotic Producers, Environmental Bacteria, and Clinical Pathogens. *Frontiers in microbiology.*, 2018, 9, 2928.

Piddock, Laura JV. Multidrug-resistance efflux pumps? Not just for resistance. *Nature Reviews Microbiology.*, 2006a, 4(8), 629-636.

Piddock, Laura JV. Clinically relevant chromosomally encoded multidrug resistance efflux pumps in bacteria. *Clinical microbiology reviews.*, 2006b, 19(2), 382-402.

Pos, Klaas M. Drug transport mechanism of the AcrB efflux pump. *Biochimica et Biophysica Acta (BBA) - Proteins and Proteomics.*, 2009, 1794(5), 782-793.

Rammelkamp, Charles H; Thelma Maxon. Resistance of Staphylococcus aureus to the Action of Penicillin. *Proceedings of the Society for Experimental Biology and Medicine.*, 1942, 51(3), 386-389.

Randall, Christopher P; Katherine R Mariner, Ian Chopra, Alex J O'Neill. The target of daptomycin is absent from *Escherichia coli* and other gram-negative pathogens. *Antimicrobial agents and chemotherapy.*, 2013, 57(1), 637-639.

Redgrave, Liam S; Sam B Sutton, Mark A Webber, Laura JV Piddock. Fluoroquinolone resistance: mechanisms, impact on bacteria, and role in evolutionary success. *Trends in microbiology.*, 2014, 22(8), 438-445.

Rodríguez-Martínez, José Manuel, María Eliecer Cano, Carmen Velasco, Luis Martínez-Martínez, Álvaro Pascual. Plasmid-mediated quinolone resistance: an update. *Journal of Infection and Chemotherapy.*, 2011, 17(2), 149-182.

Rouquette-Loughlin, Corinne E, Jacqueline T Balthazar, Stuart A Hill William M Shafer. Modulation of the mtrCDE-encoded efflux pump gene complex of Neisseria meningitidis due to a Correia element insertion sequence. *Molecular microbiology.*, 2004, 54(3), 731-741.

Shaw, William V; Daniel G Brenner, Stuart FJ LeGrice, Sarah E Skinner, Alastair R Hawkins. Chloramphenicol acetyltransferase gene of staphylococcal plasmid pC221: nucleotide sequence analysis and expression studies. *FEBS letters.*, 1985, 179 (1), 101-106.

Sköld, Ola. Sulfonamide resistance: mechanisms and trends. *Drug Resistance Updates.*, 2000, 3(3), 155-160.

Smith, CA; Baker, EN. Aminoglycoside Antibiotic Resistance by Enzymatic Deactivation. *Current Drug Targets - Infectious Disorders.*, 2002, 2(2), 143-160.

Smith, Connie L; Keith, R. Powell. Review of the Sulfonamides and Trimethoprim. *Pediatrics in Review.*, 2000, 21(11), 368-371.

Smith, Douglas R; Joseph, M Calvo. Nucleotide sequence of dihydrofolate reductase genes from trimethoprim-resistant mutants of *Escherichia coli*. *Molecular and General Genetics MGG.*, 1982, 187 (1), 72-78.

Sugawara, Etsuko, Ekaterina M Nestorovich, Sergey M Bezrukov, Hiroshi Nikaido. *Pseudomonas aeruginosa* porin OprF exists in two different conformations. *Journal of Biological Chemistry.*, 2006, 281(24), 16220-16229.

Swedberg, Göte, Signe Ringertz, Ola Sköld. Sulfonamide resistance in Streptococcus pyogenes is associated with differences in the amino acid sequence of its chromosomal dihydropteroate synthase. *Antimicrobial Agents and Chemotherapy.*, 1998, 42(5), 1062-1067.

Tamber, Sandeep, Hancock, RE. On the mechanism of solute uptake in Pseudomonas. *Front Biosci.*, 2003, 8, s472-s483.

Tran, Que-Tien, Sarah Williams, Ramy Farid, Gül Erdemli, Robert Pearlstein. The translocation kinetics of antibiotics through porin OmpC: Insights from structure-based solvation mapping using WaterMap. *Proteins: Structure, Function, and Bioinformatics.*, 2013, 81(2), 291-299.

Tsuchido, Tetsuaki, Mitsuo Takano. Sensitization by heat treatment of *Escherichia coli* K-12 cells to hydrophobic antibacterial compounds. *Antimicrobial agents and chemotherapy.*, 1988, 32(11), 1680-1683.

Ubukata, K; Itoh-Yamashita, N; Konno, M. Cloning and expression of the norA gene for fluoroquinolone resistance in *Staphylococcus aureus. Antimicrobial agents and chemotherapy.*, 1989, 33(9), 1535-1539.

Ubukata, K; Nonoguchi, R; Matsuhashi, M; Konno, M. Expression and inducibility in Staphylococcus aureus of the mecA gene, which encodes a methicillin-resistant S. aureus-specific penicillin-binding protein. *Journal of Bacteriology.*, 1989, 171(5), 2882-2885.

Vargiu, Attilio V; Hiroshi Nikaido. Multidrug binding properties of the AcrB efflux pump characterized by molecular dynamics simulations. *Proceedings of the National Academy of Sciences.*, 2012, 109(50), 20637-20642.

Ventola, C Lee. "The antibiotic resistance crisis: part 1: causes and threats." *Pharmacy and therapeutics*, 2015, 40 (4), 277-283.

Viswanathan, Rajlakshmi, Arun Kumarendu Singh, Sulagna Basu, Suparna Chatterjee, Syamal Sardar, David Isaacs. Multi-drug resistant gram negative bacilli causing early neonatal sepsis in India. *Archives of Disease in Childhood-Fetal and Neonatal Edition.*, 2012, 97(3), F182-F187.

Voulgari, Evangelia, Aggeliki Poulou, Vasiliki Koumaki, Athanassios Tsakris. Carbapenemase-producing Enterobacteriaceae: now that the storm is finally here, how will timely detection help us fight back? *Future microbiology.*, 2013, 8(1), 27-39.

Walsh, Christopher. Molecular mechanisms that confer antibacterial drug resistance. *Nature.*, 2000, 406(6797), 775-781.

Wang, Aihua, Yonghong Yang, Quan Lu, Yi Wang, Yuan Chen, Li Deng; et al. Presence of qnr gene in *Escherichia coli* and Klebsiella pneumoniae resistant to ciprofloxacin isolated from pediatric patients in China. *BMC infectious diseases.*, 2008, 8, 68-68.

Webber, Mark A; Andrew M Bailey, Jessica MA Blair, Eirwen Morgan, Mark P Stevens, Jay CD Hinton et al. The global consequence of disruption of the AcrAB-TolC efflux pump in Salmonella enterica includes reduced expression of SPI-1 and other attributes required to infect the host. *Journal of bacteriology.*, 2009, 191(13), 4276-4285.

Webber, Mark A, Ashraf Talukder, Laura JV Piddock. Contribution of mutation at amino acid 45 of AcrR to acrB expression and ciprofloxacin resistance in clinical and veterinary *Escherichia coli* isolates. *Antimicrobial agents and chemotherapy.*, 2005, 49(10), 4390-4392.

Wieczorek, Kinga, Jacek Osek. Antimicrobial resistance mechanisms among Campylobacter. *BioMed research international.*, 2013, 2013, 340605.

Wright, Gerard D. Molecular mechanisms of antibiotic resistance. *Chemical communications.*, 2011, 47(14), 4055-4061.

Xu, Ming, Yan Ning Zhou, Beth P. Goldstein, Ding Jun Jin. Cross-resistance of *Escherichia coli* RNA polymerases conferring rifampin resistance to different antibiotics. *Journal of bacteriology.*, 2005, 187(8), 2783-2792.

Yerushalmi, Hagit, Mario Lebendiker, Shimon Schuldiner. EmrE, an *Escherichia coli* 12-kDa Multidrug Transporter, Exchanges Toxic Cations and H^+ and Is Soluble in Organic Solvents. *Journal of Biological Chemistry.*, 1995, 270(12), 6856-6863.

Zaman, Sojib Bin, Muhammed Awlad Hussain, Rachel Nye, Varshil Mehta, Kazi Taib Mamun, Naznin Hossain. A Review on Antibiotic Resistance: Alarm Bells are Ringing. *Cureus.*, 2017, 9(6), e1403-e1403.

Zhang, Xuejun C; Yan Zhao, Jie Heng; Daohua Jiang. Energy coupling mechanisms of MFS transporters. *Protein science: a publication of the Protein Society.*, 2015, 24(10), 1560-1579.

Zhu, Lei, Jinshui Lin, Jincheng Ma, John E Cronan; Haihong, Wang. Triclosan resistance of *Pseudomonas aeruginosa* PAO1 is due to FabV, a triclosan-resistant enoyl-acyl carrier protein reductase. *Antimicrobial agents and chemotherapy.*, 2010, 54(2), 689-698.

Zolnerciks, Joseph K; Edward, J. Andress; Michael, Nicolaou; Kenneth, J. Linton. Structure of ABC transporters. *Essays in Biochemistry.*, 2011, 50, 43-61.

In: Antimicrobial Resistance: Opportunities and Challenges ISBN: 978-1-53617-943-9
Editors: P. Kumar Prabhakar and V. Mishra

Chapter 3

INFECTIOUS PATHOGENS, PATHOGENESIS, ANTIMICROBIAL THERAPY, ANTIMICROBIAL RESISTANCE, AND SURVEILLANCE MEASURES

B. Premkumar[1,*], Ali Zarrabi[2], Venkatesan Perumal[3], Sohail Akther[4], Abimanyu Sugumaran[5], Pranav Kumar Prabhakar[6,†] and Ponnurengam Malliappan Sivakumar[7,‡]

[1]Department of Pharmacology, K. K College of Pharmacy, Gerugambakkam, Chennai, Tamilnadu, India
[2]Sabanci University Nanotechnology Research and Application Center (SUNUM), Istanbul, Turkey
[3]Rangel College of Pharmacy, Health Science Centre, Texas A&M University, Texas, US
[4]Department of New product Development, Combination product and Devices, Global R&D, Teva Pharmaceuticals, Runcorn, UK
[5]Department of Pharmaceutics, SRM College of Pharmacy, SRM Institute of Science and Technology, Kattankulathur, India
[6]Department of Transdisciplinary Research, Lovely Professional University, Phagwara, Punjab, India
[7]Center for Molecular Biology, Institute of Research and Development, Duy Tan University, Da Nang, Vietnam

[*] Corresponding Author's E-mail: cologyprem@gmail.com.
[†] Corresponding Author's E-mail: prabhakar.iitm@gmail.com.
[‡] Corresponding Author's E-mail: sivamedchem@gmail.com.

Abstract

Antimicrobials are drugs produced by microbes or natural substances such as antibodies, or those produced through synthetic route, that cause the inhibition of microbial growth or kill the microbes at a lower concentration. A significant public health problem that healthcare providers face from dawn to dusk is antibiotic resistance. The mechanisms of antibacterial include inhibition of cell wall synthesis or leakage of cell membranes, inhibition of protein synthesis, or interference with DNA function or intermediary metabolism, and causing misreading of mRNA to affect the permeability. Antimicrobial resistance (AMR) is a worldwide public health challenge. A sound understanding of the pathogens, modes of transmission, pathogenic picture, clinical manifestations, and the emergence of resistance are necessary for the formulation of strategies to address it. The nature of different microbes such as bacteria, fungus, protozoa, and viruses are described with a note on the resistant mechanisms. To address the AMR, several surveillance measures and awareness programs are launched in India. Even though it is a difficult task to reach the broader population, technological advancements in diagnosis, practical implementation, and monitoring the treatment outcomes have reached greater heights in the mitigation of AMR.

Keywords: antimicrobial, resistance, infection, pathogen, bacteria, fungus, protozoa, virus

1. Introduction

The deadliest infectious diseases of the 21st century are cholera, plague, dengue, chikungunya, leishmaniasis, hepatitis, which are primitive diseases, and new entrants like *Ebola* virus, *Zika* virus, *Nipah* virus, etc., which are caused by different microbes. Medically important pathogenic microbes are bacteria, fungi, viruses, and parasites, causing infectious diseases in humans, whereas human microbiomes are the host microbes that are non-pathogenic and symbiotic in healthy individuals. Antimicrobials are drugs produced by microbes or natural substances such as antibodies, or those produced through synthetic route, causing inhibition of microbial growth or kill the microbes at a lower concentration. A significant public health problem that healthcare providers face from dawn to dusk is antibiotic resistance.

Antimicrobial Resistance (AMR) is a natural phenomenon occurring in the microbes, which are initiated by acquiring resistant genetic material from other microorganisms, is known as superbugs. The logical reasoning for this condition is that the antimicrobials target the susceptible strains only, and the resistant ones multiply, and transfer the resistance to other microbes, that still aggravates the clinical conditions and cause treatment complications. The AMR is distinctly different from resistance occurring in non-infectious diseases like cancer that do not transfer the resistance. So, strict surveillance programs, reduction in inappropriate use of antibiotics, and new drug discoveries could address this issue. With this view, different sections such as national healthcare programs, research in understanding the genotyping changes occurring at the molecular levels, and education to healthcare professionals incautious use of antimicrobials and surveillance at societal uses such as agricultural applications are expected to think for better control of this resistance. The current chapter delineates the essential information about emergence, transfer, molecular genetics, and other important information related to AMR.

AMR may be acquired, disseminated, or those were having intrinsic resistance. In the case of acquired resistance that occurs by mutation, the selection of antibiotics should be such that they aim at the mutants. In the case of dissemination, prevention of spread is the solution, and in the case of a virus lacking intrinsic activity, vaccine developments prove to be beneficial. Study of minimum inhibitory concentration (MIC), minimum bacterial concentration (MBC), the minimum effective concentration of the antibiotic, bacterial persisters, and quorum sensing (QS) are important laboratory tools that guide the prognosis of antimicrobial therapy. In some pathogens formation of biofilms occurring at surfaces of the blood vessels, teeth, and catheters, provide a tough barrier for the antimicrobial action compared to those pathogens without them. Bacteriostatic antimicrobials halt the growth of cell division and rely on the host immune system to kill the static microbes. Bactericidal antimicrobials are effective, but some pathogens may release toxic products on cell lysis and may pose the risk of allergic reactions. Another critical factor is the spectrum of antimicrobials, narrow-spectrum drugs are profitable in susceptible population, and broad-spectrum antimicrobials are effective for empirical therapies. Nonetheless, the broad-spectrum antimicrobials on chronic use may lead to superinfections.

The mechanisms of antibacterials include inhibition of cell wall synthesis or leakage of cell membranes, inhibition of protein synthesis, or interference with DNA function or intermediary metabolism, and causing misreading of mRNA to affect the permeability. The antifungal antimicrobials cause alteration in fungal membrane permeability, block nucleic acid synthesis, disrupt microtubule function, or inhibit cell wall synthesis. The antiviral drugs may block the viral fusion, prevent viral uncoating, prevent the release of the virus, viral packaging blockade, or their integration with the host DNA, or their transcription or inhibit reverse transcription. The antiparasitic antimicrobials also share the above-mentioned mechanisms.

The MIC ranges for each antibiotic, and the corresponding microbes are interpreted by the Clinical Laboratory Standards Institute (CLSI) and European Committee for Antimicrobial Susceptibility Testing (EUCAST) guidelines. It is easy to differentiate the susceptible wild-type strains from resistant ones. In contrast the borderline of susceptibility is tough in case of the presence of moderate strains in the same colony. The other important measures to counter antibiotic resistance are measurement of the area under the curve (AUC24), C_{max}, pharmacokinetic and pharmacodynamic (PK/PD) indices, and bioavailability (Kahlmeter and Brown, 2004).

Clinical resistance may arise in individuals exposed to the antimicrobial, and the testing of the patient isolates is decisive. The mutations occur in the microbes spontaneously due to changes in the nucleotide sequences, may be harmful or constructive to the microbe. The constructive changes at the deoxyribonucleic acid (DNA) that happen to encounter the antimicrobials are mutations that impair the drug binding target sites, excessive production of efflux-pumps that reduce the drug penetration, and up-regulation of production of drug-metabolizing enzymes. In susceptibility testing, the ratio of mutant colonies and total colony count provides data regarding the frequency of the mutation.

The drug-related factors that chaperon drug treatment and resistance are doses, dosing frequency, and duration of the treatment. The microbe related factors conferring resistance are single or multiple mutations, transfer to different patients, and resistance development in step-wise up-hill fashion. In some cases, instead of up-hill, one-step mutation is enough to afford complete protection to the microbes, such as Mycobacterium tuberculosis. Even if the MIC is maintained above mutant protective concentration (MPC), mutation may occur in the second step and still lead to resistance. The therapeutic window hypothesis is the concentration of the antimicrobial between MIC and MPC, and keeping the antimicrobial concentration within this, prevents the emergence of new resistance that occurs as the second step, and reduces the adverse effects also (Drlica and Zhao, 2007).

AMR due to induction is, *Lex*A protein controlling SOS genes, when damaged by fluoroquinolone antibiotic pressure, generates new stretches of DNA, which confer resistance to growing cells, which are non-mutant, but get induced mutations. Such induced mutations could be handled by high dose regimens (Cirz and Romeshberg, 2005). Efflux pumps can be specific or non-specific for a particular antimicrobial, contributing resistance. *Pseudomonas aeruginosa* exhibits many efflux pumps, and the most widely studied ones are *Mex*AB-*Opr*M, *Mex*CD-*Opr*J, *Mex*EF-*Opr*N, and *Mex*XY-*Opr*M (Poole, 2005).

The other important factor contributing to AMR is the development of proteins that break the pharmacophore. The typical examples are β-lactamases, which are of four types A, B, C, and D, and the mutations in these afford high degree of resistance, which occurs as Extended-Spectrum β-lactamases (ESBLs), which create serious illness (Rice, 2001). The next mechanism which causes AMR is the alteration in the protein binding targets, such as those mutations in DNA gyrase conferring fluoroquinolone resistance, and change in PBPs conferring penicillins and cephalosporins resistance. In some cases, the proteins other than the target binding could huddle resistance, such as *MfpA* and *Qnr* leading to fluoroquinolone resistance (Vetting et al., 2011).

In the case of viral diseases, the AMR is still more complicated because the virus develops more frequently and with more mutations. The emergence of resistant mutants in case of the virus, because the loss of fitness to one antimicrobial is easily encountered, and cyclic therapy with different antimicrobials is not advisable. To reduce the viral load and prevent the emergence of viral mutants, several combination strategies are followed with strict guidelines to face this situation. The transfer of resistant genetic material may be through bacterial plasmids, viral, or from the environment. The plasmids acquire the resistant genes from transposons and integrons, which may transfer from parent cell to the daughter cell through vertical transfer or from one microbe to another through horizontal transfer.

Horizontal transfer is resilient and occurs through conjugation, transduction, and transformation. Plasmids moving set of genes to bacteria that are not closely related is through conjugation. Bacteriophages induce transduction, and transformation is through the uptake of the released DNA by the new cell. The intracellular mechanisms causing DNA transfer is through, transposition, cassette integration, and recombination. The recombination may be homologous mechanism occurring through receipt of similar genetic sequences, and site-specific recombination occurs through special enzymes such as integrase and ligase.

Plasmids are molecular parasites, which contain their replication origin and protein replication gene. Plasmids are mostly circular, but linear forms also exist, may contain fewer base pairs ~3000 or more base pairs ~3000000, and have distinct mechanisms which their entry into daughter cells, when the parent cell divides. The plasmids are also important to ensure antitoxin resistance to the daughter cells, and so-called as addiction molecules. The plasmids are transferred from male cells to female cells by conjugation, e.g., F-plasmid of *Escherichia coli* (*E. coli*), which may contain *tra* genes, which confer AMR. If the conjugation is followed by mutations causing resistance, it can move to the chromosome of the bacteria. Although bacteriophages cause lysis of the bacterial cell, the inherent property of the virus to undergo mutation by transduction, exhibit resistance. Integrons present in both Gram-positive and Gram-negative organisms accumulate resistant gene sets that exhibit drug resistance to multiple drugs. Integrons have the gene *intI*, expressing integrase enzyme, and *attI* as insertion site, and a promoter region Pc, which together represent more than 70 multidrug-resistant gene cassettes. The *Salmonella enterica* is a food-borne pathogen associated with integrons, conferring multiple-drug resistance. Genomic islands are found in *Enterococcus faecalis*, which encode toxins, have a role in conjugation, and contribute to more resistance (Mazel, 2006).

2. Bacterial Infections and AMR

Acinetobacter is a Gram-negative or cocco-bacilli, anaerobe, which are normal flora in the respiratory tract and may cause hospitalized infections. The recent reports of the World Health Organization (WHO) highlight *Acinetobacter baumanii* as one of the Priority I critical pathogens. It is a non-fermenting bacterium causing multi-drug or even pan drug resistance. It is one of the ESKAPE pathogens causing nosocomial infections, and resistance is a major issue. The mechanisms of resistance include decreased permeability to antimicrobials and generation of AmpC-β-lactamases and the development of different efflux pumps leading to carbapenem resistance. It responds to sulphonamides, colistin, or carbapenems (Barman et al., 2019).

Bacillus anthracis a spore-forming Gram-positive microbe, non-motile, and spore-forming rods. It exists in three forms, such as gastrointestinal, cutaneous, pulmonary, and produces AB-type toxins, which causes severe edema. The resistance mechanism in this bacterium was found to be starvation of phosphates in the growing medium. Cutaneous anthrax develops from a papule without pain and may progress to vesicular papule with non-pitting edema. The other forms of anthrax are intestinal anthrax, which may lead to fatal diarrhoea with hemorrhage and pulmonary anthrax causing fatal pneumonia. The clinical characteristics are high fever, bacteremia, and the drug of choice is high dose of penicillin (Aggarwal et al., 2015).

Bacillus subtilis is a foodborne pathogen causing food poison due to the production of surfactin, which occurs as an opportunistic infection in immunocompromised patients. Conjugation mechanisms transfer the plasmid *pLS20* through horizontal transfer (Yadav et al., 2018).

Bacteroides fragilis is a Gram-negative anaerobe, which is a normal intestinal flora of the large intestine. The enterotoxigenic *B. fragilis* (ETBF) may cause peritonitis, intra-abdominal, or hepatic abscesses. The commonly used antimicrobials include cephalosporins, penicillins, sulphonamides, chloramphenicol, moxifloxacin, or metronidazole. The factors responsible for AMR were integrase enzymes *(intl1*, *intl2*), and the resistant genes were *cep (A)*, and *tet (Q)*. Recent reports showed the presence of new Bacteroides genomospecies, which is carbapenem and metronidazole resistant (Sarkar et al., 2015).

Brucella spp. is a zoonotic infection, spreading through infected cattle, pigs, and goats. It causes characteristic granulomas in the reticuloendothelial system, indulgent fever, which is treated by doxycycline with rifampicin or gentamicin. It is found that the drug resistance is through *Hisl* protein, revealed through computational docking and homology modeling methods (Singh et al., 2018).

Chlamydia trachomatis is a small, obligate intracellular bacteria, having a dimorphic life cycle, and they lack peptidoglycan or cell wall. It infects the eyes, causing trachoma, urogenital infection as urethritis, genital and perinatal infections, and lymphogranuloma venereum. The drugs used for treatment are tetracyclines, macrolides, or chloramphenicol, and studies have confirmed that AMR is uncommon (Deguchi et al., 2018).

Citrobacter is present as normal flora of the bowel and has less virulence. Although clinical syndromes are not common, it can cause nosocomial infections in immunocompromised patients. Quinolones, aminoglycosides, and third-generation cephalosporins are drug treatments. The resistance produced by the bacteria is through the production of ESBL, which causes multidrug-resistance (MDR), and the resistant gene was found to be $bla_{CTX\text{-}M}$ gene tested for ceftazidime (Agarwal et al., 2014).

Clostridium difficile is a Gram-positive, spore-forming, anaerobic organism, causes antibiotic-associated diarrhoea. It produces two toxins enterotoxin (toxin A), causing secretory diarrhoea and cytotoxin (toxin B), and causing the destruction of the infected tissues. The diarrhoeal symptoms vary from severe loose stools to pseudomembranous colitis, and withdrawal of the antimicrobial is useful in mild cases. Metronidazole, vancomycin, or bacitracin are drugs used for moderate to severe cases. The MDR strain of the microbe was found to be *Clostridium difficile* strain 630, and a new drug Cadazolid is under research for such cases (Lohani et al., 2017).

Clostridium perfringens is a large spore-forming, Gram-positive obligatory anaerobe found in gut and female genital tract. It releases several enterotoxins in food poisoning diarrhoea, and necrotizing enteritis, and penicillin is used in the case of histotoxic cases. The drug-resistant strains were found to be cpa^{+}, $cpb2^{+}$, and cpe^{+} (Yadav et al., 2017).

Corynebacterium diphtheriae is a facultative anaerobic Gram-positive rod bacterium, and 'triple antigen' immunization programs have made the disease outbreaks rare. Disease surveillance programs still reveal outbreaks of diphtheria and resistance, the government of India programs such as '*Indra Dhanush*' are launched to ensure effective reach of immunization programs. It causes opportunistic infections, releases diphtheria exotoxin, and causes laryngitis, pharyngitis, and cutaneous diphtheria. Penicillin is effective in combating the disease, but recent studies highlight resistance ampicillin, chloramphenicol, macrolides, and tetracyclines (Dandinarasaiah et al., 2012).

Enterococcus faecalis is microbiome of intestinal, oral, and vaginal mucosa and opportunistic nosocomial pathogen causing urinary tract infections, contains conjugative transposons and integrative conjugative elements, the strain is *Tn916*. It is resistant to penicillin, ampicillin, and ciprofloxacin, but vancomycin since *vancomycin-resistant-enterococcus faecium* (VRE) have emerged. Linezolid was found to be effective in all resistant cases (Kim and Woo, 2016).

Escherichia coli (*E.coli*) is a Gram-negative bacterium, an important pathogen implicated in urinary, wound, lung, septic, and meningeal infections and traveller's diarrhoea. The clinical diseases of *enterotoxigenic-E. coli* (ETEC) are neonatal meningitis, urinary tract infections, intraperitoneal and lung infections. Gastroenteritis or dysentery is the more common form of ETEC. Ampicillin and cotrimoxazole are the drugs used in sensitive cases, whereas gentamicin is reliable, and newer cephalosporins and penicillins are preferable. It is tough to eradicate *E. coli* due to the formation of biofilms that aggregate and live as a community in solid surfaces. The biofilm bacterial cells communicate through QS, secrete virulence factors, and an essential factor for AMR. The biofilm genes in *E. coli* were found to be *csgD*, *hha*, *bcsA*, *pgaC*, and *fimB*, and several other genes are involved in QS. The other crucial mechanisms that contribute to AMR was found to be resistance-nodulation-division (RND) efflux pumps, and New Delhi Metalloprotein-1 (NDM-1) β lactamases, conferring MDR. (WHO, 2010; Sharma et al., 2016).

Helicobacter pylori is an important pathogen, previously classified as *campylobacter*, is a motile, microaerophilic Gram-negative rod, releasing urease, and causes chronic gastritis, gastric and duodenal ulcer. Chemotherapy includes three or four drug combinations with omeprazole, amoxicillin, clarithromycin or metronidazole, and bismuth subcitrate is involved in metronidazole resistance. Eradication of *H. pylori* is difficult in resistant cases leading to failure of therapy; recent research has featured huge variation in sensitivity patterns in different parts of our country (Pandya et al., 2014).

Hemophilus influenzae is an encapsulated organism, pleomorphic, Gram-negative, coccobacillus are normal flora of the respiratory tract. Only non-encapsulated strains cause respiratory diseases like bronchitis, otitis media, and the capsulated forms that cause meningitis and epiglottitis. In the past penicillins have a therapeutic role, but most strains now produce ESBL, challenging to treat, and hence third-generation cephalosporins, cotrimoxazole, and azithromycin, ciprofloxacin, and chloramphenicol are tried. Vaccines are available for *Haemophilus influenzae B*, for children. The resistant genes and virulence were found to be fimbrial gene *hifA* in non-typable *H. influenzae* (*NTHi*) (Saikia and Das, 2012).

Klebsiella pneumoniae is a non-motile, aerobic, non-motile, Gram-negative bacillus, capsulated, and microbiome of the gut. In opportunistic cases, it causes nosocomial pneumonia, which is due to the destruction of the alveolar spaces and cavities, and the production of bloody-sputum. It also causes skin and soft tissue infections, urinary tract infections (UTIs), liver abscess, endophthalmitis, and meningitis. It is one of the ESKAPE (*Enterococcus faecium*, *Staphylococcus aureus*, *Klebsiella pneumonia*, *Acinetobacter baumannii*, *Pseudomonas aeruginosa*, and *Enterobacter*) pathogen producing metallo-β-lactamases (AmpC-β-lactamase) and carbapenemases, and so gentamicin, ciprofloxacin, and third-generation cephalosporins are tried. The resistant strains were found to be *ST1310* and *ST721* (Singh et al., 2017).

Listeria monocytogenes is a facultative anaerobe, Gram-positive rods, causes endocarditis, meningitis, bacteraemia, in immunocompromised patients and children. It is an intracellular pathogen in macrophages and monocytes; the drugs used are ampicillin, cotrimoxazole, and gentamicin. In stress conditions due to antimicrobial pressure, the *RecA* gene gets induced, expresses antimicrobial-resistant proteins as save our souls (SOS) response (Ojha and Patil, 2019).

Neisseria gonorrhoea is a Gram-negative coccus, reside in the mucosal surfaces, invade, and cause inflammation. It leads to urethritis, salpingitis, pelvic inflammatory disease, epididymo-orchitis, and other disseminated infections. It is an important pathogen of sexually transmitted infections (STI). Penicillin G is still the drug of choice, while ciprofloxacin and ceftriaxone are preferred for *penicillinase-producing N. gonorrhea* (PPNG). The WHO has launched the Gonococcal Antimicrobial Surveillance Programme (WHO GASP) to analyze the AMR trends and to design new treatment guidelines and improve public health (Wi et al., 2017).

Neisseria meningitidis is a Gram-negative capsulated that coccus only in human hosts, and releases endotoxin. It grows in the nasopharynx, invades the blood, and then invades the meninges. It causes septicaemia and may lead to Waterhouse-Friderischen syndrome, and rarely it may cause pneumonia and arthritis. Penicillin, third-generation cephalosporins, and supportive measures are the treatment measures. The resistant strains of *N. gonorrhea* were *MenA* (*ST4789*) and *MenC*, which were found to be ciprofloxacin-resistant (Tsang et al., 2017).

Nocardia asteroids are Gram-positive microbe causing pulmonary pseudotuberculosis and brain abscess, and subcutaneous tissues in and immunocompromised patients. The antimicrobials used are cotrimoxazole, amikacin, imipenem, and cephalosporins. Although in vitro studies reveal AMR, clinical isolates were found to be susceptible to most cases (Sivaprakash et al., 2007).

Proteus mirabilis is aerobic, motile, non-spore forming, Gram-negative rod, causes community-acquired, and nosocomial urinary tract infections. *Proteus mirabilis* is treated with gentamicin, third-generation cephalosporins, or meropenem. Proteus mirabilis producing ESBL and AmpC β lactamases have already been explored. Recent studies revealed Catheter-associated urinary tract infections (CAUTI) infections with MDR strains of *Proteus mirabilis*, *HI4320* (Pawar et al., 2018).

Pseudomonas aeruginosa are Gram-negative rods, anaerobic, motile, and they are normal bowel flora of healthy people. *P. aerugonisa* causes lung infections, septicaemia, wounds, ear infections, and nosocomial infections. Gentamicin with either Antipseudomonal penicillins such as ticarcillin, or third-generation cephalosporins such as ceftazidime are the treatment options. The MDR strains were associated with efflux pumps, AmpC-β-lactamases, and resistant genes were *phoP*, and *phoQ* (Miryala et al., 2019).

Salmonella typhi is motile, aerobic, Gram-positive rod, that infect humans, birds, and animals. The serotypes were based on O, H, and K antigens used to test the epidemiological outbreaks. *S. typhi* causes localized gastroenteritis, bacteraemia, and systemic enteric fever called as typhoid. The drugs of choice were cotrimoxazole, ampicillin, ciprofloxacin, and chloramphenicol. Enteric fever is a significant health issue in India, and studies have highlighted MDR to the pharmacological interventions, the MDR proteins were *EmrA*, *EmrB*, *EmrD*, *MdtB*, *MdtG*, *MdtH*, *MdtK*, *MdtL*, and *TolC*. Cadmium accumulation and relation to antibiotic resistance in *Salmonella typhi Ty2*, was also demonstrated (Oo et al., 2019).

Shigella dysenteriae is a non-motile, Gram-negative rod that spread by contaminated food or water through the fecal-oral route. It produces an enterotoxin, which leads to diarrhoea due to the invasion of the colonic mucosa. It also has a neurotoxin, which causes meningeal symptoms. Now Shigella is resistant to most drugs cotrimoxazole, tetracycline, chloramphenicol, quinolones, and cephalosporins. There are around 15 serotypes, and plasmid-mediate resistances were *bla-CTX-M-1*, *bla-OXA1*, *bla-TEM*, *dhfr1A*, *aac(6')-lb-cr*, and *qrirS* (Muthuirulandi et al., 2016).

Staphylococcus is important Gram-positive coccus causing several diseases in humans, and the different species are *S. aureus*, *S. epididermis*, *S. pyogenes*, and *S. saprophyticus*. *S. aureus* causes local purulent infections such as carbuncles, impetigo, wound infections, sinusitis, otitis media, sepsis, and pneumonia. The toxin-related infections are food poisoning, dermatitis, and toxic shock syndrome. An important strain that is widely studied is *Methicillin-Resistant Staphylococcus aureus* (MRSA), which makes the clinical condition critical and is spread by personal contact through the living and non-living things. There are two types of MRSA infections such as community-acquired and hospital-acquired, CA-MRSA and HA-MRSA respectively. The neonates, children, emergency units in hospitals, and immunocompromised patients are vulnerable populations for MRSA infection. The CA-MRSA strains were found to be ST80, USA300, and ST39 mupirocin resistance encoded by *ile*S and high-resistance through *mup*A gene, which is plasmid-mediated. The CA-MRSA is susceptible to non-β-lactams, such as macrolides, fluoroquinolones, and clindamycin. In Indian population, the CA-MRSA genotypes were *ST22* (*SCCmec IV*), *ST672* (*SCCmec type V*), and *ST722* (*SCCmec V*), and HA-MRSA were *ST239* (*SCCmec type III*) (Sharma et al., 2017).

Streptococcus is Gram-positive, non-motile, anaerobic cocci, classified into α, β, and γ, and the antigenicity is due to the Lancefield antigen present their cell walls. The β-hemolytic *Streptococci* called *S. pyogenes* causes URTI, infections of subcutaneous, and connective tissues. It may cause necrotizing fasciitis, sepsis, or septic shock. The α-hemolytic streptococcus is called as *S. pneumoniae* causing infections of the respiratory tract. It is the most frequent pathogen causing community-acquired pneumonia (CAP), and the drug-resistant strains of *S. pneumonia* (CAP-DRSP), is a threat, limiting the therapeutic efficacy. The pneumococcal conjugate vaccines *PCV10* and *PCV13* are the serotypes suggested by research for immunization programes. The erythromycin-resistant strains of *S. pyogenes* were found to be *emm49* (Aliberti et al., 2019).

Treponema pallidum is syphilis (STI) causing bacteria, is a medically important spirochaete having a spiral shape, identified by serology and morphological characteristics, and not by Gram stain or culture. The non-specific antigens are VDRL, RPR, and the specific ones are TPHA. It is a venereal disease characterized by three stages, stage I causes painless local lymphadenitis, stage II is disseminated and leads to polylymphadenopathy, stage III is subdivided into neurosyphilis, cardiovascular syphilis, and gummatous syphilis. Penicillin G is still the drug of choice, and the resistance to antimicrobials was not reported (STAMM, 2014).

Vibrio cholera is a short, curve, facultative aerobe, Gram-positive rod, and the pathogenic strains were *O1*. The cholera toxin produces massive secretory diarrhoea, where fluid and electrolyte replacements are critical management and tetracycline, ampicillin or cotrimoxazole are the preferred drugs. Recent reports highlight the biofilm induced AMR in *O139* strain of *Vibrio cholera*, and the strain *443V* was found to be MDR and XDR (Gupta et al., 2018).

Mycobacterium tuberculosis is a chronic infection in humans and animals, causing tuberculosis (TB). It affects both healthy and immunocompromised persons. It is primarily a lung disease but can spread to other organs. The disease has different stages, the primary stage is asymptomatic with mild fever and malaise, progressive primary pulmonary tuberculosis leads to fever, cough, hemostypsis, and excessive diarrhoea, post-primary pulmonary tuberculosis leads to chronic cough, sputum with hemostypsis, anorexia, and progressive wasting, and the extrapulmonary TB stage leads to spread of the disease to lymph nodes, pleura, genitourinary tract, bone and joints, and meningeal TB. The confirmatory tests include Mantoux or Quantiferon test, acid-fast bacilli staining, and chest x-ray. About 1/3rd of the population is infected with TB, as per WHO statistics 2014, in India, there are 2.2 million cases, and about 600 people die every day, because of the disease. In 2012, the Government of India had declared TB as a notified disease. The treatment regimens are quadruple for new cases with Isoniazid (H), Rifampicin (R), Pyrazinamide (Z), and Ethambutol (E) for 2 months and continuation phase of triple regimen (HRE) for 3 months in newly diagnosed cases. For previously treated cases, four-drug combinations with Streptomycin (HRZES) for 2 months, 1-month HRZE, and continuation phase with HRZE for 5 months. In MDR-TB cases, the Revised National Tuberculosis Control Programme regimen is Kanamycin, Levofloxacin, Ethionamide, Cycloserine, Pyrazinamide, and Ethambutol combination for 6-9 months intensive phase, followed by levofloxacin, ethionamide, cycloserine and ethambutol combination for 18 months in the continuation phase. The XDR-TB contains 7 drug combination capreomycin, moxifloxacin, high dose isoniazid, paraaminosalycilic acid, clofazimine, linezolid, and amoxicillin-clavulanate for 6-12 months with 6 drug combination in the continuation phase for 18 months. Poor patient compliance is an important contributor to AMR. For uninterrupted chemotherapy in TB, directly observed therapy short-course (DOTS) launched by WHO in 1995, and '*The Stop TB strategy*' was launched in 2006 for addressing MDR-TB. Mycobacterium avium causes an atypical mycobacterial infection in immunocompromised patients, particularly in AIDS (Dua et al., 2018; Lange et al., 2018).

Mycobacterium leprae causes leprosy (Hansen's disease), which is a chronic infection of nerves and skin. The clinical features are erythema nodosum leprosum, nasal cartilage destruction, food drop, neuropathic ulcer, claw hand of the ulnar nerve, and local sensory loss. There clinical forms of leprosy are tuberculoid and lepromatous leprosy, whereas the WHO classifies it as paucibacillary and multibacillary types. The multibacillary types of leprosy are infectious, and the National Leprosy Eradication Programme of India, 2009 suggests Rifampicin, Dapsone, and Clofazimine combination for multidrug therapy for 12 months in multibacillary leprosy, and rifampicin and dapsone combination for 6 months, in cases of paucibacillary leprosy. Mutation in the *rpoB* gene has been identified as a critical factor for resistance to rifampicin in leprosy cases (Vedithi et al., 2018).

3. Fungal Infections and AMR

Aspergillus is the most common invasive mold with septate hyphae, are ubiquitous saprobes, spread through an inhalation route. It causes fungal infections in skin, eyes, ears, and 'fungus ball' in the lungs causing allergic bronchopulmonary aspergillosis (ABPA), as an opportunistic infection. The triazole antifungals are the drugs of choice for the prevention and treatment of aspergillosis. Voriconazole is the preferred triazole for systemic fungal infections, and patients on ABPA can be treated with corticosteroids and itraconazole. The mutations in the *cyp51C* gene and overexpression of efflux pumps lead to the high MIC of voriconazole, resistance to polyenes had been reported, and resistance to triazoles is infrequent (Rudramurthy et al., 2019).

Candida is yeast form with pseudohyphae and causes thrush, vaginitis, esophagitis, diaper rash, and mucocutaneous candidiasis. It also causes disseminated right-sided endocarditis in intravenous drug users, bloodstream candidemia, and endophthalmitis. Infection through indwelling urinary and intravenous catheters are important. *Candida* is a normal flora of the human system, and the line of defenses are skin and mucous membranes, Th-1 cells, and neutrophils. In immunocompromised individuals, it causes chronic mucocutaneous candidiasis, causing prolonged infection of the skin, oral, genital mucosa, and nail infections. Fluconazole is the drug of choice for oropharyngeal and oesophageal thrush, and the alternates are echinocandins. Skin infections can be treated with topical clotrimazole or nystatin. Certain *candida* infections such as thrush can be treated with clotrimazole, or nystatin 'swish and swallow,' or micafungin can also be used. Alterations in drug binding proteins *UPC2* causing upregulation of *ERG11*, active efflux of *candida* through *CaCdr1* and *CaCdr2*, mutations in *ERG2*, *ERG6*, and *ERG24* genes, and *TAC1*, a transcriptional activator or *candida* resistance are the strategies reported for antifungal resistance (Prasad et al., 2019a, 2019b).

Cryptococcus neoformans causes cryptococcal meningitis, is the most common, life-threatening invasive fungal disease, an opportunistic infection in AIDS patients. It is an oval, budding yeast having a wide polysaccharide capsule, and the infection is spread through inhalation. Lung infection leading to pneumonia, dissemination to meninges, and other organs are common in immunocompromised patients. Combination therapy with amphotericin B and flucytosine are used in disseminated cases and fluconazole for long-term treatment. The less common forms, such as *Cryptococcus Lauretii*, infections have escalated in the recent scenario, and biofilm formation was identified as the contributing factor for AMR of fluconazole, amphotericin B, and itraconazole (Ajesh and Sreejith, 2012; Altamirano et al., 2017).

Blastomyces dermatitidis is a dimorphic fungus, spore-forming, mold, infects humans, dogs, and horses by inhalation. Blastomycosis is a systemic fungal infection. Pulmonary infection is asymptomatic, and treatment is initiated in cases of skin and bone infections;

itraconazole is the drug of choice and in severe cases, amphotericin B is used. Resistance to drug treatment is infrequent (Denning and Chakrabarti, 2017).

Histoplasma capsulatum is a dimorphic fungus existing as mold in soil and yeast in tissues. Histoplasmosis is a systemic fungal infection. Two forms of asexual spores tuberculate macroconidia and microconidia transmit the infection, spread through inhalation. The organism can spread to the liver and spleen, cause pneumonia, and cavitary lung lesions. In severe cases, disseminated histoplasmosis occurs in immunocompromised patients and infants. Pancytopenia and ulcerated lesions on the tongue are the clinical manifestations of the disease in AIDS cases. There is no treatment for asymptomatic or mild cases; in progressive cases, itraconazole is the drug of choice. In disseminated cases, parenteral itraconazole and amphotericin B are used; fluconazole is used in meningitis. For chronic therapy, oral itraconazole is followed in AIDS patients. The AMR has not been reported for the histoplasmosis chemotherapeutic agents (Mittal et al., 2018).

Coccidioides immitis infection causes coccidioidomycosis, is a systemic fungal disease. It is a dimorphic fungus, exists as spherule, dissemination from lungs to other organs is through cell-mediated immunity. Infected people have delayed hypersensitivity, and the treatment is amphotericin B or itraconazole for lung lesions and disseminated cases. Asymptomatic cases require no treatment, and the patients recovered from the disease have to undergo chronic therapy with fluconazole to prevent recurrent infections. Few case reports have highlighted antifungal resistance to fluconazole (Donovan et al., 2019).

Paracoccidioides brasiliensis is a dimorphic fungus, having multiple buds, causing paracoccidioidomycosis, a systemic fungal infection. It is spread through the inhalation route, causing oral mucosal lesions, and lymph node enlargement and dissemination occurs rarely. Several months of pharmacotherapy with itraconazole is essential in the infected cases. *Tetracycline resistant protein* (*TETA*) and *hemolysin iii channel protein* (*HLYiii*) were found to be associated with AMR to itraconazole, by regulation of *ERG3*, *ERG5*, *ERG6*, *ERG11*, and *ERG25*, *in vitro* studies (Neto et al., 2014).

Microsporum, *Epidermophyton*, and *Trichophyton* are examples of dermatophytes causing cutaneous infections such as *tinea capitis, tinea cruris, and tinea pedis*. These fungi are molds, which use the nutrition of the keratin. Transmission is through skin contact. The superficial keratinized layer of the skin is infected, causes an inflammatory response, moisture increases the infection frequency, and fingers and foot are mostly affected. Topical agents such as miconazole, tolnaftate, clotrimazole are used, undecylenic acid is effective against tinea pedis, and griseofulvin is the choice drug for *tinea capitis and tinea unguium*. Resistance to antifungal drugs are infrequent (Gupta et al., 2016).

Malassezia furfur causes tinea versicolor, a skin infection of superficial nature, and is of cosmetic importance only. In the hypopigmented areas, lesions are noticed in summer. Budding yeast and hyphae are found in skin lesions. The drug of choice is topical miconazole, oral antifungals such as itraconazole or fluconazole can be used in recurrent infections. Studies regarding the AMR are very less, but the lipophilic nature of the fungus warrants it (Sharma et al., 2017).

4. Protozoal Infections and AMR

Entamoeba histolytica is an amoeba causing amoebic dysentery and live abscess. The life cycle has two stages trophozoite and cyst, where the trophozoite is found within the intestine and extraintestinal lesions and diarrhoeal stools; the cyst is located in non-diarrhoeal stools. In invasive amoebiasis, antigens are formed, and corresponding antibodies are useful for serologic diagnosis. It is spread through faecal-oral route, ingested cysts differentiate to trophozoites, invade the colonic epithelium, progress to the liver, and cause dysentery and amoebic liver abscess. The drug of choice is metronidazole or tinidazole. Asymptomatic carriers are treated with iodoquinol or paromomycin. For antibiotic resistance, a database is available, Database of Diarrheal Single Nucleotide Polymorphisms (DBDiaSNP), covers mutations and resistant genes for the pathogens associated with diarrhoea (Mehla and Ramana, 2015).

Giardia lamblia causes giardiasis, which is a microaerophilic, unicellular, intestinal protozoa, existing in two stages trophozoite and the cyst. It is transmitted through the faecal-oral route. Excystation in the duodenum releases trophozoite, which attaches to the intestinal wall and causes inflammation and malabsorption. The clinical findings are non-bloody diarrhoea, and foul-smelling diarrhoea, in addition to anorexia, nausea, abdominal cramps, and flatulence that exists for weeks to months. The drug of choice is metronidazole or tinidazole. The oxidative stress conditions leading to differential gene expression, and AMR had been studied, but not clinically proven (Raj et al., 2014).

Cryptosporidium hominis causes diarrhoea in immunocompromised patients, especially AIDS, which is an intestinal protozoan. The life cycle involves oocysts, sporozoites, trophozoites, schizonts, and merozoites. The gametes differentiate into oocyst, and the route of transmission is the faecal-oral route. The clinical picture is watery, non-bloody diarrhoea, causing debilitating significant volume fluid loss. The drug of choice is nitazoxanide, and clean environmental conditions could prevent the disease. AMR was not studied, but the species was found to be the second-highest contributing factor for diarrhoea in AIDS patients (Shah et al., 2016).

Trichomonas vaginalis is a uropathogenic protozoan, causing trichomoniasis, and exists only in trophozoite form. It is transmitted only through sexual contact, and the organism is localized in the prostate and the vagina. Sexually active women have symptomatic disease, and others are usually carriers. The clinical picture is watery, foul-smelling, greenish vaginal discharge, and itching and burning accompany. In men, it can cause urethritis. The drug of choice is metronidazole or tinidazole, lowering of pH of the vagina, and the use of condoms can arrest the spread. In vitro studies had confirmed the resistance, but the major reasons are non-adherence to therapy. Thus, it plays a minor role in the possibility of treatment failure (Kissinger, 2015).

Plasmodium has four species *Plasmodium ovale*, *malariae*, *vivax*, and *falciparum* and causes malaria. *P. falciparum* and vivax are the more common causes, where vivax is widely distributed, and falciparum causes the most severe form of the disease. In Southeast Asia, another species called *Plasmodium knowlesi* is found. Female anopheles mosquito takes the gametes, the sexual cycle occurs in it; in humans, they inject sporozoites, which are taken up within 30min to the liver; it differentiates into merozoites. In *P. vivax* and *ovale*, the merozoites produce a latent form, called hypnozoite, which is responsible for relapses. Merozoites from the liver enter the general circulation, infect the red blood cells, grow into schizont filled merozoites, after they rupture, infect new erythrocytes. The periodic rupture of red blood cells causes typical chills, fever, and sweats in malaria subjects. Some of the erythrocytic merozoites differentiate into gametocytes, which are taken up by the mosquitoes, and a new cycle begins in another person when the mosquito injects. The pathogenic features include the destruction of red blood cells, enlargement of spleen, with hyperplasia of macrophages and lymphocytes. The more severe form of malaria is caused by falciparum, which may lead to cerebral malaria and renal damage, which are life-threatening. The hemoglobinuria leads to the dark colour of the urine, resulting in 'blackwater fever.' The fever cycle timing is 72 hours for *P. malariae*, and for other species, it is 48 hours. *P. falciparum* causes parasitemia of high levels, whereas others produce at lower levels. Individuals with sickle cell anemia and glucose-6-phosphate-dehydrogenase deficiency are protected against the disease, whereas those with Duffy blood group are affected. In 1958 the Indian Government launched the National Malaria Eradication Program. In the next few years, resistance was identified as a major setback, which was addressed with the launch of the National Vector Borne Diseases Control Program to improve the coverage in endemic areas. Due to improved attention, the number of malaria cases and deaths has drastically reduced. The important strategies followed for effective control measures were chemical control by indoor residual spray and malathion fogging, biological control by larvivorous fish and use of biocides, personal prophylactic measures, community participation programs, monitoring and evaluation by regular field visits and feedback measures for corrective actions by Indian Council of Medical Research. For chloroquine-sensitive *P. falciparum* oral chloroquine is followed, for chloroquine-sensitive *P. vivax* and *ovale*, chloroquine in combination with primaquine is developed, for chloroquine-resistant *P. falciparum* of uncomplicated infections are intervened with artemether and lumefantrine, and in severe complicated cases of *P. falciparum*, intravenous artesunate or quinidine is used. A non-significant correlation with chloroquine resistance was studied in *P. falciparum* chloroquine resistance transporter (*pfcrt*), *P. falciparum* multidrug resistance 1 N86Y (*pfmdr1 N86Y*), and *P. falciparum* dihydrofolate reductase (*pfdhfr*) and *P. falciparum* dihydropteroate synthase (*pfdhps*), and *P. falciparum* ATPase6 (*pfATPase6*) and *pfK13* (Mishra et al., 2017; Gowda et al., 2018; Chhibber-Goel and Sharma, 2019).

Toxoplasma is a protozoan causing toxoplasmosis that its hosts are cats. The contamination of cat feces causes ingestion of cysts that differentiate to tachyzoites and bradyzoites, and the infection is spread through uncooked meat. Congenital and transplacental infections are possible; the clinical picture is chorioretinitis, intracranial calcifications, and blindness in congenital cases. In AIDS cases, encephalitis occurs as disseminated disease. Symptomatic and asymptomatic cases are treated with pyrimethamine and sulfadiazine combination. Nuclear Factor of Activated T Cells 1 (NFAT1) regulation of $Ly6C^{hi}$was studied and reported as an important contributor of AMR (Benevides et al., 2019).

Trypanosoma includes three important pathogens *T. cruzi*, *T. gambiense*, and *T. rhodesiense*, causing trypanosomiasis, *Trypanosoma cruzi* causes Chaga's disease, where the vector is reduviid bug. The disease weakens the cardiac muscle causing congestive heart failure and causes megacolon, and megaesophagus and the drug of choice is nifurtimox. *T. gambiense and T. rhodesiense* which could cause sleeping sickness, spread by the tsetse fly, and could infecting lymph nodes and brain, and progressing to coma due to demyelination encephalitis. The drug of choice is suramin, pentamidine, or melarsoprol. Two ABC transporters, *TbMRPA*, and *TbMRPE*, are related to AMR (Pramanik et al., 2019).

Leishmania is a protozoan causing Kala-azar fever. It is spread by sand-flies and affects the reticuloendothelial system, bone marrow, spleen, and liver. The clinical picture is the enlargement of the spleen and marked increase in IgG, and prevalent in the rural and poor socioeconomic population of India. The National Kala-azar Elimination Programme by Government of India follows different strategies to combat the disease, such as early diagnosis, integrate vector management, strengthening capacity at rural health, treatment with liposomal amphotericin B, combination regimens with miltefosine and paromomycin, and also the treatment of post-Kala-azar Dermal Leishmaniasis with the combination therapies. The drug efflux transporters contributing to AMR, which are studied were, ABCA, ABCB, ABCC, and ABCG and their subfamilies (Ponte-Sucre et al., 2017).

Pneumocystis jiroveci is classified as a yeast according to molecular analysis but has many similarities to a protozoan. Pneumocystis is spread through inhalation to lungs, causes inflammatory exudation, and dyspnoea. Mostly it occurs as an opportunistic infection in immunocompromised patients, and the clinical findings include fever, cough, dyspnoea, and the chest X-ray shows 'ground-glass' pattern. The drug of choice is cotrimoxazole and can be used for prevention also. The DHFR polymorphisms at codon 55 and 57, contribute to AMR (Singh et al., 2015).

5. Viral Infections and Resistance

Influenza is a seasonal viral disease, is an important pathogen belonging to the orthomyxovirus family. It is a single piece of the RNA genome, Bird or avian influenza is caused by influenza A of H5N1 strain and influenza A of H1N1 causes swine flu. The other strains reported are H7N9. Influenza A is a segmented single-stranded RNA genome, containing RNA-dependent-RNA polymerase, enveloped in hemagglutinin, and neuraminidase, and they have nearly 16 antigenically distinct types and 9 for neuraminidase. The changes in the antigenicity, the antigenic drift, and the M2 matrix protein ion channel determine the virulence and resistance of the virus. The disease is spread through airborne droplet infection. The clinical picture after 24-48hrs incubation is fever, headache, myalgia, sore throat, sudden cough, and complications such as pneumonia due to *Streptococcus pneumoniae*. Encephalopathy is characterized as Reye's syndrome in children, encountered with aspirin. Treatment includes neuraminidase inhibitors such as oseltamivir and zanamivir; resistance is infrequent, the effectiveness depends on early initiation of the therapy. Amantadine and Rimantadine are M2 protein channel inhibitors used for prevention and treatment, although vaccines are more effective as a prophylactic measures. The substitution of serine to asparagine in the 31st position of the M2 channel is reported as an important contributor for amantadine resistance (Sood et al., 2018).

Herpes simplex virus (HSV-1 and HSV-2) is herpesvirus distinguishable by their antigenicity and location of lesions. In HSV-1, the lesions are above the waist, whereas it is above the waist in HSV-2. The virus has an icosahedral envelope, and importantly, it causes latent infections. The virus integrates with the protein VP16, activates immediate early genes, and generates DNA polymerase, followed by thymidine kinase. It is transmitted through saliva and sexual contact, and the clinical pictures are gingivostomatitis, herpes labialis, keratoconjunctivitis, encephalitis, herpes gladiatorum, eczema herpeticum, and disseminated infections in immunocompromised patients, in HSV-1 infections. HSV-2 causes many asymptomatic infections, genital herpes, neonatal herpes, and skin reactions. The treatment schedule includes acyclovir, a short-term course that reduces the shedding extent. Foscarnet is preferable in acyclovir mutants of HSV-1, and in eye infections, trifluridine is used. Recurrent infections could be controlled with valacyclovir and famciclovir. Increasing resistance to antimicrobials is evident in immunocompromised patients. Varicella-zoster is commonly called chickenpox, requires no antiviral therapy. Although acyclovir is used in moderate or severe cases to reduce the severity and duration of treatment (Chen et al., 2015).

Cytomegalovirus (CMV) is the most important cause of congenital abnormalities and the major reason for pneumonia in AIDS patients. Mother to fetal and sexual route of transmission are modes of spread. The clinical picture includes heterophil negative mononucleosis, causing fever, lethargy, and infection of the intestinal tract causing intractable colitis and diarrhoea and retinitis leading to blindness. Ganciclovir and valganciclovir are the treatment choices, and mutations in gene encoding phosphokinase which leads to the drug resistance. Foscarnet is useful in CMV retinitis. Leflunomide has also been tried as an alternative in ganciclovir resistant CMV infections (Moellering et al., 2007).

Respiratory syncytial virus (RSV) infects infants, is an important pathogen causing pneumonia and bronchiolitis. It has fusion proteins, multinucleated giant cells, transmitted through droplets, and ribavirin is the drug of choice. Due to mutation threats, it is doubtful regarding its efficacy (de Souza et al., 2015).

Ebola virus is a filovirus, causing Ebola hemorrhagic fever, named after the river Zaire. It is a single-stranded RNA genome, having RNA-dependent RNA polymerase. The reservoirs are fruit bats and rodents, transmitted through blood and body fluids. The incubation period is 5-7 days or up to 21 days, causing fever, headache, myalgia, vomiting, epigastric pain, and diarrhoea, with marked lymphopenia and hemorrhages. Serum immunoglobulins and quarantine are the only treatment and prevention options. The newer molecules tried are tetrandrine, and drug resistance has not been explored yet (Bhagya and Chandrasekhar, 2016).

Rotavirus is an important cause of viral gastroenteritis, in young children, is a double-stranded RNA genome, and contains RNA-dependent RNA polymerase. It is transmitted through the fecal-oral route, and only vaccines are the choice of prevention, live attenuated, and live reassortant one. No drug treatment and resistance are studied in rotavirus. India has launched 116E vaccine and adopted it in the nation's universal immunization program (Rose et al., 2017).

There are around five important viruses causing hepatitis, such as *hepatitis A virus* (HAV), hepatitis B virus (HBV), hepatitis C virus (HCV), hepatitis D virus (HDV), and hepatitis E virus (HEV). Other viruses causing liver infections are Epstein-Barr virus, CMV, but they infect other organs also. HAV is a picornavirus, single-stranded RNA, which is also called enterovirus 72, transmitted through the fecal-oral route. It has a short incubation period of 2-4 weeks, detected by elevation of IgM antibodies, and most cases resolve within this period. Both active and passive immunizations are available for prevention. HBV is a hepadnavirus, double-stranded DNA, has a longer incubation period of 10-12 weeks, the envelope containing surface antigen HBsAg is important for diagnosis and prognosis. Another core antigen HBeAg is found, which indicates transmissibility. HBx protein of the virus is involved in oncogenesis, which suppresses p53. The clinical manifestations include hepatitis; chronic cases that lead to cirrhosis and death. The therapeutic options are tenofovir or entecavir with peginterferon 2α. HBV polymerase chain mutations such as A181V, and N236T, confer AMR. HCV is a flavivirus, single-stranded RNA, is the most important cause of hepatocellular carcinoma. The drugs used are sofosbuvir, dasabuvir, ledipasvir, ombitasvir, daclatasvir, boceprevir, simeprevir, telaprevir, and paritaprevir. A novel HCV called HCV G18, was identified in India. Mutation in the NS3/4A proteins causes AMR. HDV is a single-stranded RNA delta virus, which has a similar infection to HBV, and could be mitigated using peg-interferon. HEV is a calicivirus, single-stranded RNA, is a more common infection than HAV, spread through the fecal-oral route, but less studied. *Hepatitis G virus* (HGV) causes hepatitis in posttransfusion hepatitis, and its role has not been established yet (Gupta et al., 2014; Borgia et al., 2018).

Human immunodeficiency virus (HIV) is a retrovirus, exists in two forms HIV-1 and HIV-2, causing AIDS. HIV-1 occurs worldwide, and HIV-2 is prevalent in West Africa. It targets the killer helper cells (CD4+) loss and develops opportunistic infections. The structural genes are gag, pol, and env, regulatory genes are tat and rev, and accessory genes are nef, vif, vpr, and vpu. The viral fusion proteins are gp41 and gp120, which mutates rapidly. Viral fusion causes entry of the virus through chemokine receptors CCR5 (delta-32 mutation), viral uncoating occurs, reverse transcriptase forms viral DNA, viral integrase integrates it to host DNA, transcribed into viral proteins along with host protein transcriptions, the viral proteins are cleaved to functional ones by viral protease, then the virus gets packed, and new virions release and infect new CD4+ cells. Transmission is through sexual route, blood transfusion, mother to fetus, and the latent period for the viral detection is 4-11 days. The viral antibodies to p24 antigen (detection kits use to detect anti-p24 antibodies), anti-gp 120 antibodies, and CD4 lymphocytes cause acute symptoms within 6 months of the infection. Chronic cases reveal acquired immune deficiency syndrome (AIDS) in 3-10 years, as opportunistic infections. The opportunistic infections are pneumonia and tuberculosis in lungs, thrush, and ulcers in mouth, thrush, and esophagitis in the esophagus, diarrhoea of intestinal tract, meningitis, brain abscess in the central nervous system, retinitis in the eye, Kaposi's sarcoma and zoster infections of the skin, and lymphadenopathy and splenomegaly of reticuloendothelial system. Intensive treatment regimens are followed to avoid the emergence of resistance, such as highly active antiretroviral therapy (HAART). The antiretroviral drugs include nucleoside/nucleotide reverse transcriptase inhibitors (NRTI), non-nucleoside reverse transcriptase inhibitors (NNRTI), viral fusion inhibitors, protease inhibitors (PIs), and integrase inhibitors. The coinfection of HIV with TB, is an increasing risk factor for the emergence of resistance and failure of chemotherapy. The mutations related to antiviral resistance, in the case of NRTIs and NNRTIs were found to be K219R, L74V, K219N, Y181C, E157Q, L74M, Q95K, and T97K. In Indian studies, HIV-1 subtype C was characterized for the gene mutations related to AMR (Sinha et al., 2018; Nesakumar et al., 2019; Karade et al., 2019; Chauhan et al., 2019).

Dengue (DENV) is caused by the bite of Aedes mosquito, humans are reservoirs, prevalent in tropical areas, and no vaccine is available for prevention; four serotypes exist DENV-1, DENV-2, DENV-3, and DENV-4. Classical dengue fever, called breakbone fever has influenza-like onset, fatigue, headache, and retro-orbital pain. Myalgia, arthralgia, flushing, maculopapular rash, and leukopenia occur. It has few sequelae, and the symptoms regress. Dengue hemorrhagic fever, in contrast, is more severe form of the disease, causing high mortality. The initial picture is similar to the classic dengue, but shock and hemorrhage, especially in the gastrointestinal tract, and skin occurs. Neither antiviral therapy nor vaccines are available for dengue, only a polyherbal formulation, *nilavembukudineer*, is successful in DENV-2 viral cases. From the past decade, the number of dengue cases is rising in India, due to improper drainage systems and lack of awareness in household practices of stagnated water management (Jain et al., 2019).

Chikungunya (CHIKV) is caused by the bite of Aedes mosquito, humans are reservoirs, prevalent in tropical areas, and no vaccine is available for prevention. It causes sudden high fever in onset, and wrist and ankle joint pain. Encephalitis and maculopapular rash may occur. The treatment strategies include supportive measures, paracetamol, weaker opioids, ibuprofen, oral prednisolone, intraarticular steroids, and tricyclic antidepressants. In resistant cases, TNFα inhibitors are used (Pathak et al., 2019).

Zika virus was reported in India in 2017, an infection spread through Aedes mosquito. The clinical manifestations mimic DENV, and CHIKV, maculopapular exanthema, at the onset, and other features such as hematomas, aphthous ulcers, low-grade fever, non-purulent conjunctivitis, and myalgia. Supportive measures are the treatment choices, and the resistance has not been studied (Hasan et al., 2019).

Nipah virus is a paramyxovirus, first reported in Kerala, India, in 1998. It is transmitted from human-human, having an incubation period of 4days to 2 months, and the clinical features of the disease are headache, dizziness, vomiting, doll's eye reflex, seizures, myoclonic jerks, and other neurological manifestations. Only supportive measures are treatment options (Plowright et al., 2019).

Human T-Cell lymphoma virus (HTCL), exists as HTCL-1 and HTCL-2, both are associated with lymphomas and leukemia. The retroviral genes are gag, pol, env, tax, and rex. Human papillomavirus (HPV) is a DNA tumor virus, begins as warts, and proceed to carcinomas. HPV genes E7 and E6 interrupt the tumor-suppressing genes, such as p53 and Rb. Epstein-Barr virus (EBV) is a herpes virus causing Burkitt's lymphoma, causing translocation of the c-myc gene to chromosome 14, from chromosome 8. Human Herpes Virus (HHV-8) causes Kaposi's sarcoma in AIDS, a protein encoded by the virus, impairs the function of tumor suppressor genes p53 and Rb (Taylor et al., 2015; Pinto et al., 2019).

6. Antimicrobial Surveillance and National Programmes to Control AMR

Infectious diseases are a major cause of mortality globally, and in India, it causes 30% deaths approximately. Unnecessary use of antimicrobials, the emergence of multidrug resistance, pan-drug resistance, therapeutic interventions without susceptibility testing, the incremental cost of antimicrobials, poses a challenge to modern science, in the post-antibiotic era. Several measures are taken by the Indian Government, and National Antimicrobial Containment Programme is launched during the period 2012-2017. Currently, around 20 medical colleges of different states are running this program effectively. The main aims of the program are the establishment of countrywide laboratories to document the antimicrobial usage data, make rational use of antimicrobials, and conducting nationwide awareness programs among the healthcare community. National AMR surveillance prioritizes the following pathogens, *Enterococcus*spp, *Staphylococcus aureus*, *E. coli*, *Klebsiella*, *Acinetobacter*, *Pseudomonas*spp, and *Salmonella enterica*, *typhi*, and *paratyphi* for antimicrobial stewardship program. The clinical specimen guidelines govern the collection of blood, urine, pus aspirate, stool, and other sterile body fluids. Standard antimicrobial susceptibility testing methods such as automated cultures and advanced machines such as Vitek2, Phoenix, and Microscan are adopted. Mandatory clinical and epidemiological details of each isolate are documented, such as hospital code, patient data, specimen data, and region, specimen collection data, pathogen isolated, AST results, and the data file is in WHONET format. The alert forms for reporting of AMR resistance is launched, which makes reporting convenience, and effective surveillance. List of antibiotics under the AMR surveillance, for shipping of isolates to national center for disease control (NCDC) is, amikacin, ampicillin azithromycin, cefepime, cefotaxime, cefoxitin, ceftazidime, ceftriaxone, erythromycin, gentamicin, imipenem, linezolid, minocycline, nitrofurantoin, piperacillin/tazobactam, vancomycin, and colistin (NCDC, India, 2019).

The National Vector Borne Diseases Control Programme identified *P. falciparum* resistance to chloroquine as a major issue, and monitoring centers concentrate in this context. Several control strategies, tools for monitoring were initiated, and the new protocol classifies three categories of malaria: Adequate Clinical and Parasitological Response (ACPR), Early Treatment Failure (ETF), and Late Treatment Failure (LTF). The program aims at providing a complete cure in malaria, prevention of progression of uncomplicated disease to severe stages, prevention of relapses, and prevention of drug resistance by rational therapeutic interventions. *P. falciparum* is treated with artesunate, sulfadoxine, and pyrimethamine (ACT) combinations, or lumefantrine as an add-on in North-Eastern states (ACT-AL). Different dosage charts with different colors (containing artesunate, sulfadoxine, and pyrimethamine at different dose levels) are issued for different age groups, include: Pink blister for 0-1yrs of age, yellow blister for 1-4yrs, green blister for 5-8yrs, red blister for 9-14yrs, and white blister for ≥15yrs of age. For the treatment of chloroquine-resistant falciparum malaria cases, artesunate, sulfadoxine, mefloquine, quinine, doxycycline, and clindamycin combinations are used. For severe cases, artesunate, artemether, arteether, and quinine combinations are used (NHP, 2019).

National guidelines for clinical management of dengue in India are laid to address *dengue*, the major public health concern. A recent trend suggests a steep increase in the spread of the disease from urban to rural areas. The national course of dengue infection has three phases, febrile phase, critical phase (leakage phase), and convalescent-phase (recovery phase). Diagnosis is graded as DF, DHFII, DHFIII, and DHFIV, and IgM ELISA kits are supplied to all the laboratories by the National Institute of Virology. Several algorithm guidelines for pharmacotherapeutic interventions and supportive care are followed for the different stages of the disease, specially cases with coinfections, pregnancy, and neonatal cases (NHP, 2019).

The NDC has laid guidelines on the categorization of seasonal *influenza* for quarantine, testing, hospitalization, and treatment. The categorization segregates as A for uncomplicated and mild influenza, B for uncomplicated but with severe symptoms as high-risk groups, and C for complicated cases. In infection control, the practices, disinfection of contact surfaces, avoiding reuse of disposable masks, and home care measures are educated. Clinical measures of early infection control, minimization of nosocomial infection spread, early identification of risk persons, and instituting oseltamivir is recommended. Guidelines are laid for the collection, storage, and transportation of human clinical samples for the diagnosis of *influenza*. There are around 36 approved labs for testing the clinical samples for *H1N1* in different parts of our country (NCDC, 2019).

The National AIDS Control Program (NACP), launched in 1992, is a successful program in the prevention and control of HIV/AIDS in India. NACP-I focused on reducing the spread of the disease, NACP-II increased the capacity of the country to encounter the disease on a long-term basis, NACP-III was launched to halt and reverse the epidemic. The current phase, NACP-IV aims at prevention of new infections, prevention of parent to child transmission (PPTCT), reducing the stigma for people living with HIV (PLHIV), and ensuring effective use of strategies at all levels of the program. The integrated counseling and testing centers (ICTC) conduct HIV diagnostic tests, provides essential information on modes of transmission, and link people together for HIV prevention, care, and treatment services. The PPTCT was launched in the year 2002, offers services to pregnant women. The Antiretroviral Treatment (ART) centers across the country, dispense medications at free of cost, and monitor strict adherence to the treatment, and particular emphasis is given to seropositive children and women. There are around 578 approved ART centres across the country, taking care of the HIV/AIDS treatment (NACO, 2019).

Conclusion

India, the fastest growing economy, has burgeoning population demands on healthcare, has made enormous efforts to address the high prevalence of AMR. The necessary actions to be taken are improving awareness and understand AMR through effective communication, training, and education, strengthen the evidence-based surveillance, reduce the incidence of AMR, promote optimal use of antimicrobials in health, food, and animals, promote research activities and funding, and strengthen the AMR stewardship.

Establishment of AMR surveillance through NDC, national action plan and strategic network, vector control programs, influenza control programs and NACP with advanced diagnostic facilities, building awareness at health care providers levels and consumer levels, will hopefully address the infectious disease burden and AMR in our country.

REFERENCES

[1] Drlica, K. and Perlin, D. *Antibiotic resistance.* Upper Saddle River, N. J.: FT Press. 2011.

[2] Kahlmeter, G. and Brown, D. Harmonization of antimicrobial breakpoints in Europe — Can it be achieved? *Clinical Microbiology Newsletter.* 2004; 26(24), 187.

[3] Drlica, K. and Zhao, X. Mutant Selection Window Hypothesis Updated. *Clinical Infectious Diseases.* 2007; 44(5), 681-8.

[4] Cirz, R. and Romesberg, F. Induction and Inhibition of Ciprofloxacin Resistance-Conferring Mutations in Hypermutator Bacteria. *Antimicrobial Agents and Chemotherapy.* 2005; 50(1), 220-5.

[5] Poole, K. Efflux-mediated antimicrobial resistance. *Journal of Antimicrobial Chemotherapy.* 2005; 56(1), 20-51.

[6] Rice, L. Evolution and Clinical Importance of Extended-Spectrum β-Lactamases. *Chest.* 2001; 119(2), 391S-6S.

[7] Vetting M, Hegde, S., Wang, M., Jacoby, G., Hooper, D. and Blanchard, J. (2011). Structure of QnrB1, a Plasmid-mediated Fluoroquinolone Resistance Factor. *Journal of Biological Chemistry.* 2011; 286(28), 25265-73.

[8] Mazel, D. Integrons: agents of bacterial evolution. *Nature Reviews Microbiology.* 2006; 4(8), 608-20.

[9] Barman, S., Konai, M., Samaddar, S. and Haldar, J. Amino Acid Conjugated Polymers: Antibacterial Agents Effective against Drug-Resistant Acinetobacter baumannii with No Detectable Resistance. *ACS Applied Materials & Interfaces.* 2019; 11(37), 33559-72.

[10] Aggarwal, S., Somani, V. and Bhatnagar, R. Phosphate starvation enhances the pathogenesis of Bacillus anthracis. *International Journal of Medical Microbiology.* 2015; 305(6), 523-31.

[11] Yadav, A., Sirohi, P., Saraswat, S., Rani, M., Singh, M., Srivastava, S. and Singh, N. Inhibitory Mechanism on Combination of Phytic Acid with Methanolic Seed Extract of *Syzygium cumini* and Sodium Chloride over Bacillus subtilis. *Current Microbiology.* 2018; 75(7), 849-56.

[12] Sarkar, A., Pazhani, G., Dharanidharan, R., Ghosh, A. and Ramamurthy, T. Detection of integron-associated gene cassettes and other antimicrobial resistance genes in enterotoxigenic *Bacteroides fragilis. Anaerobe.* 2015; 33, 18-24.

[13] Singh, D., Kumar, B., Shrinet, G., Singh, R., Das, A., Mantur, B., et al., Draft genome sequence of field isolate *Brucella melitensis* strain 2007BM/1 from India. *Journal of Global Antimicrobial Resistance*. 2018; 13, 152-53.

[14] Deguchi, T., Hatazaki, K., Ito, S., Kondo, H., Horie, K., Nakane, K., et al., Macrolide and fluoroquinolone resistance is uncommon in clinical strains of Chlamydia trachomatis. *Journal of Infection and Chemotherapy*. 2018; 24(8), 610-4.

[15] Agarwal, J., Mohan, S., Srivastava, R. and Singh, M. Observations on Citrobacter species from a tertiary care health center with special reference to multi-drug resistance and presence of CTX-M gene. *Indian Journal of Pathology and Microbiology*. 2014; 57(3), 439.

[16] Lohani, M., Dhasmana, A., Haque, S., Wahid, M., Jawed, A., Dar, S., et al., Proteome mining for the identification and in-silico characterization of putative drug targets of multi-drug resistant *Clostridium difficile* strain 630. *Journal of Microbiological Methods*. 2017; 136, 6-10.

[17] Yadav, J., Das, S., Dhaka, P., Vijay, D., Kumar, M., Mukhopadhyay, A., et al., Molecular characterization and antimicrobial resistance profile of Clostridium perfringens type A isolates from humans, animals, fish and their environment. *Anaerobe*. 2017; 47, 120-4.

[18] Dandinarasaiah, M., Vikram, B., Krishnamurthy, N., Chetan, A. and Jain, A. Diphtheria Re-emergence: Problems Faced by Developing Countries. *Indian Journal of Otolaryngology and Head & Neck Surgery*. 2012; 65(4), 314-8.

[19] Kim, M. and Woo, G. Characterization of antimicrobial resistance and quinolone resistance factors in high-level ciprofloxacin-resistant *Enterococcus faecalis* and *Enterococcus faecium* isolates obtained from fresh produce and fecal samples of patients. *Journal of the Science of Food and Agriculture*. 2016; 97(9), 2858-64.

[20] Sharma, G., Sharma, S., Sharma, P., Chandola, D., Dang, S., Gupta, S. et al., *Escherichia colibiofilm:* development and therapeutic strategies. *Journal of Applied Microbiology*. 2016; 121(2), 309-19.

[21] Antimicrobial resistance: revisiting the "tragedy of the commons." *Bulletin of the World Health Organization*. 2010; 88(11), 805-6.

[22] Pandya, H., Agravat, H., Patel, J. and Sodagar, N. Emerging antimicrobial resistance pattern of *Helicobacter pylori* in central Gujarat. *Indian Journal of Medical Microbiology*. 2014; 32(4), 408.

[23] Saikia, K. and Das, B. Multi locus sequence type comparison of invasive and commensal *Haemophilus influenzae* isolates from Delhi. *International Journal of Infectious Diseases*. 2012; 16, e238.

[24] Singh, S., Mishra, M., Sahoo, M., Patole, S., Sahu, S., Misra, S. et al., Antibiotic resistance determinants and clonal relationships among multidrug-resistant isolates of *Klebsiella pneumoniae*. *Microbial Pathogenesis*. 2017; 110, 31-6.

[25] Ojha, D. and Patil, K. p-Coumaric acid inhibits the *Listeria monocytogenes* RecA protein functions and SOS response: An antimicrobial target. *Biochemical and Biophysical Research Communications*. 2019; 517(4), 655-61.

[26] Wi, T., Lahra, M., Ndowa, F., Bala, M., Dillon, J., Ramon-Pardo, P., et al., Antimicrobial resistance in *Neisseria gonorrhoeae*: Global surveillance and a call for international collaborative action. *PLOS Medicine*. 2017; 14(7), e1002344.

[27] Tsang, R., Law, D., Deng, S. and Hoang, L. Ciprofloxacin-resistant *Neisseria meningitidis* in Canada: likely imported strains. *Canadian Journal of Microbiology*. 2017; 63(3), 265-8.

[28] Shivaprakash, M., Rao, P., Mandal, J., Biswal, M., Gupta, S., Ray, P. et al., Nocardiosis in a tertiary care hospital in North India and review of patients reported from India. *Mycopathologia*. 2007; 163(5), 267-74.

[29] Pawar, S., Ashraf, M., Mujawar, S., Mishra, R. and Lahiri, C. In silico Identification of the Indispensable Quorum Sensing Proteins of Multidrug Resistant Proteus mirabilis. *Frontiers in Cellular and Infection Microbiology*. 2018; 8, 269.

[30] Miryala, S., Anbarasu, A. and Ramaiah, S. Systems biology studies in *Pseudomonas aeruginosa* PA01 to understand their role in biofilm formation and multidrug efflux pumps. *Microbial Pathogenesis*. 2019; 136, 103668.

[31] Oo, K., Myat, T., Htike, W., Biswas, A., Hannaway, R., Murdoch, D., et al., Molecular mechanisms of antimicrobial resistance and phylogenetic relationships of *Salmonella enterica* isolates from febrile patients in Yangon, Myanmar. *Transactions of The Royal Society of Tropical Medicine and Hygiene*. 2019; 113(10), 641-8.

[32] Muthuirulandi Sethuvel, D., Devanga Ragupathi, N., Anandan, S. and Veeraraghavan, B. Update on: Shigella new serogroups/serotypes and their antimicrobial resistance. *Letters in Applied Microbiology*. 2016; 64(1), 8-18.

[33] Sharma, V., Sharma, S., Dahiya, D., Khan, A., Mathur, M. and Sharma, A. Coagulase gene polymorphism, enterotoxigenecity, biofilm production, and antibiotic resistance in Staphylococcus aureus isolated from bovine raw milk in North West India. *Annals of Clinical Microbiology and Antimicrobials.* 2017; 16(1), 65.

[34] Aliberti, S., Cook, G., Babu, B., Reyes, L., H. Rodriguez, A., Sanz, F., et al., International prevalence and risk factors evaluation for drug-resistant Streptococcus pneumoniae pneumonia. *Journal of Infection*. 2019; 79(4), 300-11.

[35] STAMM, L. Syphilis: antibiotic treatment and resistance. *Epidemiology and Infection*. 2014; 143(8), 1567-74.

[36] Gupta, P., Mankere, B., Chekkoora Keloth, S., Tuteja, U., Pandey, P. and Chelvam, K. Increased antibiotic resistance exhibited by the biofilm of Vibrio cholerae O139. *Journal of Antimicrobial Chemotherapy*. 2018; 73(7), 1841-7.

[37] Dua, K., Rapalli, V., Shukla, S., Singhvi, G., Shastri, M., Chellappan, D., et al., Multi-drug resistant Mycobacterium tuberculosis & oxidative stress complexity:

Emerging need for novel drug delivery approaches. *Biomedicine & Pharmacotherapy*. 2018; 107, 1218-29.

[38] Lange, C., Chesov, D., Heyckendorf, J., Leung, C., Udwadia, Z. and Dheda, K. Drug-resistant tuberculosis: An update on disease burden, diagnosis and treatment. *Respirology*. 2018; 23(7), 656-73.

[39] Vedithi, S., Malhotra, S., Das, M., Daniel, S., Kishore, N., George, A., et al., Structural Implications of Mutations Conferring Rifampin Resistance in Mycobacterium leprae. *Scientific Reports*. 2018; 8(1), 5016.

[40] Rudramurthy, S., Paul, R., Chakrabarti, A., Mouton, J. and Meis, J. Invasive Aspergillosis by Aspergillus flavus: Epidemiology, Diagnosis, Antifungal Resistance, and Management. *Journal of Fungi*. 2019; 5(3), 55.

[41] Prasad, R., Nair, R. and Banerjee, A. Emerging Mechanisms of Drug Resistance in Candida albicans. *Yeasts in Biotechnology and Human Health*. 2019a; 135-153.

[42] Prasad, R., Nair, R. and Banerjee, A. Multidrug transporters of Candida species in clinical azole resistance. *Fungal Genetics and Biology*. 2019b; 132, 103252.

[43] Altamirano, S., Fang, D., Simmons, C., Sridhar, S., Wu, P., Sanyal, K. et al., Fluconazole-Induced Ploidy Change in *Cryptococcus neoformans* Results from the Uncoupling of Cell Growth and Nuclear Division. *mSphere*. 2017; 2(3).

[44] Ajesh, K. and Sreejith, K. *Cryptococcus laurentii* Biofilms: Structure, Development and Antifungal Drug Resistance. *Mycopathologia*. 2012; 174(5-6), 409-419.

[45] Denning, D. and Chakrabarti, A. Pulmonary and sinus fungal diseases in non-immunocompromised patients. *The Lancet Infectious Diseases*. 2017; 17(11), e357-e366.

[46] Mittal, J., Ponce, M., Gendlina, I. and Nosanchuk, J. Histoplasma Capsulatum: Mechanisms for Pathogenesis. *Fungal Physiology and Immunopathogenesis*. 2018; 157-191.

[47] Donovan, F., Shubitz, L., Powell, D., Orbach, M., Frelinger, J. and Galgiani, J. Early Events in Coccidioidomycosis. *Clinical Microbiology Reviews*. 2019; 33(1).

[48] Neto, B., Carvalho, P., Bailão, A., Martins, W., de Almeida Soares, C. and Pereira, M. Transcriptional profile of *Paracoccidioides* spp. in response to itraconazole. *BMC Genomics*. 2014; 15(1), 254.

[49] Gupta, A., Foley, K. and Versteeg, S. New Antifungal Agents and New Formulations against Dermatophytes. *Mycopathologia*. 2016; 182(1-2), 127-41.

[50] Sharma, A., Rabha, D. and Ahmed, G. In vitro antifungal susceptibility of Malassezia isolates from pityriasis versicolor lesions. *Indian Journal of Dermatology, Venereology, and Leprology*. 2017; 83(2), 249.

[51] Mehla, K. and Ramana, J. DBDiaSNP: An Open-Source Knowledgebase of Genetic Polymorphisms and Resistance Genes Related to Diarrheal Pathogens. *OMICS: A Journal of Integrative Biology*. 2015; 19(6), 354-60.

[52] Raj, D., Ghosh, E., Mukherjee, A., Nozaki, T. and Ganguly, S. Differential gene expression in Giardia lamblia under oxidative stress: Significance in eukaryotic evolution. *Gene*. 2014; 535(2), 131-39.

[53] Shah, S., Kongre, V., Kumar, V. and Bharadwaj, R. A Study of Parasitic and Bacterial Pathogens Associated with Diarrhea in HIV-Positive Patients. *Cureus*. 2016.

[54] Kissinger, P. Trichomonas vaginalis: a review of epidemiologic, clinical and treatment issues. *BMC Infectious Diseases*. 2015; 15(1).

[55] Gowda, D., Dayananda, K. and Achur, R. Epidemiology, drug resistance, and pathophysiology of *Plasmodium vivaxmalaria*. *Journal of Vector Borne Diseases*. 2018; 55(1), 1.

[56] Mishra, N., Sharma, S., Bharti, R., Bhardwaj, N., Anvikar, A. and Valecha, N. Correlation of in vitro sensitivity of chloroquine and other antimalarials with the partner drug resistance to plasmodium falciparum malaria in selected sites of India. *Indian Journal of Medical Microbiology*. 2017; 35(4), p. 485.

[57] Chhibber-Goel, J. and Sharma, A. Profiles of Kelch mutations in Plasmodium falciparum across South Asia and their implications for tracking drug resistance. *International Journal for Parasitology: Drugs and Drug Resistance*. 2019; 11, 49-58.

[58] Benevides, L., Saltarelli, V., Pioto, F., Sacramento, L., Dias, M., Rodríguez, G., et al., NFAT1 Regulates Ly6Chi Monocyte Recruitment to the CNS and Plays an Essential Role in Resistance to *Toxoplasma gondii* Infection. *Frontiers in Immunology*. 2019; 10.

[59] Pramanik, P., Alam, M., Roy Chowdhury, D. and Chakraborti, T. Drug Resistance in Protozoan Parasites: An Incessant Wrestle for Survival. *Journal of Global Antimicrobial Resistance*. 2019; 18, 1-11.

[60] Ponte-Sucre, A., Gamarro, F., Dujardin, J., Barrett, M., López-Vélez, R., García-Hernández, R., et al., Drug resistance and treatment failure in leishmaniasis: A 21st century challenge. *PLOS Neglected Tropical Diseases*. 2017; 11(12), e0006052.

[61] Singh, Y., Mirdha, B., Guleria, R., Khalil, S., Panda, A., Chaudhry, R., et al., Molecular detection of DHFR gene polymorphisms in Pneumocystis jirovecii isolates from Indian patients. *The Journal of Infection in Developing Countries*. 2015; 9(11), 1250-1256.

[62] Sood, R., Kumar, N., Bhatia, S., Chanu, K., Gupta, C., Pateriya, A., et al., Neuraminidase inhibitors susceptibility profiles of highly pathogenic influenza A (H5N1) viruses isolated from avian species in India (2006–2015). *Antiviral Research*. 2018; 158, 143-6.

[63] Chen, D., Su, A., Fu, Y., Wang, X., Lv, X., Xu, W., et al., Harmine blocks herpes simplex virus infection through downregulating cellular NF-κB and MAPK pathways induced by oxidative stress. *Antiviral Research*. 2015; 123, 27-38.

[64] Moellering, R., Graybill, J., McGowan, J. and Corey, L. Antimicrobial resistance prevention initiative—an update: Proceedings of an expert panel on resistance. *American Journal of Infection Control*. 2007. 35(9), S1-S23.

[65] de Souza, A., de Freitas, D., Antuntes Fernandes, K., D'Avila da Cunha, M., Antunes Fernandes, J., Benetti Gassen, R., et al., Respiratory syncytial virus induces phosphorylation of mTOR at ser2448 in CD8 T cells from nasal washes of infected infants. *Clinical & Experimental Immunology*. 2015; 183(2), 248-57.

[66] Bhagya, N. and Chandrashekar, K. Tetrandrine – A molecule of wide bioactivity. *Phytochemistry*. 2016; 125, 5-13.

[67] Rose, J., Homa, L., Meropol, S., Debanne, S., Bielefeld, R., Hoyen, C. et al., Health impact and cost-effectiveness of a domestically-produced rotavirus vaccine in India: A model based analysis. *PLOS ONE*. 2017; 12(11), e0187446.

[68] Gupta, N. Goyal, M. Wu, CH. Wu, GY. The Molecular and Structural Basis of HBV-resistance to Nucleos(t)ide Analogs. *Journal of Clinical and Translational Hepatology*. 2014; 2(3), 202-11.

[69] Borgia, S., Hedskog, C., Parhy, B., Hyland, R., Stamm, L., Brainard, D., et al., Identification of a Novel Hepatitis C Virus Genotype From Punjab, India: Expanding Classification of Hepatitis C Virus Into 8 Genotypes. *The Journal of Infectious Diseases*. 2018; 218(11), 1722-9.

[70] Sinha, S., Gupta, K., Khan, N., Mandal, D., Kohli, M., Das, B. et al., Higher Frequency of HIV-1 Drug Resistance and Increased Nucleoside Reverse Transcriptase Inhibitor Mutations among the HIV-1 Positive Antiretroviral Therapy–Naïve patients Coinfected With Mycobacterium tuberculosis Compared With Only HIV Infection in India. *Infectious Diseases: Research and Treatment*. 2018; 11, p. 117863371878887.

[71] Nesakumar, M., Haribabu, H., Cheedarla, N., Karunaianantham, R., Kailasam, N., Sathyamurthi, P., et al., Transmitted HIV-1 Drug Resistance in a Treatment-Naive Cohort of Recently Infected Individuals from Chennai, India. *AIDS Research and Human Retroviruses*. 2019; 35(8), 775-779.

[72] Karade, S., Sen, S. and Sashindran, V. Absence of Integrase Strand Transfer Inhibitor Associated Resistance in Antiretroviral Therapy Naïve and Experienced Individuals from Western India. *AIDS Research and Human Retroviruses*. 2019; 35(6), 567-71.

[73] Chauhan, C., Lakshmi, P., Sagar, V., Sharma, A., Arora, S. and Kumar, R. Primary HIV Drug Resistance among Recently Infected Cases of HIV in North-West India. *AIDS Research and Treatment*, 2019, 1-8.

[74] Jain, J., Kumar, A., Narayanan, V., Ramaswamy, R., Sathiyarajeswaran, P., Shree Devi, M., et al., Antiviral activity of ethanolic extract of Nilavembu Kudineer against dengue and chikungunya virus through *in vitro* evaluation. *Journal of Ayurveda and Integrative Medicine*. 2019 (https://doi.org/10.1016/j.jaim.2018. 05.006).

[75] Pathak, H., Mohan, M. and Ravindran, V. Chikungunya arthritis. *Clinical Medicine.* 2019; 19(5), 381-85.

[76] Hasan, S., Saeed, S., Panigrahi, R. and Choudhary, P. Zika virus: A global public health menace: A comprehensive update. *Journal of International Society of Preventive and Community Dentistry.* 2019; 9(4), 316.

[77] Plowright, R., Becker, D., Crowley, D., Washburne, A., Huang, T., Nameer, P., et al., Prioritizing surveillance of Nipah virus in India. *PLOS Neglected Tropical Diseases.* 2019; 13(6), .e0007393.

[78] Pinto, D., DeMarino, C., Pleet, M., Cowen, M., Branscome, H., Al Sharif, S., et al., HTLV-1 Extracellular Vesicles Promote Cell-to-Cell Contact. *Frontiers in Microbiology.* 2019; 10.

[79] Taylor, G., Long, H., Brooks, J., Rickinson, A. and Hislop, A. The Immunology of Epstein-Barr Virus–Induced Disease. *Annual Review of Immunology.* 2015; 33(1), 787-821.

[80] India, M. Antimicrobial Resistance (AMR) *Containment: National Centre for Disease Control (NCDC).* [online] Ncdc.gov.in. Available at: https://ncdc.gov.in/index1.php?lang=1&level=1&sublinkid=145 &lid=74 [Accessed 22 Oct. 2019].

[81] Nhp.gov.in. *National Vector Borne Disease Control Programme | National Health Portal of India.* [online] Available at: https://www. nhp.gov.in/national-vector-borne-disease-control-programme_pg [Accessed 22 Oct. 2019].

[82] India, M. *Technical Guidelines for H1N1:: National Centre for Disease Control (NCDC).* [online] Ncdc.gov.in. Available at: https://ncdc.gov.in/ index1.php?lang=1&level=1&sublinkid=409&lid=359 [Accessed 22 Oct. 2019].

[83] Naco.gov.in. *National AIDS Control Organization* | MoHFW | GoI. [online] Available at: http://naco.gov.in/ [Accessed 22 Oct. 2019].

In: Antimicrobial Resistance: Opportunities and Challenges ISBN: 978-1-53617-943-9
Editors: P. Kumar Prabhakar and V. Mishra

Chapter 4

FUNGAL RESISTANCE TO ANTIBIOTICS

Yachana Mishra[1,*], Sujit Bose[2,†], Pallavi Nayak[2], Namita Giri[3] and Chaudhery Mustansar Hussain[4]

[1]Department of Zoology, Shri Shakti Degree College, Shankhahari, Ghatampur, Kanpur Nagar, UP, India
[2]School of Pharmaceutical Sciences, Lovely Professional University, Phagwara, Punjab, India
[3]School of Pharmacy, Ferris State University, Big Rapids, Michigan, US
[4]Department of Chemistry and Environmental Science, New Jersey Institute of Technology, Newark, New Jersey, US

ABSTRACT

The utilization of antifungal agents in the treatment of fungal infections can prompt the commencement of antifungal resistance. Resistance has been outlined among all antifungal agents in various species, including *Candida* and *Aspergillus*. Regardless of the progress made so far in antifungal treatments, the conjecture of resistance is still of a significant concern in clinical practice. Fluconazole resistance is also causing specific worry in *non-Candida albicans* species because of the expanded rate of diseases brought about by these species worldwide. As of now different investigational antifungal agents are in preclinical or clinical stage. A few of them have the potential to overcome the resistance observed against the azoles and echinocandins. These classes involve agents that target ergosterol biosynthesis.

Keywords: fungal infection, fungal resistance, antifungal, antibiotics

---

* Corresponding Author's E-mail: yachanamishra@gmail.com.
† Corresponding Author's E-mail: sjtbose308@gmail.com.

1. FUNGAL DISEASES: AN INTRODUCTION

Fungi are uni- or multi-cell eukaryotic organisms that can survive in any environmental condition around the world. Fungi can exist in any size or number of structure such as large mushrooms to minute yeasts and molds. While most fungi don't cause human infections, there are a few hundred species that can cause infection or disease. Fungal infections (mycoses) can range from mild infections like itching to severe life-threatening infections such as cryptococcal meningitis. The term 'antifungal' referred to any single synthetic or natural compound with pharmacological properties that are utilized to treat infections of fungal origin. Clinically, fungal infections are classified based on the site and degree of the disease, followed by how they were acquired and lastly by the destructiveness of the fungal source. Mycoses can be local (superficial cutaneous, subcutaneous) or systemic (deep, blood borne). The source of the fungal disease is of either exogenous (airborne/inward breath, cutaneous exposure, percutaneous inoculation) or endogenous origin (normal flora or reactivated contamination). The destructiveness of the life form arranges as either an essential infection (ailment emerging in a sound host), or shrewd disease (sickness emerging in population with weak immune response) (Grover, 2010).

Antifungal medications speak to a pharmacologically assorted gathering of medications that are important parts in the advanced restorative regimen of mycoses. While antimycotic pharmacology has progressed notably, over the past three decades, regular harmful fungal infections still convey a high mortality: *Candida albicans* (~20 to 40% mortality), *Aspergillus fumigatus* (~50 to 90%), *Cryptococcus neoformans* (~20 to 70%) (Birnbaum, 1990; Walsh and Dixon, 1996). Amphotericin B deoxycholate, a polyene was the primary antimycotic substance found in the year 1958 for the treatment of systemic mycoses. Although, Amphotericin B deoxycholate was a viable specialist, its interest for other topical, oral, and intravenous use was also evident. Griseofulvin was discovered in 1959, representing second category of the antifungals. The resulting presentation did not occur until 1971 when an antimetabolite drug flucytosine entered the market. Azoles first became accessible in 1973 with the appearance of clotrimazole; with the extra azoles that were removed in last five decades: miconazole (1979), ketoconazole (1981), fluconazole (1990), itraconazole (1992), voriconazole (2002), posaconazole (2006), and most recently isavuconazonium (2015) (Niewerth et al., 2003). Terbinafine, an allylamine antifungal was approved by the United States Food and Drug Administration (USFDA) in 1996, for the treatment of local fungal infections. The further advancement in systemic treatment had a premise in amphotericin B lipid preparations, which have increasingly good symptoms. Following lipid preparations of azoles, the other class of antifungal specialists that was exceptionally compelling in treating some of the systemic mycoses, were the echinocandins class. The echinocandins displayed less renal toxicity than amphotericin B, however they still have noteworthy hepatotoxicity and were more costly than the azoles; this viably consigns this class to a second or third line drugs (Prasad et al., 2016). Due to disturbing and quick increment in drug resistant systemic fungal infections, the need of new agents is ever growing (Pappas et al., 1995; Mohr et al., 2008; Reeder et al., 2011; Pappas et al., 2009).

2. Pathogens and Characteristics of Diseases

Various species are responsible for variety of fungal diseases. Table 1 represents different fungal diseases, their causative agents and associated symptoms (de Pauw, 2011).

Table 1. Invasive fungal infections, pathogens and characteristics of disease

Opportunistic Pathogens		
Disease type	Causative agent	Clinical signs and symptoms
Candidiasis	*Candida spp*	Acute disseminated: fever, chills, polymyalgia, polyarthralgia, not tender pinkish skin lesions, and retinal exudates.
Aspergillosis	*Aspergillus spp*	Unremitting fever and pulmonary infiltrates during antibiotic therapy. Chest pain, pleural rubs, pleural effusion, haemoptysis. Halo and air crescent sign on chest radiograph and CT scan. Clinical and radiology sinusitis
Cryptococcosis	*Cryptococcus neoformans*	Flu like symptoms; skin lesions, headache without meningismus
Zygomycosis	*Rhizopus spp* *Absidia spp* *Mucor spp*	Like aspergillosis, more outspoken rhino-cerebral form with serosanguinous nasal discharge
Others	*Malassezia furfur* *Trichosporon spp* *Fusarium spp* *Pseudallescheria boydii* *Scedosporium spp* *Alternaria spp*	Often catheter-associated; pneumonia Skin and lung lesions, Often positive blood cultures. Skin lesions, severe myalgia. Abscess formation with symptoms depending on organ involved. Like aspergillosis; wound infections
Endemic Pathogens		
Blastomyscosis	*Blastomyces dermatitidis*	Ulcerative lesions; Skin and Urogenital tract, Central nervous system
Histoplasmosis	*Histoplasma capsulatum*	Pulmonary infiltrates; mucocutaneous ulcers, Hepatosplenomegaly
Coccidioido mycosis	*Coccidiodis immitis*	Pulmonary infection, Dissemination with osteomyelitis, arthritis, meningitis
Para-coccidioido mycosis	*Paracoccidiodis brasiliensis*	Pulmonary infection, Dissemination to skin, mucosa and lymphnodes.
Penicilliosis	*Penicillium marneffei*	Skin and subcutaneous lesions, lung, lymphadenitis, splenomegaly.

3. Types of Fungal Infections

Different fungal infections and responsible causative agents are represented in Figure 1 (Kauffman et al., 2007; Wheat et al., 2007; Chapman et al., 2008; Perfect et al., 2010).

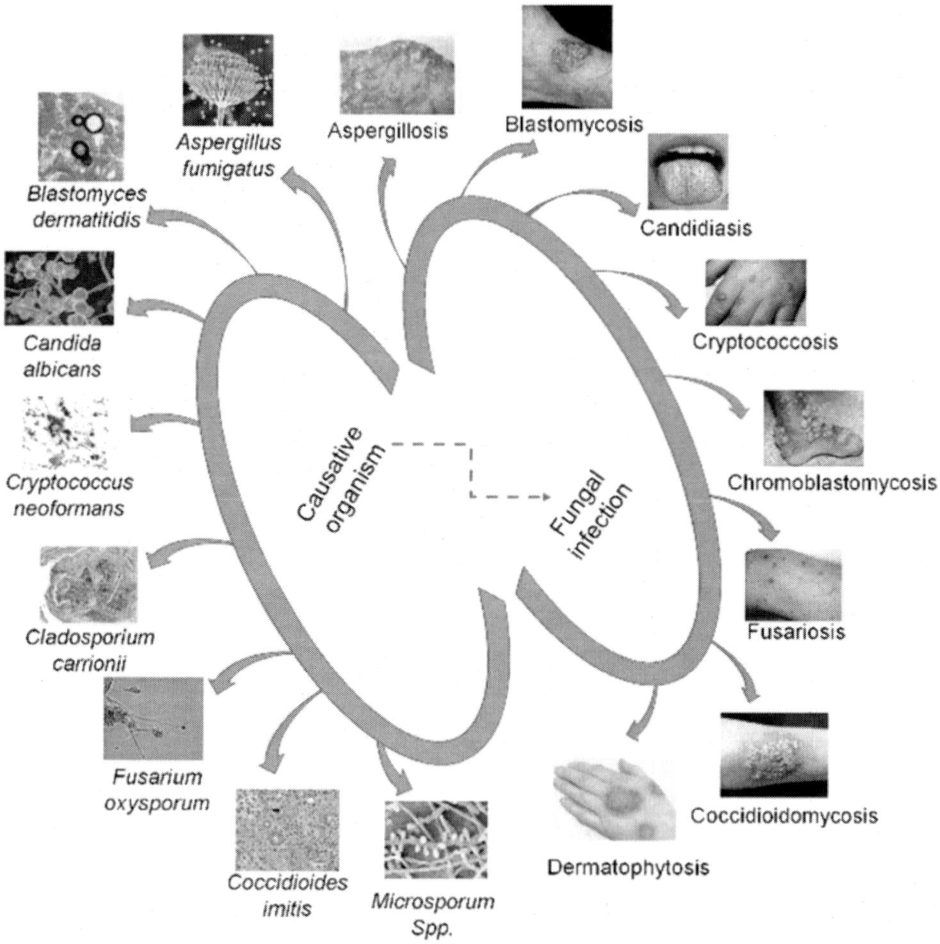

Figure 1. Types of fungal infections and their cusative organisms.

4. Types of Antifungal Drugs

Various antifungal agents (Figure 2) available for the treatment of variety of fungal diseases are discussed below:

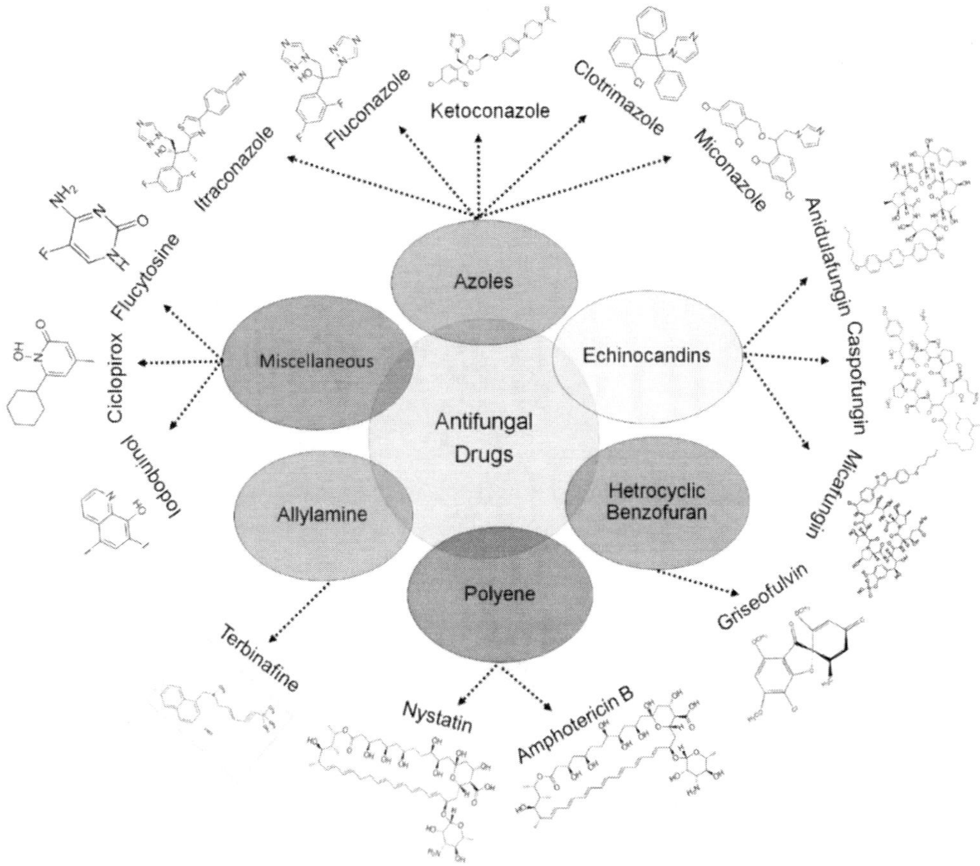

Figure 2. Different category of antifungal drugs.

4.1. Amphotericin B Deoxycholate (AMB-d)

It is indicated for the treatment of life-threatening or potentially life-threatening fungal infections such as aspergillosis, cryptococcosis, blastomycosis, systemic candidiasis, coccidioidomycosis, histoplasmosis, and mucormycosis. AMB-d is also approved for treating the parasitic diseases such as American mucocutaneous leishmaniasis. AMB-d has an off-label use for esophageal candidiasis (both Human Immunodeficiency Virus-HIV infected and non-HIV infected adults and adolescents; HIV-exposed and or infected infants and children), fluconazole-refractory oropharyngeal candidiasis, candidal endophthalmitis, candidal urinary tract infections, visceral leishmaniasis, and ophthalmic aspergillosis.

4.2. Liposomal Amphotericin B (L-AMB)

It has been approved to treat systemic aspergillosis, candidiasis, and cryptococcosis in patients with impaired renal function, and patient's refractory to AMB-d therapy. Additionally, L-AMB is used in empiric antifungal therapy in febrile neutropenic patients, and HIV infected patients with cryptococcal meningitis. L-AMB is also indicated for visceral leishmaniasis, which is a parasitic infection. L-AMB has extensive off-label usage for patients infected with candidiasis, coccidioidomycosis, cryptococcosis, and histoplasmosis or exposed to HIV (Fosses and Waymack, 2018; Miceli and Krishnamurthy, 2019; Karray and McKinney, 2019; Akram and Koirala, 2019).

4.3. Amphotericin B Lipid Complex (ABLC)

Like L-AMB, ABLC is indicated to treat invasive mycoses in patients who are unable to tolerate AMB-d. Off-label use of ABLC is indicated for HIV positive patients with coccidioidomycosis, cryptococcal meningitis, and histoplasmosis; empiric therapy for candidiasis and neutropenic fever; and also in the treatment of the visceral leishmaniasis of parasitic origin (Marques, 2013; Nucci et al., 2015; Akram and Koirala, 2019).

4.4. Nystatin

Nystatin is approved as an oral antifungal agent in the form of "swish-and-swallow" suspension for the treatment of cutaneous, mucocutaneous, and oral Candida infections. Topical use of nystatin is approved for the treatment of mucocutaneous and cutaneous infections with Candida (most commonly *C. albicans*).

4.5. Ketoconazole

Topical application of Ketoconazole is approved for treating tinea species including *T. corporis*, *T. cruris*, *T. pedis*, *T. versicolor*, cutaneous candidiasis, and seborrheic dermatitis. Off-label use of topical ketoconazole is common to treat several oral candidal pathologies including chronic mucocutaneous candidiasis and oral thrush. Ketoconazole can also be used systemically and is approved for treating blastomycosis, coccidioidomycosis, chromomycosis, histoplasmosis, and paracoccidioidomycosis.

4.6. Miconazole

Topical miconazole is used to treat cutaneous and mucocutaneous mycoses, particularly vulvovaginal candidiasis. Oral formulations of miconazole are indicated for oropharyngeal candidiasis.

4.7. Clotrimazole

In topical forms, it is approved to treat tinea corporis, tinea pedis, tinea versicolor, cutaneous candidiasis, and vaginal yeast infections. Oral miconazole is recommended for the treatment of oropharyngeal candidiasis.

4.8. Itraconazole

It is given orally and is approved to treat aspergillosis (pulmonary and extrapulmonary), blastomycosis (pulmonary and extrapulmonary), and histoplasmosis (systemic/disseminated not involving the CNS, cavitary pulmonary histoplasmosis) in both immunocompromised and immunocompetent patients. Itraconazole is also approved to treat oropharyngeal and esophageal candidiasis, and onychomycosis (toe nail or finger nail) in immunocompetent patients.

4.9. Fluconazole

It is indicated in the treatment of esophageal, oropharyngeal, peritoneal, urinary tract and vaginal candidiasis- additionally, fluconazole also treats systemic fungal infections including candidemia, candida pneumonia, and cryptococcal meningitis. Fluconazole is used as a first line prophylactic agent for mycosis in allogeneic hematopoietic stem cell transplant patients. Fluconazole has a variety of off label applications including blastomycosis, empiric antifungal therapy in non-neutropenic ICU patients, candida prophylaxis (ICU with a high risk of invasive Candida spp, transplant patients), and tinea.

4.10. Voriconazole

Voriconazole is indicated for invasive aspergillosis, candidemia in non-neutropenic patients, esophageal candidiasis, and disseminated candidiasis. This drug also treats life-threatening mycoses from fungi like Fusarium spp. Off-label uses for voriconazole are mostly aimed at prophylactic and suppression therapy of fungal infections, including but not limited to aspergillosis, candidiasis, coccidioidomycosis, hematopoietic stem cell transplant patients with or without graft versus host disease, acute myelogenous leukemia, empiric therapy in neutropenic fever, and myelodysplastic syndrome.

4.11. Posaconazole

It is approved for the prophylaxis of both invasive aspergillosis as well as invasive candidiasis. Additionally, posaconazole is used to treat oropharyngeal candidiasis, typically in patients refractory to treatment with fluconazole and itraconazole.

4.12. Terbinafine

Terbinafine is approved for both topical as well as systemic use. Topical terbinafine is approved to treat tinea (pedis, cruris, and corporis). Oral therapy is indicated in the systemic treatment of onychomycosis *T. unguium* and *T. capitis*. The most common off-label use of oral formulations includes the treatment of tinea (cruris, corporis, penis, and manuum) as well as lymphocutaneous and cutaneous sporotrichosis. The echinocandin anidulafungin is only given intravenously and has approval for the treatment of candida infections (esophageal candidiasis, candidemia, Candida peritonitis, and intra-abdominal abscesses when candida is grown in culture or the suspected organism).

4.13. Caspofungin

Caspofungin is approved for the treatment of invasive aspergillosis in patients refractory to Amphotericin B and Itraconazole administered intravenously. It is also approved for the treatment of candida infections (candidemia, esophageal candidiasis, intra-abdominal abscess, peritonitis, and empiric therapy in neutropenic patients). Off-label use includes use with other agents in treating severe candida infections.

4.14. Micafungin

Micafungin is only approved for the intravenous administration for the treatment of esophageal candidiasis, prophylaxis of candida infections, candidemia, candida peritonitis, Candida abscesses, and disseminated candidiasis.

4.15. Griseofulvin

Griseofulvin is only approved as a systemic agent for the treatment of dermatophytoses of the skin, hair, and nails which is severe or refractory to the topical therapy. Specifically, it treats tinea (corporis, pedis, cruris, barbae, capitis, and unguium).

4.16. Flucytosine

It is approved as an adjunct antifungal agent for the treatment of systemic candida or cryptococcus infections. Off-label, flucytosine is also utilized to treat paediatric endocarditis caused by *Aspergillus spp.*

4.17. Ciclopirox

It is approved for the treatment of *T. corporis, T. pedis, T. cruris, T. unguium* (onychomycosis), *T. versicolor* (pityriasis), and the candida infection, moniliasis.

4.18. Iodoquinol

It is a topical agent that is approved in the treatment of *T. corporis, T. pedis, T. cruris, T. unguium, T. capitis*, and candidal intertrigo.

4.19. Clioquinol

It is approved as a combined agent (clioquinol-hydrocortisone) in the United States for the treatment of same spectrum of dermatoses as iodoquinol: *T. corporis, T. pedis, T. cruris,* moniliasis, and candidal intertrigo.

4.20. Potassium Iodide

It is formulated as a saturated solution of potassium iodide (SSKI) has no official anti-fungal approvals; but is used as an off-label treatment for both cutaneous and lymphocutaneous sporotrichosis.

4.21. Zinc Pyrithione

It is not officially approved for its independent anti-fungal use but can be utilized as primary or adjunct therapy in the treatment of mycoses leading to hyperkeratotic skin conditions. A common off-label use is the treatment of tinea (pityriasis) versicolor.

5. Mechanisms of Antifungal Drug Resistance

5.1. Polyenes

Amphotericin B and Nystatin are the classic polyenes (Kurien, 2019). Amphotericin B attaches to sterol the primary fungal plasma membrane and modifies the membrane penetrability and causes the cell death. Amphotericin B additionally causes oxidative damage to fungal cells. Since the 1950s until the discovery of the azoles, polyene class of antifungals was used routinely in the treatment of systemic fungal infections (Masia and Gutierrez, 2002).

Amphotericin B and other polyenes, including Nystatin, act basically by binding to the ergosterol. Interaction of Amphotericin B with sterols brings about the development of aqueous pores (Holz, 1974; Sugar, 1986) that disturbs the membrane structure and its polarity. Such modification enhances the permeability of the fungal membrane to the protons and monovalent cations, and eventually causes the cell death (de Krujff and Demel, 1974; Kerridge, 1980). Although, Amphotericin B has a high affinity towards ergosterol as compared to the other sterols, the host virulence shown by this drug is principally the after effect of its attachment to the cholesterol. Amphotericin B has been involved as an inducer of phagocytic cells. Regardless of its clinical use for over 30 years resistance to the polyene anti-infection agents is uncommon, with safe detaches kept generally to the less common types of candida, for example, *Candida lusitaniae, C. glabrata*, and *C. guillermondii* (Kerridge, 1985; Martins, 1985).

5.2. Fluoropyrimidines

Flucytosine, or 5-fluorocytosine (5-FC), is a synthetic fluorinated pyrimidine. It is transported into susceptible fungal cells by the enzymatic activity of cytosine permease and eventually converted by cytosine deaminase to the active form, fluorouracil. Fluorouracil inserts itself into the RNA backbone instead of uracil and also obstructs the thymidylate amalgamations, a basic compound required for the DNA synthesis.

Flucytosine has inhibitory action against numerous fungi (Hamilton-Miller, 1973; Pappagianis et al., 1979; Guinet et al., 1983; Barchiesi et al., 2006).

5.3. Azoles

The azole antifungals in clinical use contain either imidazoles (ketoconazole, miconazole, clotrimazole, econazole, and butoconazole) or triazoles (itraconazole, fluconazole, terconazole). The newer azoles include voriconazole, posaconazole, ravuconazole, and albaconazole. These azoles prevent ergosterol union in the fungal cell wall through their activity on the cytochrome P450-dependent enzyme lanosterol 14α-demethylase (Kartsonis et al., 2005). Inhibition of 14α-demethylase depletes ergosterol, which conducts the arrangement of membranes with altered structure and capacity of sterol precursors, particularly 14α-methyl fecosterol and 14α-methyl-ergosta-8, 24 (28)- dien-3β, 6α-diol (Nozawa and Morita, 1986; Kelly et al., 1995; Kelly et al., 1997). Although azoles are in general strongly against different *Candida* species, they are less active against the rising pathogen *Candida krusei* (although the new triazoles, such as voriconazole, have intense activity against *C. krusei*). Mammalian cholesterol biosynthesis is also interrupted by the azoles at 14α-demethylation phase; in any case, the dose required to deliver a same response of restraint is a lot higher than that required for fungi (Vanden, 1982; Hitchcock et al., 1990; Vanden, 1991). Human sterol biosynthesis is most noticeably affected by ketoconazole.

5.4. Echinocandins

This new class of echinocandins comprises of parenteral caspofungin, micafungin, and anidulafungin. These antifungals prevent fungal cell wall synthesis of through enzyme 1, 3-β-d-glucan synthase, anticipating the arrangement of 1,3-β-d-glucan, a fundamental segment of the fungal cell wall. These antifungals create a debilitated cell wall bringing about fungal cell lysis and are considered candidacidal (Vanden, 1978).

Echinocandin like substances have powerful antifungal action against *C. albican*, *A. fumigatus*, and *Pneumocystis carinii*. Conversely, this class of antifungals has limited action against *C. neoformans*. This lack of efficient activity is proposed to be expected due to absence of 1, 3-β-D-glucan in the cryptococcal cell wall. A report by Feldmesser et al., demonstrated that albeit 1, 3-β-D-glucan is available in the cryptococcal cell wall, yet to a lesser degree than in candida species (Feldmesser et al., 2000). In this way, deficiency of 1, 3-β-D-glucanperse does not clarify the overall absence of adequacy of echinocandins and recommends extra system (Bartizal et al., 1997) suggested that moderately low adequacy of caspofungin against *C. neoformans* may result from diminished action against

C. neoformans glucan synthase or from yet unfamiliar component of activity employable in other fungal pathogens but yet not in *C. neoformans*.

6. Clinical Implications of Antifungal Resistance

Antifungal resistance is now emerging as one of the clinically significant issue in patients requiring long term prophylactic or who have a background marked by antifungal use. Also, resistance advancement is affecting the study of disease transmission of *Candida* contamination evidenced by a change from the *albicans* to the *non-albicans Candida* (Feldmesser et al., 2000). Increasing incidences proposes that resistance to azoles originated from prolonged treatment with prophylactic use of these drugs, particularly fluconazole. Out of need, HIV-contaminated patients keep on being regulated support dosages to control intermittent fungal diseases, especially cryptococcal meningitis and oropharyngeal candidiasis (Feldmesser et al., 2000). Most of the AIDS patients experience at least one fungal infection during the disease, with oropharyngeal candidiasis.

In spite of the fact that fluconazole is generally successful in treating oropharyngeal candidiasis, numerous patients experience recurrent infection because of an inadequate annihilation of the infecting yeasts (White et al., 1998; Georgopapadakou and Walsh, 1994; Gale et al., 1975). Unlike the circumstance with candida circulation system contaminations, with high occurrence, the frequency of oesophageal candidiasis in the HIV/AIDS infection is reduced. This abatement could be credited to the host invulnerable reconstitution achieved by the emergence of an exceptionally dynamic antiretroviral treatment i.e., Highly Active Antiretroviral Therapy (HAART). A reduced occurrence in oesophageal candidiasis in the both adult and kids is recorded in the study of Kaplan et al., (Fothergill et al., 1996) who studied disease transmission of HIV-related artful infection in the United States during HAART. The occurrence of oesophageal candidiasis dropped significantly somewhere in the range of 1996 and 1997; in any case, an exasperating pattern of expanding rate was observed during the year 1997 and 1998. This perception shows that it is judicious to follow the study of disease transmission and other infections, and that it's too soon to conclude any firm inferences with respect to the observed patterns. The abatement in the frequency of contaminations, including candidiasis, has been credited to the reclamation of explicit resistance (Kalpan et al., 2000; Autran et al., 1997). It is basic to explain the perception that occurrence of an astute diseases, including candidiasis, is diminishing in the patients contaminated with HIV is restricted to developed nations (basically the US and Western Europe); in any case, the frequency is still high in developing nations, especially sub Saharan Africa where half of the population is tainted with the HIV infection.

7. Development of New Antifungal Agents

Amphotericin B, has been used for over 30 years, and fluconazole was not discovered until the 1980s (Angel et al., 1998; Neely and Ghannoum, 2000). The fast quest for the next generation antifungals represents the unmistakable requirement for the new or modified derivatives of currently used antifungals that keep risk of resistance advancement away and furthermore successfully treat the infections brought about by strains previously evident of resistance phenotype. Notwithstanding, the wide substrate specificity shown by the efflux siphons requires that all the new investigational agents be tried as potential substrates. Correspondingly, the investigational medications ought to likewise be observed for their capability to make cross-obstruction among the polyenes, particularly amphotericin B. The revelation of fungal efflux siphons raises an important question whether the antifungal drugs are joined with an efflux pumps inhibitor to build viability? This circumstance is undifferentiated from the blend of a β-lactam anti-toxin with a β-lactamase inhibitor. MC-510027 is an efflux inhibitor that has been appeared to decrease the minimum inhibitory concentrations (MIC) estimations of three azoles and terbinafine by 1 to 3 sets of extent against a few types of candida (Neely and Ghannoum, 2000). Despite the absence of clinical information in animal studies or human trials, this methodology unmistakably has the potential, and may emerge as promising antifungal candidate.

8. Prevention and Control of Antifungal Resistance

In spite of the fact that the development of antifungal resistance affected fluconazole prophylactic use up to some extent, specific measures to reduce or avoid the stifle rise of antifungal resistance have not been yet characterized. In any case, approaches undifferentiated from those prescribed for antibacterial could be recommended (Hossain and Ghannoum, 2000; Chamberland et al., 1999; Levy, 1990). These measures include: (i) reasonable utilization of antifungals; (ii) treatment with the appropriate antifungal (in situations where the etiological agent is known); and (iii) directing observation to decide the genuine recurrence of antifungal resistance. Most of the mechanism involved in resistance advancement presently can't seem to be distinguished makes it unreasonable to portray clinical disengages at the sub-atomic level. Nonetheless, an alluring methodology proposed by White et al. is to utilize molecular procedures to assess the general soundness of a cell in the presence and absence of a drug. Also, enhancements to the vulnerability testing techniques might be useful in reducing the use of agents that have known or reported resistance to the infecting fungi (White et al., 1998).

9. Adverse Effects of Antifungals

The systemic polyene antifungal Amphotericin B (formulated as AMB-d, L-AMB, ABLC, and ABCD), has severe adverse reactions. AMB-d therapy has reported side effects such as hypotension, chills, headache, hypokalemia, hypomagnesemia, anemia, renal insufficiency, renal function abnormalities, injection site pain, nausea, vomiting, rigors, and fever (Cohen, 1992). L-AMB therapy has lowered the incidence of renal function abnormalities as compared to the AMB-d, but still carries a risk of nephrotoxicity. The most common adverse events caused by L-AMB therapy includes hypertension, hypotension, tachycardia, localized phlebitis, chills, headache, skin rash, electrolyte abnormalities (hypokalemia, hypomagnesemia, hyponatremia, hyponatremia), hyperglycemia, abnormal liver function tests. ABLC also carries a risk of nephrotoxicity leading to increased serum creatinine, fever related to infusion, rigors, and chills; but these risks are less than the treatment regimens including AMB-d.

9.1. Nystatin

It is approved only for the topical and oral "swish-and-swallow" applications because of severe systemic side effects. Adverse events related to topical nystatin include mild contact dermatitis, with the most severe side effect being Stevens-Johnson syndrome. Oral "swish-and swallow" nystatin carries a lower risk of hypersensitivity reactions than topical formulations; there are also reports of diarrhoea, nausea, vomiting, and abdominal pain.

9.2. Azoles

Azoles are typically well tolerated, but frequently cause nausea, vomiting, diarrhoea, and abdominal pain. Hepatotoxicity (elevated liver functions tests, hepatitis, cholestasis, and or fulminant liver failure) is a common adverse reaction associated with all azoles. Each of the azole drugs has unique as well as general adverse events. Some of the common side effects are listed below:

1. Ketoconazole reported orthostatic hypotension, thrombocytopenia, pruritis, rash, myalgias, and a rare suppression of glucocorticoid production in the adrenal glands. Of note, ketoconazole also correlates with a significantly higher gastrointestinal distress than other azoles.
2. Fluconazole has been shown to cause mild headaches, dizziness, and alopecia in high doses (Lai et al., 2008).

3. Itraconazole has a triad of heart failure-like symptoms, hypertension, peripheral edema, and hypokalemia. An increased risk of herpes zoster activation or reactivation, headache, dizziness, and fatigue has been reported.
4. Voriconazole has the most numerous and unique of all the side effects in the azole class. These include vision changes, auditory and/or visual hallucinations, neurotoxicity, photosensitivity rash, photophobia, periostitis, cardiotoxicity, and alopecia.
5. Posaconazole is most commonly link to thrombophlebitis secondary to peripheral intravenous catheters, hypertension, hypotension, headache, rash, hypokalemia, and thrombocytopenia. Another reported adverse event is a rare prolongation of the QT interval (Shlaes et al., 1997; Park et al., 2009).
6. Isavuconazole has more severe gastrointestinal side effects than most of the other azoles. Other reported adverse events include a headache, hypokalemia, dyspnea, cough, and peripheral edema.
7. Miconazole has no reported serious adverse events but commonly causes contact dermatitis, burning/stinging, and pruritis at the application site.
8. Clotrimazole has no severe adverse reactions, but commonly causes irritation, burning/stinging, pruritis, urticaria, and possible blistering at the application site.

9.3. Terbinafine

This allylamine most commonly results in central nervous system (CNS) side effects, with a headache being the most frequently reported symptom. Other adverse events include skin rashes, diarrhoea, dyspepsia, and upper respiratory inflammation or infections.

9.4. Echinocandins

Like many other antifungal agents, these can result in hepatotoxicity.

1. Anidulafungin is associated with hypotension, peripheral edema, insomnia, hypokalemia, hypomagnesemia, increased risk of urinary tract infections, dyspnea, and fever.
2. Caspofungin has links to hypotension, peripheral edema, tachycardia, chills, headache, rash, anemia, localized phlebitis, respiratory failure, and infusion-related reactions.
3. Micafungin correlates with phlebitis, anemia, transaminitis, hyperbilirubinemia, renal failure, and fever.

9.5. Griseofulvin

It has numerous potential adverse events, the most commonly reported adverse events includes skin rash and urticaria. More severe complications can occur and include an erythema multiforme-like drug reaction, skin photosensitivity, leukopenia (rare), granulocytopenia, and hepatotoxicity.

9.6. Flucytosine

It was wide range of side effects affecting almost all physiological systems. The most commonly reported adverse effects are: cardiovascular (cardiotoxicity, chest pain), central and peripheral nervous systems (confusion, headache hallucination, parkinsonian-like syndrome, peripheral neuropathy), integumentary system (pruritis, urticaria, rash), gastrointestinal (abdominal pain, nausea, vomiting, GI hemorrhage), circulatory system (agranulocytosis, aplastic anemia, pancytopenia, eosinophilia), hepatic (acute hepatic injury / insufficiency / necrosis), and renal (acute kidney injury, renal failure).

9.7. Ciclopirox

These have no major reported serious adverse reactions, but common being mild reaction includes skin irritation/burning, contact dermatitis, and pruritis.

9.8. Quinolines

Quinolines (like clioquinol and iodoquinol) reported dry skin, contact dermatitis, allergic reaction, rapid hair growth in areas the agent is applied, and folliculitis as adverse reactions.

9.9. Potassium Iodide

A saturated solution of potassium iodide, has reported several serious adverse reactions, which includes arrhythmia, GI bleeding, angioedema, parotitis, thyroid adenoma, and goitre. More common but mild reactions include possible metallic taste, urticaria, acne, cutaneous haemorrhage, numbness, and paraesthesia.

9.10. Zinc Pyrithione

No reported serious adverse reactions are reported but it is more commonly correlates with mild skin irritation.

10. Contraindications of Antifungals

10.1. Amphotericin B

All formulations of Amphotericin B (AMB-d, L-AMB, ABLC, and ABCD) are contraindicated in patients with known or possible hypersensitivity to Amphotericin B or any components of the LAMB, ABLC or ABCD formulations.

1. AMB-d carries two sincere warnings first, only to be used for invasive and potentially life-threatening mycoses and must be avoided in non-invasive fungal infections (oral thrush, esophageal candidiasis, and vaginal candidiasis in patients with neutrophil counts within normal limits) and second exercising extreme caution in patients with renal impairment and or electrolyte abnormalities due to the risk of accidental overdose
2. L-AMB, ABLC, and ABCD do not carry sincere warnings but require caution in patients with renal impairment.

10.2. Nystatin

It is contraindicated in patients with hypersensitivity to the drug or any additional components in a combination formulation.

10.3. Azole

All azoles should be avoided in patients with hypersensitivities to azole drugs or components and must be used with caution in patients with impaired renal or hepatic functions

1. Fluconazole should be cautiously administered in patients with electrolyte abnormalities; torsades de pointes; and or medical history, family history, and or current QT prolongation.

2. Itraconazole has a sincere warning against its use in treating onychomycosis in patients with congestive heart failure (CHF). Itraconazole is contraindicated in pregnancy, left ventricular dysfunction, and current or active congestive heart failure. This drug should be used cautiously in patients with cystic fibrosis, cardiovascular diseases, pulmonary diseases, and or in geriatric patients.
3. Ketoconazole carries several sincere warnings such as (i) used only in situtaions when another effective antifungal, cannot be tolerated or unavailable, (b) has significant risk of hepatotoxicity, even in patients without predisposing factors and thus any treatment with ketoconazole should include close liver function monitoring, (c) has several contraindicated drug interactions that may cause QT prolongation by increasing levels of cisapride, disopyramide, dofetilide, dronedarone, methadone, quinidine, or ranolazine.
4. Voriconazole is contraindicated in galactose malabsorption/ intolerance, Lapp lactase deficiency, glucose malabsorption, uncorrected electrolyte abnormalities, and pregnancy. This agent should be utilized with caution in patients with a medical or family history of QT prolongation, torsades de pointes, and or hematologic malignancy.
5. Isavuconazole is contraindicated in patients with familial short QT syndrome and should be used with caution in patients with hematologic malignancies.
6. Posaconazole is contraindicated in pregnancy. Caution is advisable in patients with electrolyte abnormalities, renal insufficiency, cardiomyopathy, torsades de points, or medical history/family history/congenital prolonged QT interval.

10.4. Terbinafine

It should be utilized with caution or avoided in patients with hyper-sensitivity reactions, depression, gastrointestinal issues, liver failure, and immune suppression secondary to hematologic effects.

10.5. Echinocandins

All of the Echinocandins are contraindicated in patients with hypersensitivities to any of the echinocandin drug class, or components. Caspofungin should be used with caution in patients with impaired hepatic functions.

10.6. Griseofulvin

Treatment with griseofulvin should include considerations for potential adverse events in susceptible patients, and those with existing disease states; particularly patients with a hypersensitivity to griseofulvin or penicillins (there is a possible cross-reaction between penicillins and griseofulvin), hepatic failure, patients with known porphyrias, and patients that are pregnant or nursing.

10.7. Flucytosine

It carries a black box warning that agent should be used with extreme caution in renal impairment and that hematologic, hepatic and renal function should be closely monitored. This agent is contraindicated in patients with hypersensitivities to this drug/components, first trimester pregnancies, and women that are breastfeeding. Caution is advisable with this agent in patients with renal impairment, hepatic impairment, bone marrow depression, and pregnant patients in their second or third trimester.

10.8. Quinolones

The quinolones (iodoquinol and clioquinol) are contraindicated in patients with hypersensitivities to the antifungal agents.

10.9. Others

Antifungals used only as topical agents including ciclopirox, potassium iodide, and zinc pyrithione should be avoided in patients with hypersensitivities to these agents.

11. Overcoming Antifungal Resistance

The development of antimicrobial drug resistance is a practically unavoidable procedure that is universal in the microbial world. In spite of the fact that, the advancement of antifungal drug resistance has not paralleled their antibacterial anti-infection partners, the monetary features related with fungal infections remains unsuitably high. Hence, beating antifungal resistance can be considered as the backbone for improving restorative methodologies to treat mycoses (Waldorf and Polak, 1983). One of the main reasons leading to antifungal-drug resistance is the wrong and inappropriate use of antifungals. Various examinations have explained the relationship of ill-advised antifungal drug introduction and subsequent development of resistance. Along these lines antifungal treatment should utilize powerful operators to ease the advancement of drug resistance. Moreover, the portion and range of activity of the antimicrobial agent has significant repercussions due to their potential consequences for the human microbiome. It has been distinguished that the lop-sidedness in the host microbial assorted variety gives pernicious impact and once in a while advancement of resistance in less vulnerable organisms. Thinking about these variables, the potential systems to set out to beat antifungal resistance will be examined further.

Conclusion

The declaration of protection from antifungal resistance is the sensible and unavoidable outcome of utilizing these agents to treat human infections. With expanded use and accessibility of various classes of antifungal agents, it is foreseen that see an expanding number and assortment of fungal species impervious to these agents will be observed. Proceeded with endeavours to think about the instruments of antifungal resistance and the advancement of exploratory frameworks, in which individual resistance systems can be contemplated will be the significant segments of a procedure to confine the rise of resistance for these agents and to create more secure and increasingly intense mixes for what's to come.

References

[1] Akram SM, Koirala J. Stat Pearls. Stat Pearls Publishing; Treasure Island (FL): 2019. *Coccidioidomycosis.*

[2] Akram SM, Koirala J. Stat Pearls. Stat Pearls Publishing; Treasure Island (FL): 2019. *Histoplasmosis.*

[3] Angel JB, Kumar A, Parato K, et al. Improvement in cell-mediated immune functions during potent anti-human immunodeficiency virus therapy with ritonavir plus saquivir. *J Infect Dis* 1998; 177: 898-904.

[4] Autran B, Carcelain G, Li TS, et al. Positive effect of combined antiretroviral therapy on CD4 T-cell homeostasis and function in advanced HIV disease. *Science* 1997; 277: 112-6.

[5] Barchiesi F, Spreghini E, Tomassetti S, Della Vittoria A, Arzeni D, Manso E, et al. Effects of caspofungin against Candida guilliermondii and Candida parapsilosis. *Antimicrob Agents Chemother.* 2006;50(8):2719-27.

[6] Bartizal K, Gill CJ, Abruzzo GK, Flattery AM, Kong L, Scott PM, et al. In vitro preclinical evaluation studies with the echinocandin antifungal MK-0991 (L-743,872). *Antimicrob Agents Chemother.* 1997;41(11):2326-32.

[7] Birnbaum JE. Pharmacology of the allylamines. *J. Am. Acad. Dermatol.* 1990; 23(4 Pt 2):782-5.

[8] Chamberland S, Blais J, Cotter DP, et al. Impact of MC-510,027, a fungal efflux pump inhibitor, on the susceptibility of clinical isolates of Candida spp. to antifungal agents [abstract no. 1270]. *Proceedings of the 39th Interscience Conference on Antimicrobial Agents and Chemotherapy*; 1999 Sept 26-29; San Francisco.

[9] Chapman SW, Dismukes WE, Proia LA, Bradsher RW, Pappas PG, Threlkeld MG, et al. Infectious Diseases Society of America. Clinical practice guidelines for the management of blastomycosis: 2008 update by the Infectious Diseases Society of America. *Clin. Infect. Dis.* 2008; 46(12):1801-12.

[10] Cohen ML. Epidemiology of drug resistance: implications for a post-antimicrobial era. *Science* 1992; 257: 1050-5.

[11] de Kruijff B, Demel RA. Polyene antibiotic-sterol interactions in membranes of Acholeplasma laidlawaii cells and lecithin liposomes. III. Molecular structure of the polyene antibioticcholesterol complexes. *Biochim Biophys Acta.* 1974; 339: 57-70.

[12] de Pauw BE. What are fungal infections?. *Mediterranean journal of hematology and infectious diseases.* 2011;3(1), e2011003.

[13] Feldmesser M, Yvonne K, Mednick A, et al. The effect of the echinocandin analogue caspofungin on cell wall glucan synthesis by Cryptococcus neoformans. *J Infect Dis* 2000; 182:1791-5.

[14] Fosses Vuong M, Waymack JR. Stat Pearls. Stat Pearls Publishing; Treasure Island (FL): 2018. *Aspergillosis.*

[15] Fothergill WA, Sutton DA, Rinaldi MG. An in vitro head-to-head comparison of Schering 5692, amphotericin B, fluconazole, and itraconazole against a spectrum of filamentous fungi [abstract no. F89]. *Proceedings of the 36th Interscience Conference on Antimicrobial Agents and Chemotherapy;* 1996 Sep 15-18; New Orleans, Louisiana: American Society for Microbiology, 115.

[16] Gale EF, Johnson AM, Kerridge D, et al. Factors affecting the changes in amphotericin B sensitivity of Candida albicans during growth. *J Gen Microbiol* 1975; 87: 20-36.

[17] Georgopapadakou NH, Walsh TJ. Human mycoses: drugs and targets for emerging pathogens. *Science* 1994; 264: 371-3.

[18] Grover ND. Echinocandins: A ray of hope in antifungal drug therapy. *Indian J Pharmacol.* 2010; 42 (1):9-11.

[19] Guinet R, Chanas J, Goullier A, Bonnefoy G, Ambroise-Thomas P. Fatal septicemia due to amphotericin B-resistant Candida lusitaniae. *J Clin Microbiol.* 1983;18(2):443-4.

[20] Hamilton-Miller JMT. Chemistry and biology of the polyene macrolide antibiotics. *Bacteriol Rev.* 1973; 37: 166-96.

[21] Hitchcock C, Dickinson K, Brown SB, et al. Interaction of azole antifungal antibiotics with cytochrome P450-dependent 14α-sterol demethylase purified from *Candida albicans. J Biochem (Tokyo)* 1990; 266: 475-80.

[22] Holz RW. The effects of the polyene antibiotics nystatin and amphotericin B on thin lipid membranes. *Ann N Y Acad Sci.* 1974; 235: 469-79.

[23] Hossain MA, Ghannoum MA. New investigational antifungal agents for treating invasive fungal infections. *Expert Opin Investig Drugs* 2000; 9: 1797-813.

[24] Kaplan JE, Hanson D, Dworkin MS, et al. Epidemiology of human immunodeficiency virus: associated opportunistic infections in the United States in the era of highly active antiretroviral therapy. *Clin Infect Dis* 2000; 30: s5-s14.

[25] Karray M, McKinney WP. Stat Pearls. Stat Pearls Publishing; Treasure Island (FL): 2019. *Tinea (Pityriasis) Versicolor.*

[26] Kartsonis N, Killar J, Mixson L, Hoe CM, Sable C, Bartizal K, et al. Caspofungin susceptibility testing of isolates from patients with esophageal candidiasis or invasive candidiasis: relationship of MIC to treatment outcome. *Antimicrob Agents Chemother.* 2005; 49(9):3616-23.

[27] Kauffman CA, Bustamante B, Chapman SW, Pappas PG., Infectious Diseases Society of America. Clinical practice guidelines for the management of sporotrichosis: 2007 update by the Infectious Diseases Society of America. *Clin. Infect. Dis.* 2007; 45(10):1255-65.

[28] Kelly SL, Lamb DC, Corran AJ, et al. Mode of action and resistance to azole antifungals associated with the formation of 14α-methyl-ergosta-8,24(28)-dien-3β,6α-diol. *Biochem Biophys Res Commun* 1995; 207: 910-5.

[29] Kelly SL, Lamb DC, Kelly DE, et al. Resistance to fluconazole and cross-resistance to amphotericin B in Candida albicans from AIDS patients caused by defective sterol Delta (5,6) desaturation. *FEBS Lett* 1997; 400: 80-2.

[30] Kerridge D. The plasma membrane of *Candida albicans* and its role in the action of antifungal drugs. In: Gooday GW, Lloyd D, Trinci APJ, editors. *The Eukaryotic Microbial Cell.* Cambridge: Cambridge University Press, 1980: 103.

[31] Kerridge D. The protoplast membrane and antifungal drugs. In: Peberdy JF, Ferenczy L, editors. *Fungal Protoplasts, Applications in Biochemistry and Genetics.* New York: Marcel Dekker, 1985: 135.

[32] Kurien G, Sugumar K, Chandran V. Stat Pearls. Stat Pearls Publishing; Treasure Island (FL): 2019. *Chromoblastomycosis (Chromomycosis).*

[33] Lai CC, Tan CK, Huang YT, Shao PL, Hsueh PR. Current challenges in the management of invasive fungal infections. *J. Infect. Chemother.* 2008; 14(2):77-85.

[34] Levy SB. Starting life resistance-free. *N Engl J Med* 1990; 323: 335-7.

[35] Marques SA. Paracoccidioidomycosis: epidemiological, clinical, diagnostic and treatment up-dating. *An Bras Dermatol.* 2013; 88(5):700-11.

[36] Martins MD, Rex JH. Resistance to antifungal agents in the critical care setting: Problems and perspectives. *New Horizons.* 1996; 4: 338-44.

[37] Masia Canuto M, Gutierrez Rodero F. Antifungal drug resistance to azoles and polyenes. *Lancet Infect Dis.* 2002; 2(9):550-63.

[38] Miceli A, Krishnamurthy K. Stat Pearls. Stat Pearls Publishing; Treasure Island (FL): 2019. *Blastomycosis.*

[39] Mohr J, Johnson M, Cooper T, Lewis JS, Ostrosky-Zeichner L. Current options in antifungal pharmacotherapy. *Pharmacotherapy.* 2008; 28 (5):614-45.

[40] Neely MN, Ghannoum MA. The exciting future of antifungal therapy. *Eur J Clin Microbiol Infect Dis* 2000; 19: 897-914.

[41] Niewerth M, Kunze D, Seibold M, Schaller M, Korting HC, Hube B. Ciclopirox olamine treatment affects the expression pattern of Candida albicans genes encoding virulence factors, iron metabolism proteins, and drug resistance factors. *Antimicrob. Agents Chemother.* 2003; 47(6):1805-17.

[42] Nozawa Y, Morita T. Molecular mechanisms of antifungal agents associated with membrane ergosterol: dysfunction of membrane ergosterol and inhibition of ergosterol biosynthesis. In: Iwata K, Vanden Bossche H, editors. *In vitro and in vivo evaluation of antifungal agents.* Amsterdam: Elsevier Science Publishers, 1986: 111.

[43] Nucci F, Nouér SA, Capone D, Anaissie E, Nucci M. Fusariosis. *Semin Respir Crit Care Med.* 2015; 36(5):706-14.

[44] Pappagianis D, Collins MS, Hector R, Remington J. Development of resistance to amphotericin B in *Candida lusitaniae* infecting a human. *Antimicrob Agents Chemother.* 1979; 16(2):123-6.

[45] Pappas PG, Kauffman CA, Andes D, Benjamin DK, Calandra TF, Edwards JE, et al. Infectious Diseases Society of America. Clinical practice guidelines for the management of candidiasis: 2009 update by the Infectious Diseases Society of America. *Clin. Infect. Dis.* 2009;48(5):503-35.

[46] Pappas PG, Kauffman CA, Perfect J, Johnson PC, McKinsey DS, Bamberger DM, et al. Alopecia associated with fluconazole therapy. *Ann. Intern. Med.* 1995; 123(5):354-7.

[47] Park BJ, Wannemuehler KA, Marston BJ, Govender N, Pappas PG, Chiller TM. Estimation of the current global burden of cryptococcal meningitis among persons living with HIV/AIDS. *AIDS.* 2009;23(4): 525-30.

[48] Perfect JR, Dismukes WE, Dromer F, Goldman DL, Graybill JR, Hamill RJ, et al. Clinical practice guidelines for the management of cryptococcal disease: 2010 update by the infectious diseases society of America. *Clin. Infect. Dis.* 2010; 50(3):291-322.

[49] Prasad R, Shah AH, Rawal MK. Antifungals: Mechanism of Action and Drug Resistance. *Adv. Exp. Med. Biol.* 2016; 892:327-349.

[50] Reeder NL, Xu J, Youngquist RS, Schwartz JR, Rust RC, Saunders CW. The antifungal mechanism of action of zinc pyrithione. *Br. J. Dermatol.* 2011;165 Suppl 2:9-12.

[51] Shlaes DM, Gerding DN, John Jr JF, et al. Society for Healthcare Epidemiology of America and Infectious Diseases Society of America Joint Committee on the Prevention of Antimicrobial Resistance: guidelines for the prevention of antimicrobial resistance in hospitals. *Clin Infect Dis* 1997; 25: 584-99.

[52] Sugar AM. The polyene macrolide antifungal drugs. In: Peterson PK, Verhoef J, editors. *Antimicrobial Agents* Vol. 1. Amsterdam, The Netherlands: Elsevier Science Publishers, 1986:229-244.

[53] Vanden Bossche H, Willemsens G, Cools W, et al. Biochemical effect of miconazole on fungi. II. Inhibition of ergosterol biosynthesis in *Candida albicans*. *Chem Biol Interact* 1978; 21:59-78

[54] Vanden Bossche H, Willemsens G. Effect of the antimycotics, miconazole and ketoconazole on cytochrome P450 in yeast microsomes and rat liver microsomes. *Arch Int Physiol Biochim* 1982; 90: B218-9.

[55] Vanden Bossche H. Ergosterol biosynthesis inhibitors. In: Prasad R, editor. *Candida albicans.* Berlin, Germany: Springer-Verlag KG, 1991: 239-257.

[56] Waldorf AR, Polak A. Mechanisms of action of 5-fluorocytosine. Antimicrob. *Agents Chemother.* 1983; 23(1):79-85.

[57] Walsh TJ, Dixon DM. Spectrum of Mycoses. In: Baron S, editor. *Medical Microbiology*. 4th ed. University of Texas Medical Branch at Galveston; Galveston (TX): 1996.

[58] Wheat LJ, Freifeld AG, Kleiman MB, Baddley JW, McKinsey DS, Loyd JE, et al. Infectious Diseases Society of America. Clinical practice guidelines for the management of patients with histoplasmosis: 2007 update by the Infectious Diseases Society of America. *Clin. Infect. Dis.* 2007; 45(7):807-25.

[59] White TC, Marr KA, Bowden RA. Clinical, cellular and molecular factors that contribute to antifungal drug resistance. *Clin Microbiol Rev* 1998; 11: 382-402.

In: Antimicrobial Resistance: Opportunities and Challenges ISBN: 978-1-53617-943-9
Editors: P. Kumar Prabhakar and V. Mishra

Chapter 5

ANTIBIOTIC RESISTANCE AND TOLERANCE IN BACTERIAL BIOFILM

***Vijay Mishra*[1,*], *Yachana Mishra*[2] *and Manvendra Singh*[1]**
[1]School of Pharmaceutical Sciences, Lovely Professional University, Phagwara, Punjab, India
[2]Department of Zoology, Shri Shakti Degree College, Sankhahari, Ghatampur, India

ABSTRACT

As multi-cell communities known as biofilms, bacteria can grow, and this style of life is distinct from plankton. While the natural and industrial ecosystems are omnipresent of bacterial biofilms, their relevance in the case of human infections has been fully recognized in recent decades. Usually biofilm-related bacteria cause chronic and subacute infections. A significant biofilm model of many bacterial pathogens, such as Staphylococcus aureus, and *Pseudomonas aeruginosa* that causes chronic infections of the airway in CF patients are essential biofilm studies model species. Its continuance despite ongoing therapies with antimicrobials and effective host protection was clinically significant. Highly immune to a wide range of antimicrobial compounds and disinfectants, the bio-film bacteria have multifactoral functions. This section discusses cell cycle and mechanisms for the sensitivity of bacterial biofilms and antibiotic resistance.

1. INTRODUCTION

The sessilian lifestyle is distinct from planktonic development and bacteria may evolve as multiple cell communities called biofills. Throughout the natural and industrial

* Corresponding Author's E-mail: vijaymishra2@gmail.com.

environment, microbial biofilms, however, have become fully acknowledged in recent decades in the context of human infections only (Potera 1999; Costerton et al, 1999; Parsek and Singh 2003). Biofilm-like bacteria cause numerous subacute and persistent infections spanning from periodontitis to airway infections linked to cystic fibroids, catheter infections, medical devices, recurrent infections of the urinary tract, osteomyelitis, chronic otitis media, endocarditis, chronic wound infections. Their persistence, despite sustained antimicrobial treatments and proper host defenses, makes them clinically significant. For biofilmed studies, many bacterial pathogens, like Staphylococcus aureus, readily form biofilms and *Pseudomonas aeruginosa*, which causes chronic airway infection in CF patients. Biofilms are also often polymicrobial, multiple bacterial species communities (McKay et al., 2014).

2. Role of Biofilm in Antibiotic Resistance

Biofilm played an important role during the time during which antibiotic resistance developed naturally. The environment for horizontal propagation of antibiotic resistance genes and virulence factors has been described as a privileged environment (Savage, 2013) and conjugation between cells within biofilm has been reported as 700 times more effective than free (planktonic), free-living bacterial cells (Flemming et al., 2016). Although the use of antibiotic resistance as a vehicle for antibiotic resistance and a tolerance mechanism for bacteria is widely recognized as a worldwide health crisis, biofilm is relatively unappreciated. Currently, it is a significant threat to human safety, as a consequence of an increasingly common and highly deterrence genetically resistant bacteria present in biofilm that establish an additional physical/physiological resistance to antimicrobial agents (Stewart, 2015).

Current inflammation and tissue damage are causing chronic (Costerton et al., 2003) infections (Bjarnsholt et al., 2009). Infections that survive I through antibiotic therapy and host endogenous, adaptive and pro-inflammatory activity, as well as (ii) invasion, are distinguished by immune response and persistent pathology, including infections from outside of the corpora. Chronic infections Abiofilm is a structured bacterial consortium incorporating polysaccharide, protein and DNA into a self-produced polymer matrix. Chronic infection caused by bacterial biofilms has increased tolerance for antibiotics and disinfectants and to resist phagocytosis and other body-defense components. Biofilm production is responsible for, for example, the survival of staphylococcal infections linked to alien bodies. Similarly, chronic *Pseudomonas aeruginosa* lung infection is a result of biofilm-growing mucoid strains in patients with cystic fibrosis. Characterized by nutrients gradient and an increased doubling in times for bacterial cells, these gradients exist from top to bottom of biofilms, and are the cause of some of their antibiotic tolerance. These are the more or less dormancy cells. Growth from biofilms is linked to an increased level of mutations and to mechanisms regulating quorum sensing. The persistence of biofilms are also aided by traditional strength factors such as chromosomal lactamases, upregulated efflux pumps, and mutations in antibiotic target molecules in bacteria. An early aggressive antibiotic prevention or therapy can be used to prevent biofilms and chronically suppressive therapy can be used for treatment. A promising strategy could be the use of enzymes which can dissolve the biofilm matrix (e.g., DNAse and alginate lyase) and inhibitors which improve the sensitivity of the biofilm to antibiotics (Høiby et al., 2010).

3. Mechanism of Antibiotic Resestance in Bacterial Biofilm

Antibiotics avoid the killing of bacteria which bind to the surface and expand like a biofilm. Reduced susceptibility to antibiotics contributes to the persistence of biofilm infections, like implanted equipment. Within active biofilm, the defense processes tend to be different from those responsible for typical resistance to antibiotics. Low antibiotic absorption, nutritional limits and slow growth are thought to represent a multi-layered protection for biofilm, adaptive stress responses and development of permanent cells. Only now are genetic and biochemical details of these biofilm defenses emerging. The development of new chemotherapy agents can be targeted at each gene and gene material that leads to this resistance. Dismissing tolerance to biofilms may improve the ability of existing antibiotics to remove biofilm-refractory infections (Stewart and Costerton, 2001).

Once bacteria adhere themselves to the ground and develop like a biofilm, toxins, biocides and other chemical or physical threats are shielded from being destroyed (Lewis 2001; Mah and O'toole 2001; Stewart and Costerton 2001). The persistence of diverse infections is increasingly identified as a key factor (Costerton, 1999). Chronic diseases that biofilm has shown include periodontitis, cystic pneumonia and many illnesses related to occupants such as catheters, heart valves and protheses. Bacterial antibiotic resistance is easily reported in vitro in biofilms, which shows that host factors are not necessary for these biofilm defense demonstrations. While Biofilm Staphylococcus aeruginosa is known for its antibiotic resistance to the popular opportunistic pathogenes Staphylococcus aeruginozes, several other microorganisms are also biofilms which provide defense against a large number of antibiotics. The yeast of *Candida albicans* and the obligatory anaerobic *Porphyromonas gingivalis* (Hansen et al., 2000), for example, was shown to be less vulnerable to antibiotics when grown in biofilms than free-floating cells. Several kinds of antibiotics were tested for bacterial biofilms and bacteria were shown to be less likely to kill when they were growing in free aqueous suspension in almost every biometric growth mode than the same strain. Such findings warn us that the capacity to shape defensive biofilms is widely distributed among microbes and that the resistance mechanisms that function in biofilms represent a broad spectrum of defenses.

However, these familiar protective methods ± target mutations (low cell permeability) and efflux pumps and modifying enzymes ± do not seem to root the reduced antimicrobial susceptibility in biofilms. A lot is known about the genetic and molecular base of antibiotic resistance in bacteria. For instance, when grown in biofilm condition, bacteria that are devoid of protective mutations or lack of plasmids or other mobile genetic elements carrying resistance genes become less susceptible (Anderl et al., 2000). Once pathogens are removed from a biofilm, antibiotic tolerance is usually quickly regained. The rapid reversal of resistance when a biofilm is dispersed suggests a adaptive and non-genetic resistance mechanism. The intersection between conventional antibiotic resistance and a lowered susceptibility to biofilm is interesting and relatively unexplored. Spontaneous mutants can be reproduced on the basis of the natural protection provided by a biofilm. In order to speed up plasmid transmission, bacterial cells are also closely connected with a biofilm (Hausner and Wuertz, 1999).

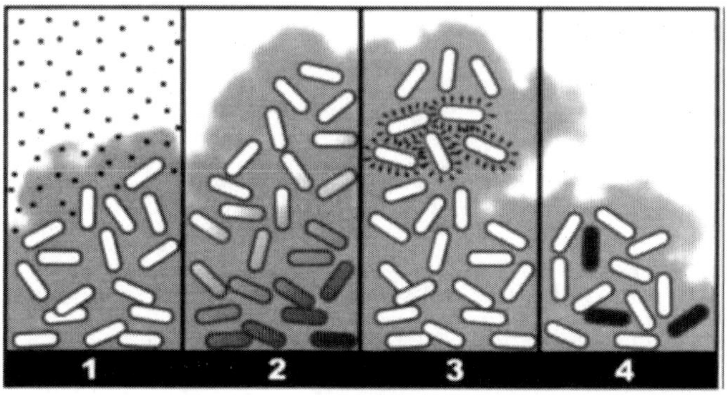

Figure 1. Four biofilm resistance pathways. (1) The antibiotic (squares) penetrates gradually or incompletely; (2) A concentration gradient of the metabolic substrate or material contributes to areas of sluggish or non-growing bacteria (shaded cells); (3) An adaptive stress response is displayed by some of the cells (marked cells); (4) A small fraction of the cells discriminate between elevated and safe persistence (dark cells).

The reduced susceptibility of bacteria to antibiotics in biofilms is represented in Figure 1. Although there is virtually no work on the genetic and molecular basis of each of these processes, in view of the theoretical mechanisms.

Biofilms are mainly water, and antibiotic solutes spread easily throughout the matrix of biofilms. Measurements of effective solute diffusion coefficients of the biofilm, weighing 100 to 1000 moleculars, average about 40% of the respective pure water diffusion coefficient (Stewart 1998). This is consistent with measurements of antibiosphere diffusion coefficients in pure water (Cheema et al., 1986; Gordon et al., 1988) in glycoprotein or multi-saccharid gels. Decreased antibiotic mobility in biofilms by a factor of perhaps 2-3 is inadequate to explain the resistance of aggregate bacteria to murder. The physical mobility of biofilm antibiotics does not, however, ensure that the antibiotic penetrates the biofilm. When the antibiotic is inactivated with a response or binding as it spreads into the biofilm, it can be deeply delayed in its delivery to the depths of the film. Such an association between a reaction and diffusion is enough to avoid penicillin antibiotics from reaching a biofilm made up of a bacterium that is positive for betalactamase (Anderl et al., 2000). Ampicillin has been shown to quickly penetrate a biofilm of a beta-lactamase negative intermediate in this same analysis. Evidences suggest that linking the aminoglycosides with the negative loaded matrix polymers of the biofilm is slowing their penetration (Gordon et al., 1988; Nichols et al. (1988), Kumon et al. (1994), etc.). But in most cases the antibiotic does not occur quickly or rapidly neutralize and antibiotics have been found to penetrate biofilms in in-vitro experimental test systems (Stewart, 1996) (Table 1). Only beta lactams and aminoglycosides have significant limitations on biofilm penetration, and these limitations have not been found in every study. If an antibiotic slowly penetrates the biofilm, enzymes which inactivate or modify the antibiotic will be vital, otherwise the biofilm will eventually be saturated unless it is totally impermeable. These include beta-lactama enzymes, aminoglycoside-modeling enzymes, or acetyltransferases of chloramphenicol, for example. The researchers report survival of the sample microorganism in most of the cases in which successful antibiotic penetration was experimentally seen (Dunne et al., 1993; Darouiche et al., 1994; Anderl et al., 2000). This proves that biofilms should be protective mechanisms other than the failure to penetrate.

Table 1. Mechanisms of tolerance to antibiotics in biofilms (Ciofu et al., 2017)

Antibiotics	Physiological tolerance	Physical Biofilm matrix-related tolerance	Adaptive tolerance
Aminoglycosides	Impaired activity on non-dividing cells	LPS modification and impaired uptake of antibiotics. eDNA creates cation-limited conditions and induces. Diffusion impaired by alginate.	ndv-dependent glucan production in the periplasm and sequestration of AG in this compartment. Subinhibitory concentrations induce biofilm formation. Upregulation of MexXY efflux pumps.
Quinolones	Impaired activity in anaerobic conditions	No impact	Up-regulation of PA1874-1877 efflux pump in combination with ndv (PA14) Up-regulation of efflux pumps as stress response.
Beta-lactam	No activity on non dividing cells	Diffusion partially impaired	Up-regulation of efflux pumps as stress response. Sub-inhibitory concentrations induce alginate genes. Induction of beta-lactamase that by hydrolyzing the beta-lactam molecules can impair the diffusion in biofilm.
Azithromycin	-	-	Up-regulation of efflux pump (mexCD-oprJ)
Antimicrobial peptides and other membrane active biocides	-	LPS modification and impaired uptake of antibiotics. eDNA creates cation-limited conditions and induces.	Upregulation of efflux pumps as stress response. Non-specific: up-regulation of efflux pumps. Specific: Induction of arn operon and LPS modification.

4. Antibiotic Resistance and Biofilm Tolerance

The first live inhabitants of Earth were bacteria. Bacteria will also likely be the last remaining organisms on Earth, supported and supported by their evolutionary adaptivity to produce and overcome antimicrobial molecules and tolerate extreme conditions of the environment through the biofilm way of life. During the last few years we have discovered that the natural and inextricably linked antibiotic resistance and bio film tolerances create an enormous threat and catalyst for chronic infections. The new know-how allows us to reflect differently on how we develop and deploy antibiotic resistance and biofilm tolerance antimicrobial strategy. Combination treatments using both antibiotic film as well as antimicrobial medications have already been effective for treatment of chronic infections, such as those involved with non-healing wounds.

After antibiotic resistance has been introduced into human medicine at an alarming rate in the 1940s and now represents a major public health risk. This problem is exacerbated by pathogenic bacteria, which are most frequently found in biofilm form, and therefore create additional tolerance of antimicrobials. The primary cause for chronic infections is now considered biofilm, with antibiotic-resistant bacteria in biofilm forms predominant. Chronic non-healing injuries, in particular, usually harbour, in polymicrobial complex pathogens which are tolerant of systemic and topical therapies. Globally, antibiotic management systems have evolved to strengthen antibiotic prescribing procedures and to prevent bacterial resistance growth and spread. In this respect, the use of antibiofilm/antimicrobial combinations to interrupt biofilm, thus facilitating the efficiency of antifilm agents, and reduce the possibility for transfers within biofilms of antibiotic resistance genes, must be considered in new antimicrobial strategies. In several medical situations, like persistent nonhealing wounds, antibiotics are often used unnecessarily and indiscriminately, this Strategy is tested. An antimicrobial-antibiofilm injury combination has been shown to facilitate healing in previous non-healing biofilm impaired injuries. This approach must be considered in antibiotic management programs to reduce antibiotic therapy use and impacts and improve outcomes for chronic infections (Bowler, 2018).

4.1. Tolerance, Adaptive Resistance and Efflux Pumps

Active colistin is only in antimicrobials against the central part of the biofilms *P. aeruginosa* in vitro, which do not divide. Due to the upregulation of the two-component regulatory system PmrA-Pmran, which is concerned with the adaptive resistance to cationic peptides, leading to the addition of aminoarabinosa to LPS lipid A (Anwar and coster) (Anwar and Coster) In addition to the sleeping central part of the biofilm, combined therapy with ciprofloxacin and colistin has been able to kill all cells in biofilm in Vitro since the metabolically active surface layer of the biofilm is responsive towards ciprofloxacin (Pamp et al., 2008). The combination therapy for early eradication of *P. aeruginosa* in CF patients has demonstrated clinical efficacy (Hansen et al., 2008)

Biofilm tolerance to tobramycin also mediates low metabalic activity, but high cell density leading to the accumulation of extracellular signaling molecules is likely significant because *P. aeruginosa* strain biofilm tolerance of tobramycin was shown to be quorum sensing-mediated in *P. aeruginosa* strain PAO1 (Bjarnsholt et al., 2005). Furthermore, the MexAB-OprM effluss pump upregulation was shown to be a non-specific tolerance mechanism for metabolically active colistin cells (Pamp et al., 2008). In fact, decreased efflux pump behavior due to mutations has shown that *P. aeruginosa* in CF patients is a significant resistance mechanism against aminologlycoside antibiotics and fluoroquinolones (Jalal et al., 2000; Islam et al., 2009).

4.2. Model of Antibiotic-Tolerant Cell Formation in *P. Aeruginosa* Biofilms

In development of *P. aeruginosa* biofilm, a link between TYPE IV pili and the colistin-tolerant population was shown previously (Haagensen et al., 2007), whereas a position of importance for development of the colistin-tolerant populations in more mature *P. aeruginosa* bio-films appears to be important in the nutrient-rich micro-environment (Pamp et al., 2008). We monitored the development of antibiotic-tolerant subpopulations in *P. aeruginosa* biofilms and selectively mark the proteome of SILAC as the product of the discovery of the multiple genes/proteins essential for antibiotic-tolerant biofilm in new protein abundance subpopulations. In the development of *P. aeruginosa* biofilm, we propose the model for colistin tolerant subpopulations linked to the phoenix effect.' (a); remaining cell aggregates for the type IV overexpress for selective migration to the top of the dead biofilm, (b) generate QofC-regulated factors to facilitate the formation of new microcolonies of antibiotic-tolerant, and (c) most of the biofilm will be destroyed by colistin therapy. Microcolony width scale is about 50 mm in our in vitro *P. aeruginosa* biofilm, which suits in vivo (Bjarnsholt et al. 2009) and ex vivo (Moreau et al. 2008, and Moreauet al. 2009) and microcolony measurements (Moreau et al. 2008). Biofilms with aeruginosa. This suggests that in vitro *P. aeruginosa* biofilms could have similar effects as those found on in vitro *P. aeruginosa* biofilms. Our study also shows that concentrating on essential pathways of antibiotic resistant cells significantly improved traditional biofilm management strategies. This stressed the importance of developing QS and motility inhibitors for chronically infected patients in order to create functional anti-biofilm chemotherapy. It might also be used in other dynamic segregated cultures such as leukemia, as a recent study has shown that mobile drug resistant cells are found in chemotherapy care populations of cancer cells (Wu et al., A 2013).

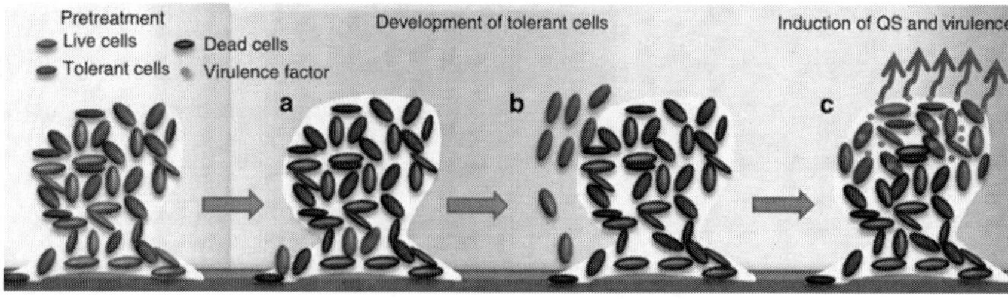

Figure 2. Model of antibiotic-tolerant cell formation in *P. aeruginosa* biofilms. (a) antibiotics like colistin therapy destroy most biofilm cells, but leave at the bottom of the biofilm a handful of antibiotic-tolerant cells. (b) Cell tolerant of antibiotics increase in numbers and migrate with pilus-mediated motility to the top of the biofilm. (c) Antibiotic cell assemblies trigger QS, which triggers the development of virulence factors linked to QS such as elastase or pyocyanine. There is a new biofilm antibiotic resistant.

4.3. Eradication of Antibiotic Tolerant Bacterial Populations in *P. Aeruginosa* Biofilms

The therapeutic potential of antibiotics against diseases is greatly reduced by the resistance and tolerance to medicines. The resistance to antibiotics by bacterial biofilms also contributes to continuing infections. We use here a proteomic method to quantify newly developed proteins in colistin-tolerant subpopulations of *Pseudomonas aeruginosa* biofilms (colistin, the last resort antibiotic against multi-drug-resistant gram-negative pathogens) with pulsed, stable isotope labeling of amino acids (puls-SILAC). To order to form colistin-tolerant biopopopulations, movement is important, with colistin-tolerant cells use type IV pili moving to the top of the biofilm with the colistin-killed. Quorum sensing (QS) is used for the formation of new subpopulations of colistin-tolerants, highlighting multicellular behavior in the development of antibiotic tolerance. The previously proved macrolide erythromycin inhibits *P. aeruginosa* motility and QS, boosts colistin biofilm eradication. Our work offers insights into the mechanisms underlying the development of antibiotic tolerant groups in bacterial biofilms and outlines avenues for research into more effective treatment against biofilms-associated infections two major bacterial defense strategies can severely damage the therapeutic effectiveness of antibiotics. The' resistance' occurs when genetic mutations or mobile genetic elements are developed in bacteria, allowing growth in constant antibiotic exposure conditions (Neu, 1992), Tolerance ' mediates transient phenotypes that allow bacteria to survive at the most exposed doses of antibiotics (Lewis, 2007) based on their expression in protein. Although the current view is that tolerance formation is inherently linked to the biofilm lifestyle, early studies in planctonic bacteria show that tolerance can grow to a variety of strains, including antibiotics (Harrison et al., 2005). But biofilms are sessile, highly populated, multicellular communities, embedded in biopolymeric matrix components, in contrast to their mode of life in planktonic (Costerton et al., 1995). One major benefit in biofilm aggregation is the fact that a variety of matrix components and physiological conditions provide additional levels of immune and antibiotic protection that cannot be achieved in planktonic growth mode (Colvin et al., 2011; Chua et al., 2015). Pathogenic bacteria are widely acknowledged to play an important role in the production of nosocomial infections in biofilm form, as they colonize hospital and recurrent infections locations, and pose a serious public health threat. Bacterial cells are distinguished within a biofilm by several distinct biological groups. Antibiotic treatment often eradicates susceptible species, but it lies behind limited antibiotic-tolerant subpopulations, resulting in recurrent infections after the termination of antibiotic therapy (Parsek and Tolker, 2008). The bacterial cells of antibiotic-tolerants are variations of wild-type cells, which may return to the wild if antimicrobial therapy is discontinued. Its transient nature and low abundance make it difficult to investigate its physiology linked to tolerance. In just a handful (Conlon et al., 2013, Vega et al., 2012), the mechanism underlies the formation of antibiotic tolerance subpopulation in biofiles have not been sufficiently studied by a number of studies that have pre-isolated planktonic antibiotic tolerant cells by means of flow. In contrast, these antibiotic resistance subpopulations in biofilms can and cannot be identified as damaging and preisolation-based strategies. This prompted us to create an innovative high resolution "omics" method that characterizes simultaneously the physiologies of the subpopulations of susceptible and antibiotic resistant films.

P. aeruginosa forms biofilms with extreme antibiotic tolerance for nosocomial infections such as pneumonia and surgical site infections which lead to *P. aeruginosa* being classified as' Serious' Threat Level by the Centers of Disease Control and Prevention and is declared to belong to six "Hazardous, Drug-Resistant Microbes" by the Infective Diseases Society of America. Colistin is a new antibiotic polymyxin used for the management of drug-resistant Gram-negative bacteria infections (Li et al., 2006). In flow chambers, *P. aeruginosa* biofilms quickly develop colistin-tolerant cells with high levels of pmr operone expression (Haagensen et al., 2007) and the other cells. In *P. aeruginosa* biofilms following colistine treatment we observe a dynamic development of drug-toleran subpopulations. To gain an awareness of the physiology of biofilm colistin-tolerant cells in order to selectively measure the new proteins which incorporate, and so label with heavy C13 lysine, we apply a pulsed stable isotope label using amino acid (pulsed-SILAC) proteomics to promote an understanding of colistin-tolerant subpopulation physiology. By using type IV pili-dependent motility, the colistin-tolerant populations can migrate to the top of the dead biofilm and initiate the formation of the new biofilm by quorumsensing group activity (QS). Synergistic treatment using erythromycin, which can inhibit QS and colistin motility, can boost *P. aeruginosa* biofilm elimination. Therefore, our study offers key insights into new biofilm-associated treatments that otherwise prove difficult to eradicate.

4.4. Chemical Approach to Combat Colistin-Tolerant Subpopulations

Basically in the establishments of antibiotic tolerant subpopulations in the development of *P. aeruginosa* biofilm require both type IV migration pili mediated and QS, we use a chemical biology approach to disable these activities. In tandem with colistin, macrolides used to treat *P. aeruginosa* biofilms were used to suppress type IV pil and QS attachment in *P. aeruginosa* (Perez and Haas 2011)..

Erythromycin therapy had a mild killing effect on the biofilm cells and planktonic cells and therefore could not kill the biofilm. The complete eradication and repression of the development of antibiotic tolerance subpopulations are caused by combination processes using erythromycin and colistin. We did not see the synergistic effect of colistin and erythromycin against planctonic *P. aeruginosa* at concentrations equivalent to that of the biofilm treatment, in support of the specific treatment of colistin-tolerant cells in the biofilm way of life. We used a murine model of implant-associated infection (Christensen et al. 2007), to check the role of the eradication of antibiotic and resistant cell by both antibiotics in vivo. Because opportunistic host infections may occur in devices such as heart catheters and valves once biofilm is established it is important to manage the increase in antimicrobial sensitive or resistant populations (Schierholz and Beuth, 2001; Costerton et al., 2005). Surgically inserted into the mice peritoneum implants covered in *P. aeruginosa* biofilms. Locally the mouse was used for the treatment of patients with device diseases with antibiotic medications. For every mouse the antibiotic rate (1 mg/kg and 10 mg/erythromycin) was much smaller, guaranteeing non-lethal treatment of *P. aeruginosa* biofilm in the mouse than that used on each average lethal dose (86 mg/kg colitin and 280 mg/erythromycin). A 24-hour incubation with the mouse revealed that the administration of colistin or erythromycin alone did not eliminate after c.f.u. per ml count of *P. aeruginosa* disease from implant and spleen. However the biofilm was eradicated by c.f.u. per ml through synergistic therapies of colistin and erythromycin. We found that the dissemination of pathogens has also been prevented because we were unable to detector live bacterial cells in spleen specimens. The supplementation of conventional antibiotic therapy by a tolerant agent therefore seems to be an excellent treatment for the eradication of biofilm-related infections.

4.5. Development of Colistin-Tolerant Subpopulations in Biofilms

After exposure in real time to colistin, we followed developments in living and dead *P. aeruginosa* sub-populations using confocal image time series and listing viaible cells with colony forming units (c.f.u. per ml). We monitored live and dead cells. The sudden decline in cell viability of propidium-yodide (PI) dead and c.f.u. per ml count, caused by the exposure to colistin of *P. aeruginosa* wild biofilms with the Green Fluorescent Protein

(GFP). However, we found that even after the 24 hours of colistin treatment, several *P. aeruginosa* cells located on the substratum of the biofilms remain alive, which may be the result of the heterogenous structures of the exopolymer matrix (Walters et al., 2003; Stewart, 2002). Other components of an exopolymeric matrix that serve as physical barriers and decrease colistin penetration into a deep section of organic films (Hengzhuang et al., 2011), enabling other substratum biofilm cells to survive and develop colistin tolerance. We observed, interestingly, that colistin subpopulations of dead *P. aeruginosa* organic film cells could migrate towards the top of the large microcolonies. Follow-up on the colistin-tolerant cell aggregates near major microcolonies showed that the aggregates move in direction of dead microcolonies-a migration that is not related to the culture medium's flow direction. On the other hand, individual tolerant to the colistin appears randomly distributed far removed from the major microcolonies. These results indicated that a coordinated process of migration of colistin tolerant cell aggregates to large biofolm cell microcolonies was possible with certain signals. Colistin-tolerant cells obtained from biofilms displayed only the temporary expression of the pmr operon and lost their resistance to colistin after overnight culture in ABTGC media containing no colistin, which confirms our opinion that colistin-tolerant cells were the product of phenotypic variation and not per se resistant cells. The sequencing of biofilm DNA with and without colistin treatment revealed that, as compared to the control biofilm populations, no convergent non synonymous mutation was produced from the three colistin-treated biofilm populations. The development of mobile subpopulations in biofilms with colistin tolerant substances has significant clinical consequences as they can produce persistent infections.

5. Biofilm-Specific Mechanisms of Resistance

5.1. Extracellular Proteins

Many enzymes are involved with the deterioration of components of EPS while other nonenzymatic proteins, such as carbohydrate-binding proteins (lectins) are involved in the stamina and structure of EPS (Lynch et al. 2007; Diggle et al. 2006). Periplasmic beta lactamases can release into the PES after cell death or through membrane vesicles, but their biofilm resistance contribution to the beta lactamases is still uncertain (Ciofu et al., 2000).

5.2. Matrix and Diffusion Barriers

Biofilm bacteria, typically composed of the extracellular polysaccharides, DNA and proteins, are integrated into a self-producing extracellular (also called extracellular polymer substances or EPS). While its exact composition is not often well-designed, it may vary significantly depending on strains, species, and growth (e.g., the nutrient source and sheer strengths) in up to 90 percent of the biofilm biomass (Zogaj et al., 2001; Branda et al., 2005). The EPS establishes the multicellular structure and facilitates the external accumulation and adhesion of the substrate. The EPS acts as a molecular sieve because of its physicochemical properties, which can provide a protective barrier to some antimicrobials, including desinfectants and antibiotics. While the diffusion barrier and restricted penetration of antimicrobials were originally thought to play a major role in resistance to biofilms, more recent research suggests that it contributes more limitedly to the overall biofilm resistance.

5.3. Diffusion of Antimicrobial Compounds

The EPS biofilm comprises a mixture of charged or hydrophobic polymers, which can strengthen or restrict molecular penetration. While the protective barrier function of the EPS was initially proposed as a major biofilm resistance mechanism, many studies have now suggested that this diffusion restriction plays only a minor role. The diffusion of the soft anionic piperacillin beta-lactam, which was hindered by dialysis membranes, while the levels of strong Ca2 + tended to be 208 G. And D. and McKay. Nguyen has cut its spread (Hoyle et al., 1992). *Pseudomonas aeruginosa* biofilms sequestered and delayed the release of tobramycin (a positively charged aminoglycoside) but not ciprofloxacin (a neutral fluoroquinolone) (Walters et al., 2003; Shigeta et al., 1997; Yasuda et al., 1993; Beloin et al., 2004). The diffusion of antimicrobials by alginate was investigated by Kumon et al., who showed limited diffusion only aminoglycasides and polypeptides, whereas beta lactamia, fluoroquinolones and macrolides did not (Kumon et al., 1994). There are also significant diffusion limitations in biocides such as hypochlorite (Stewart et al., 2001). These studies combined show that the spread rate and chemical structures of particular compounds are not correlated with their biofilm efficacy. Between antibiotics there is minimal penetration of aminoglycosides only and this possibly does not fully account for the strong biofilm resistance for these compounds (Vrany et al., 1997; Stewart, 1996).

5.4. Extracellular DNA

Extracellular DNA (eDNA) is released by bacteria passively or actively and may constitute a significant EPS element for certain types of bacteria. eDNA primarily plays a structural role and improves biofilm aggregation and stability. Exogenous eDNA can interfere interestingly with the effect of cationic antimicrobial peptides, and aminoglycosides in *P. aeruginosa* through sequestration of cations and the induction of LPS modifications mediated by PhoPQ/PmrAB (Mulcahy et al., 2008). If eDNA is appropriate in the biofilm EPS for the induction of PhoPQ/PmrAB systems remains unclear (Pamp et al., 2008).

5.5. Extracellular Polymeric Substances (EPS)

In several bacterial species like *Bacillus subtilis*, *P. aeruginosa* and *S. aureus*, exopolysaccharides constitute a major component of EPS. (Branda et al, 2005). These polymers consist of several polysaccharides, including sucroses, cellulose or heteropoly saccharides loaded with glucose and fructans. Through their chemical composition and replacements, they decide the physical and chemical properties of the EPS. Polyanionic are the most common exopolysaccharids, although polycationic exopolysaccharides such as B-1,6-linked n-acetyl glucosamines have been reported in staphylococcal biofilms, for example. (Spanish only, 2002). The structure of exopolysaccharides varies significantly between organisms and even between plant strains. In natural biofilms, labo-experimental systems, and human infection-associated biofilms, exopolysaccharides have been identified. Mutants without EPS components produce biofilms with altered morphology or less adherence to the surface. While exopolyaccharides have an obvious part to play in biofilm adhesion and structure, they are still poorly defined in antibiotic resistance. Exopolysaccharides may either sequester loaded molecules or alter the structure of the EPS indirectly. Exopolysaccharides can sequester loaded molecules or modify the structure of the EPS indirectly. In several studies comparing biofilm resistance to various antibiotics, the aminoglycosides which are positively charged appear most vulnerable to exopolyzaccharide sequestering action.

For its EPS, *Pseudomonas aeruginosa* produces at least three different chemical polymers: Pel and Psl. Alginate is one of the most widely studied exopolysaccharides and uroninic acid polymers. Alginate is not needed to make biofilms in all *P. aeruginosa* strains (Wozniak et al., 2003), but is overproduced in the cystic fibrosis patient *P. aeruginosa* isolated forms. Alginate strain overproducing biofilms have developed biofilms more resistant to treatments for aminoglycosides than wild strains in several laboratory biofilm modeling methods (Alkawash et al., 2006) with alginates which may inhale the release of aminoglycosides (Gordon et al., 1988). Testing of Pel-deficient or Pel-super-expressive strains indicates a limited role in biofilm tolerance to aminoglycosides and aztreonam but the involvement of Pel in the Psl-negative strain PA14 is possibly strain-specific and most important (Colvin et al., 2011; Khan et al., 2010; Yu et al., 2012). Psl also may involve detergent resistance, as psl-deficient mutants form polysorbate-80 susceptible biofilms, whereas Psl overproduction gives resistance (Zegans et al., 2012). The resistance mechanism is still unknown.

5.6. Other Biofilm-Specific Mechanisms

BrlR, a Sea-like transcriptional regulating authority involved in both antibiotic tolerance and multiple antibiotic resistance in *P. aeruginosa*, was reported by Liao et al. BrlR's inactivation decreases biofilm resistance without affecting planktonic bacteria sensitivity (Liao and Sauer, 2012). This is presumably because of its strong production during biofilm, but not because it is planktonic. Although the BrlR mechanism is still unclear, members of the MerR family may activate multidrug efflux pump expression. Since brlR has been induced under the conditions of oxygen limit (Trunk et al. 2010) and this can lead to high brlR expression in biofilms. Periplasmic glucans are other examples of biofilm-specific antibiotic resistance mechanisms. NdvB encodes a glucosylated transferrase necessary for the synthesis of periplasmic glucans in *P. aeruginosa*, which is best expressed during biofilm development and can lead to biofilm tolerance by antibiotic sequestration (Mah et al. 2003).

5.7. Prophylaxis and Treatment of *Pseudomonas aeruginosa* Biofilms in Cystic Fibrosis Lungs

The currently used methods for preventing chronic *P. aeruginosa* biofilms in CF lungs are (i) early aggressive eradication therapy of intermittent colonisation by means of oral ciprofloxacin and nebulise colistin for 3 weeks or, even better, for 3 months or by using nebulised tobramycin as monotherapy (Döringand Hoiby,2004), (ii) daily nebulized DNase (Pulmozyme®) (Frederiksen et al., 2006) and (iii) prevention of cross-infection from other already chronically infected CF patients by isolation techniques and hygienic measures (Høibyand Pedersen,1989), . Such three approaches, which are integrated in most CF clinics, are active and cost-effective and have completely changed the epidemiology of persistent *P. aeruginosa* lung infection in CF patients from being very prevalent in CF children to being mainly adult patients, and no antibiotic resistance issues have been reported (Baumann et al., 2003; Frederiksen et al., 1999). In an animal model of *P. aeruginosa* disease in C FC, an early effective eradication treatment was also shown to be preferable (van et al., 2009). In addition, while *P. aeruginosa* vaccinations have been developed and medically tested, the effectiveness of early active eradication therapy was not further developed (Döring and Pier, 2008). Persistent suppressive antibiotic therapy (Doring et al., 2000) is the preferred way to treat a persistent *P. aeruginosa* iofilm infection and is triggered by evaluation of chronic infections (continuous bacteriologically colonised regularly for six month, and/or improved anticorps against *P. aeruginosa*) (Høiby et al., 1977), (Proesmans et al., 2006). Chronic therapy is a daily nebulised colistin or tobramycin, combined with a regular2-week I.V. every 3 months for the rest of the patient's life. Antibiotic antipseudomonal (combination therapy with two antibiotics: tobramycin, colistin + ceftazidime; or pipperazellin/tazobactam) or ad hoc i.v. Clinical disorder treatment takes place. DNase is also inhaled every day to reduce the viscosity of the sputum that contains DNA, and patients are administered oral azithromycin as stated above daily. Chronic suppression therapy has succeeded or slowed declines in lung function, prolonged patients ' lives for many years (Frederiksen et al., 1996). (Frederiksen et al., 1994). In the persistent strains and with high level of allergy to –lactam antibiotics are the side effects of maintenance therapy (Koch atr al. 1991). Anything close, such as (a) Early aggressive eradication therapy for planktonic growth eradication or early biofilm formation (b) Hronical antibiotic suppressant therapy to keep an inserted foreign body functioning (Antibiotic systemic or local prophylactic use to prevent formation of biofilms, In other regions of biofilm contamination, they are being added slowly (Stensballe et al., 2007). Perhaps other approaches using QSIs can further enhance biofilm infection management.

5.8. Modulation of Antibiotic Sensitivity and Biofilm Formation in *P. Aeruginosa* by Interspecies Signal Analogues *P. Aeruginosa*

Signs of cis-2-unsaturated fatty acids of the diffusible signal factor (DSF) family may engage in interspecies communication. Sensing these signals leads to changing the formation of biofilms and an increased tolerance to different antibiotics, requiring histidine kinase PA1396. We show that PA1396 has five transmembrane helices in the membrane-associated sensory input domain, two of which are needed for sensoring the DSF. Improved autophosphorylation PA1396 embedded in liposomes is consistent with DSF binding. We also examined the possibility for the modulation or inhibition of PA1396 activity of synthetic DSF analogs. Several of these analogs prevent the DSF from triggering autofosphorylation and gene expression, while others act as reverse agonists reducing in vitro and in murine-infection models the formation of biofilms and antibiotic tolerance. Such analogs can thus be used to develop new adjuvants to improve the effectiveness of current antibiotics. (An et al., 2019).

A symptom of public health is antibiotic resistance, coupled with minimal development of new antibiotic drugs, that highlights the need to fictitiously tackle the global threat of alternative strategies for the war against infection1-4. Such strategies address stimuli that control the development and approaches to the enhancement of current antibiotic efficacy in microbial virulences. These strategies involve The suggested potential targets for small molecular interaction are bacterial cell contact (quorum sensing), biofilm development and cyclic di-GMP coverage (Bolitho et al., 2011; Curtis et al., 2014; Ng et al., 2012; Sperandio, 2007). The influences of anti-virulence are desirable because they do not affect bacterial growth that may reduce selective resistance. The research investigates the regulation by inter-species signaling of *Pseudomonas aeruginosa*. The typically occasionistic pathogen of man, the *P. aeruginosa*, uses the quorum-sensing molecules that are responsible for the production of virular factors (PQS) as two lactones, N-acyl homoserine and the quinolone signal, the PQS-signal (2-heptyl-3-hydroxy-4(1H)-quinolone) (Bjarnsholt et al., 2010; Chugani and Greenberg, 2014; Williams and Cámara,2009). There are several reports that these pathways inhibit small molecules, which affect synthesis factor virulence (Jakobsenet al., 2012; Kim et al., 2015; Ng et al., 2011). *P. aeruginosa* may take part in interspecies signaling, mediated by diffusable signal factor (DSF) mollecules, that are cis-2-unsaturated fatty acids as well as intra-species signaling.

P. aeruginosa senses in specific cis-11-methyl-2-dodecenoic acid (DSF) and cis-2-dodecenoic acid (BDSF) formed in bacterias other than *P. aeruginosa*, such as *Burkholderia*, *Stenotrophomonas maltophilia* (Ryan et al., 2008). It is sensible to several antibiotics, including polymyxin B (Ryan et al., 2008; Twomey et al., 2012) and causes alterations in the formation of biofilm. The elimination of PA1396 from the template strain PAO1 and of other medical isolates also contributes to an improvement in sensitivity to polymyxins B and E. (Ryan et al., 2008; Twomey et al., 2012). Detection of DSF group molecules in polybacterial infections, such as those diagnosed with cystic fibrosis (CF) where *P. aeruginosa* is present along with other bacterial species, indicates that interspecies signaling exists in vivo and may therefore contribute to decreased effectiveness of antibiotic therapy (Ryan et al., 2008; Twomey et al., 2012). In this background, the function of PA1396 in sensing the family DSF signals and the capacity to modulate the PA1396 behavior of Structural Analogs were discussed in greater detail. We concentrate in particular on the effects of DSF analogs on in vitro and murine-infection system PA1396-regulated features of biofilm formation and antibiotic sensitivity. It has been shown that for DSF sensing two of five PA1366 input domain transmembrane helices are requisite and for enhanced PA1396 automotive phosphorylation, DSF connection is associated. We also found that some DSF analogs block its ability to trigger autophosphorylation, which coincides with specific changes in *P. aeruginosa* sub-set expression, reduced biofilm formation and antibiosphere tolerance with in vitro and in vivo models of infection. All in all, our findings indicate that associations with DSF mediated organisms may be pharmaceutically affected to enhance the systemic diagnosis of *P. aeruginosa*.

5.9. Selective Proteomic Analysis of Antibiotic-Tolerant Cellular Subpopulations in *P. Aeruginosa* Biofilms

Antibiotic therapy resistance for biofilm infections is high. The biofilm analysis is complicated by the phenotypic heterogeneity; the physiological behavior and the responses to antibiotics vary in biofilm subpopulations. There the use of the BONCAT system for the systematic proteomic study of the *Pseudomonas aeruginosa* sub-population is defined.

They discuss the use of bioorthogonal non canonical amino acid labeling. Through the regulated speech, we targeted BONCAT labeling in biofilm microcolonies areas that shows increased tolerance to antibiotics by means of the mutant methionyltRNA synthesis. We in these regions have enriched and established cell synthesized proteins. The marked subpopulation has a lower concentration of ribosomal proteins in comparison with the overall biofilm proteome, and is supplemented with unidentified proteins. We have conducted a pulse labelling experiment with the fluoroquinolone antibiotic ciprofloxacin to determine the dynamic proteomic response from the tolerant subspecies to super-MIC treatment. The evolutionary reaction involved up-regulation of proteins needed to recognize and repair damage to DNA and significant changes in the expression of the main carbon metabolism enzymes. We distinguished from the longer adaptive strategy which involves upregulating purine synthesis the immediate proteomic reaction defined by an improvement in flagellal motility. This focused, selective analysis of a bacterial subspecies shows how the study of proteomic dynamics can help us understand biofilm heterogeneity and antibiotic tolerance in particular (Babin et al., 2017).

Bacteria living in surface based biofilms are more tolerant of different stresses than their planktonic equivalents (Fux et al., 2005). The improved antibiotic resistance in biofilms is of therapeutic significance, but avoids persistent infection removal. The sensitivity of phenotypes is distinct from genotypical resistance, while cell persistence following treatment may encourage resistance (Levin and Rozen, 2006). *Pseudomonas aeruginosa* is an opportunistic pathogen that serves as a model for studying both biofilm physiology and antibiotic tolerant infections. Biofilms that re-alkalcitrant the host immune system and the treatment of antimicrobials are produced by *P. aeruginosa*, a principal contributor to chronic lung fibrosis infections. In host and in vitro bio-film studies (Høiby et al. 2010), antimicrobial resistance for these bio-film infections has been reported. Research on *P. aeruginosa* biofilms grown in vitro has revelated in its antibiotic answer the significance of spatial diversity, while other subpopulations do not). For example, drug groups (for instance fluoroquinolones and aminoglycosides) are known to kill forming cells with biofilm regions with greater access to Exogenous Nutrients, which control operation (DNA replication, transcription and peptoglycan synthesis respectively). These antibiotic responses include decreased penetration of antibiotic molecules, decreased metabolic rates, and modified metabolism (Stewart and Franklin, 2008). Conversely, polymyxins and detergents, which disrupt cellular membranes, preferentially kill dormant cells in the interiors of biofilm microstructures. Measurements of mRNA or protein abundance can offer comprehensive and unbiased views of a physiological response to antibiotics (Wu et al., 2015; Park et al., 2016), but experimental challenges limit the investigation of tolerant biofilm subpopulations. Because only a subset of cells exhibit tolerance, any analysis must distinguish tolerant cells from those that do not survive treatment. Laser capture microdissection was employed to isolate biofilm cells from spatially distinct biofilm regions and to quantify differences in the abundance of transcript mRNA between these regions with quantitative PCR (qPCR) and DNA microarray analyses (Williamson et al., 2012). This method was used to investigate the general heterogeneity of biofilm, but not in the analysis of antibiotic responses in subpopulations. The use of pulsated stable isotope labeling with Amino acids in the culture of cells (pSILAC) to measure improvements in protein expressions after the adaption of biofilm cells to the task of antibiotics has been commonly used to better understand biofilm physiology (Khemiri et al., 2016), but the directed select methods have been limited. The pSILAC can be a way of discriminating between proteins synthesized both before and during pulse based on size, through the application of an amino acid isotopology (Chua et al., 2016). Chua et al has described the long term proteomic reactions of sensitive cells with biofilms for 8 hour, enabling nontolerant cells to die, and labelling newer synthesized proteins with prolonged amino acid (48-h) isotopoly pulses with medical polymyxin colistine. This method meant that the responsive subpopulation of interest was used to synthesize marked and established proteins over a 2-day cycle. The results of this study show the importance of Type IV motility in *P. aeruginosa* biofilms ' tolerance to colistin. We used the bio-orthogonal noncanonic amino acid tagging (BONCAT) approach for selective proteomic analysis in response to the challenges posed by complex and heterogeneous responses (Dieterich et al., 2007). A non-canonical amino acid (ncAA) which bears a bio orthogonal chemicals handle is used in BONCAT's cellular proteins. The named proteins can be paired with an association mark upon incorporation.

Importance the development of bacteria is often characterized by behavioral diversity at the level of one cell. The physiology of biofilms in which different cell subpopulations may react differently to stress, including subpopulations of pathogenic biofilms that are more antibiotically tolerant, shows a particular heterogeneity. GPA gives information into the biology of cells, but cannot classify proteins found in a particular interest subpopulation. Here we present a chemical biology system for the selective naming, enrichment and classification of proteins of cells inside specific biofilm microcolonies regions. This method was used to research improvements in protein synthesis throughout the procedure by the subpopulation of antibioticolerant cells. The initial answer and the long-term evolutionary approach of the biofilm cells found significant variations.

6. Antibiotic Treatment Strategies for Combating Biofilm Infections

Bacterial biofilms are associated with a broad range of infections, from those relating to exogeneous devices such as catheters and prothetic joints to infections of the chronic tissue, for instance those in the lungs of patients with cystic fibrosis. This leads to the persistence of biofilm infections despite exposure to antibiotic resistance (genetic resistance). To optimize antibiotic treatment of biofilm-related infections, an understanding of the interrelationship between phenotypic and genetic resistance mechanisms in film and the diversification of environmental conditions in biofilm infections that influence antibiotic effects is needed. The optimization of PK/PD parameters is important in biofilms, topical treatments of high doses, combination and sequential/alternative treatments or the use of antibiotic adjuvants (Ciofu et al., 2017).

Providing high levels of antibiotics by topical administration, an improved sensitivity to antimicrobial agents is one of the main features of biofilms. As a proper therapy focused on antimicrobial susceptibility testing recorded by the medical microbiology lab cannot always be attributed to therapeutic effectiveness, a variety of approaches to resolve the inherent antimicrobial resistance of biofilms should be tackled. Topical administration provides high local levels by selectively administering antibiotics at lower or even undetectable serum concentrations at the location of infection and avoids systemic side effects. For example, the concentration of serum after inhalation for tobramycin nebulized is below 1 mg/L, while for sputum it exceeds 1200 mg/L (yu et al., 2012). One of the benefits of topical antibiotics is that the target of diagnosis may sometimes be in anavascular areas that are not easily reached by parenterally administered antibiotics. In fact, the risks of acquiring antimicrobial resistance may be decreased by this routes of administration and also have an impact on bacteria normally considered to be immune, as the amount of antibiotics stay well beyond MIC. This antibiotic therapy may, in this sense, play a role in the treatment and even prophylaxes for infection prevention in certain circumstances of proven bio-film-associated infections (Table 2). In the case of exacerbations, nebulized antibiotics are used in CF suppression or repair treatment. Inhalation application of antibiotics increases respiratory effects to effectively reduce and withstand the microbial pressure in the sputum (Schuster et al., 2013; Stass et al., 2015). A recent work explores the impact of treatment with *P. aeruginosa* biofilms at high concentrations with aztreonam and tobramycin on filamenting, flattening or disturbance of biofilm structures and their ability to reduce bacterial strain and biomass (Rojo et al. 2016). In addition, inhaled antibiotics may be an active suppressive antibiotic drug with an appropriate safety profile in clinicians with healthy non-CF bronchiectasis (Brodt et al., 2014). Nebulized antibiotics appear to decrease the risk of acute aggravation in these patients rather than placebo or symptomatic treatment, but without increase the likelihood of developing resistance after treatment. On the other side, the efficacy of inhaled pneumonia therapy (VAP) is still unclear. Several studies found that the symptoms of respiratory infection have reduced considerably compared to placebo (Palmer and Smaldone, 2014), whereas other tests have shown no change in contrast to IVA antibiotics to aeroso-lize antibiotics (Rattanaumpawan et al., 2010). Nevertheless, the effects of nebulized sanamicin on VAP prevention by the most common pathogens have been evaluated in a single clinical study, which shows its efficiency in the prevention of endotracheal tube (ETTs). Other VAP prevention efforts are focused on the use of antimicrobial-coated ETTs, especially in silver, that can impede the forming of biofilms and have antimicrobial impact on the nearby airways. In the manually ventilated cases, silver-coated ETTs even decreased mortality (Afessa et al., 2010). The use of coated catheters or antibiotic lock technology (ALT) can be used to prevent formation of biofilm on central vein catheters (CVCs) with topical antibiotics. For example, the use of minoclycine/rifampin coated CVCs for catheter exchange. Bloodstream infection (CR-BSI) associated with the S aureus catheter may be associated with a lower overall mortality rate. ALT is used for prophylaxis using a small volume of distilled antimicrobials, which live within the catheter lumen for 12–24 hours.

Table 2. Summary of current topical antibiotic treatment regimens according to the site of biofilm infection (Ciofu et al., 2017)

Biofilm site of infection	Route of administration	Duration	Antibiotic regimen
Orthopedic procedures	Intraoperative (beads)		2 mg/mL gentamicin aqueous solution. 40 mg/mL tobramycin + 1 g vancomycin + 10 mL packet of calcium sulfate
Endotracheal tubes	Inhalation	14 days	80 mg gentamicin + 2 mL saline, three times daily. 120 mg vancomycin HCl + 2 mL saline, three times daily.
Rhinosinusitis	Nasal drop	28 days	3 drops ofloxacin 0.3%, three times daily 28 days Nasal drops.
Catheters	Catheter lumen	12–24 h	1 mg/mL gentamicin + 2500 U/mL heparin. 10 mg/mL cotrimoxazole + 2500 U/mL heparin. 3 mg/mL minocycline + 30 mg/mL EDTA.
Lung infection in CF	Inhalation	Continuous 28 days on/off cycles	
bronchiestasis	Inhalation	Continuous	80 mg gentamicin, twice daily.
Wounds	Cutaneous	7 days	Silver sulfadiazine 1% cream
Lung infection in non-CF	Inhalation	28 days	300 mg tobramycin, twice daily

In hemodialysis patients especially, ALT is effective in preventing CR-BSI. Such approaches should however be limited to CR-BSI clients, although these techniques are useful. In the prevention of surgical infections, topical antibiotics were used. In sternotomy closure procedures the occurrence of sternal wound infection was minimized using genemicine-containing collagen implants. The empirical evidence also shows that the occurrence of prothetic joint infections correlated to biofilms with Vancomycin, Tbramycin and gentamicin (Marschall et al., 2013)

As far as orthopedic trauma therapies for proven infection are concerned there should be often antibiotic therapy paired with surgical debridement and subsequent death area that can be colonized by bacteria quickly. The prevention for chronic infections is another use, but sadly the data available render the efficacy of topical antibiotics impossible in certain cases to be accepted. Nevertheless, it is assisted in ischemic wounds and in the burn of patients to avoid septic infection (not met with systemic antibiotics) (Neely et al., 2009).

7. Development of Antibiotic Resistance in Biofilms

P. aerauginosa is also considered to be a major cause of chronic infections because of its ability to form biofilms, where it is difficult or impossible to remove antibiotic treatment in aggregates comprising an extracellular matrix of a self-produced material. *P. aeruginosa* is triggering chronic pulmonary disease, chronic urinary tract infects and ventilator-associated pneumonia in intubated patients, as well as an important pathogen in chronic wounds in patients with continuous bladder catheters. Therefore, because of its inherent antibiotic susceptibility and the creation of mutational antibiotic resistance, antibiotic treatment cannot remove such biofilmine infections. Biofilms are multifactorial tolerances of structural, biological, and genetic determinants, whereas biofilm antibiotic resistance is induced by mutation and triggered by continuously exposed bacteria to elevated antibiotic rates. The study discusses the antibiotic sensitivity in *P. aeruginosa* biofilms, as well as the production of antibiotic resistance. Possible therapeutic strategies are also discussed centered on an appreciation of the resistance and antibiotic protection processes of the biofilms (Ciofu et al., 2019).

The tolerance mechanisms mentioned all lead to the survival of biofilms, a favorable base for the evolution of antibiotic-resistant mutants. Antibiotic resistance implies genetic determinant mutations that cause increased antibiotics with minimal inhibitory concentrations in biofilm bacterial cells and it is accepted as a side-effect of prolonged maintenance therapy. The tolerance of the resistance is preceding the occurrence of resistance (Levin et al., 2017). In comparison to planktonic fast-divided cell common for the analysis of antibiotic resistance in cultures that shake, biofilm-grown bacteria have gradients of nutrients and oxygen which contribute to heterogeneous community of bacteria, including slow-growing or non-dividing cells. Knowledge that mutagenesis is not divisive and that the mutation rate is higher and is represented with different types of mutation in biofilms compared to planktonic cultures cells with a stationary or adaptive mutagenesis (Maharjan and Ferenci, 2017). Oxidative stress, SOS response to DNA damage and RpoS dependent responses are mechanisms responsible for the increased mutagenesis in the biofilm. The space-structured and heterogeneous environment of bio-films can be regarded, compared to planctonic homogenous populations, as a group of various niches, with varying selective pressures that provide the opportunity to persist in the biofilm of a broad range of resistant mutants. The diversification of the bacterial species to higher levels than seen in developmental research utilizing planktonic cultures is a common aspect observed in biofilm evolution experiments (Steenackers et al. 2013). The frequent insulation of bacteria with high rates of mutilation, e. g. hypermutators due to mutations in their DNA-repair mechanisms, is another characteristic of biofilm development, particularly in case of chronic infections. This is despite the fact that mutator phenotypes are generally accepted as self-limiting because of the high risk of deleterious mutations. This suggests that mutators are maintained in some niches and are likely to make up for the slow growth of bacteria by ensuring an abundance of spontaneous mutants which enable the biofilm to be adapted and the resistance to biofilms to develop. Previous evidence has shown that selection prefers strains with increased rate of mutation due to mutational inactivation of the same-repair system when the adaptation is restricted by the availability of mutations. In competitive experiments in flow-cell biofilm, the mutators were improving adaptability over the wildlife type. In vitro experimental studies have recently shown that antibiotic resistance in plankton culture has been developed through different pathways and by biopaths exposed to ciprofloxacin inhibitory rates (Ahmed et al., 2018). During the time ciprofloxacin was larger, resistance levels (MICs) are lower than those of planktonically evolved cultures that correlate to mutations in efflux pumps in biofilms and mutations in inplanctonic cultures of target genes. The MexCD-OprJ efflux pump in biofilm was also revealed a fast proliferation of nfxB mutants over-expressed to low doses of ciprofloxacin, with mutants comprising 80% to 90% of the biofilm community, after four days of treatment. Suggest that the production and maintenance of low resistance mutants promotes biofilm growth, represents initial mutations with possible accumulation of mutations leading to increased MIC at subsequent stages (Ciofuet al., 2019).

8. In Vivo Tolerance of Biofilms

For the survival of biofilms in various environments, various antibiotic tolerance mechanisms may be critical (Hall & Mah, 2017). In vitro, according to the place of disease, the local biofilm condition is quite complex. The antibiotic tolerance of biofilms in lung, high dose, and/or prolonged treatment required for example have been demonstrated in pharmacycinematic/pharmacodynamic studies. Such high doses are not systemically given to clinicians in the hospital due to exposure, but can be received by local administration (inhalation treatment *in vivo*, the size of biofilms in most in vitro models is smaller and the locally based biofilm environment, depending on the location of the infection, is quite complex. Thus the underlying cause of bacterial persistence of various biofilm infections may be specific antibiotic resistance mechanisms. *P. aeruginosa* biofilms, for example, are not-surface-attached in patients with cystic fibrosis, but present as clusters of airway cells surrounded by inflammatory polymorphone cells that consume much of the oxygen that is presently available (Jensen et al., 2017).

It has therefore been shown that bacteria have anaerobic environments in these biofilms. The P is the reverse. Aeruginosa-biofilms which are associated in patients in aerobic environments with nosocomial ventilator-associated pneumonia in endotracheal tubes suggesting that bacteria in the periphery of such biofilms are under aerobic conditions. According to clonal development and adaptation across persistent inflammations including *P. aeruginosa*ic cf cases, in vivo, the difficulty of resistance for the biofilm can be enhanced by the diversity of the biofilm-forming clinical strains with different gene-expression pathologies (Rossy et al. 2018). In vivo.

Multispecies biofilms are commonly found in nature, but their meanings in chronic infections are unclear since in vivo biofilms, with the exception of oral and intestinal biofilms, have in many cases demonstrated monospecies bacterial aggregates. Biofilms in in vitro, including *P. aeruginosa*, have shown increased tolerance to antibiotics compared to biofilms of monospecies (Tavernier et al., 2017; Ciofu et al., 2019).

Human beings have had acute bacterial infections for many centuries, during which planctonic cells have been attacking our bodies for life with the help of a special pathogens. We countered these acute diseases by vaccines and antibiotics, which we control to some extent. However, for millions of years organisms have been successful in the environment and are engineering successful attacks on human health, utilizing the biofilm strategy which has so well protected their indigenous habitats (Fux et al., 2005). Biofilm-associated microorganisms for many infections in advanced countries, including a variety of nosocomial infections, middle ear infections, intraocular fibrosis, periodontitis (PPI) in chronic lung infection, endocarditis, chronic prostatitis, osteomyelitis, urinary tract infections, sinusitis (Del and Patel, 2007). This medical entity's burden (that is, biofilm-disease) is huge, as it is an important problem for public health. Biofilm illness can be divided into three main types: biofilm-related clinical impairment, persistent biofilm disorder and biofilm -related device malfunction.

8.1. Biofilm-Related Device Malfunction

The external body response consisting of macrophages and foreign body giant cells is the final stage reactions after implantation of the inflammatory and wound healing responses. These events in the foreign body reaction include adsorption of proteins, monocyte/macrophage adhesion of the body, macrophage fusion forming foreign body giant cells and the effects of foreign body responses of biomaterials. In the first two to four weeks after the implant of a medical device, the surface of biomaterials plays an important role during module of the foreign body reaction although the foreign body's reaction at the tissue/material interface is present for the lifespan of the medical device (Glage et al., 2017; Del et al., 2009).

Different clinical presentations can be observed that may occasionally change from each other, depending on the balance of the interactions between the host, implant, microorganisms and their byproducts. Implant dysfunction can exhibit mild clinical signs/symptoms of negative infections/inflammation predictor, such as light pain and/or soft tissue contracture or even functional impairment. Diagnosis of subclinical disease also requires extended sampling, complex microbiological laboratory techniques (sonication) and, finally, genomic tests to identify biofilm pathogens (Del et al., 2009; Romano et al., 2017). The clinical manifestations of the malfunction of biofilms may be devastating. Prothetic joint failure, capsule breast implant contraction, dysfunction of the intravascular catheter, obstruction of the bile tube, crystalline urinary stent encrustations, brain spinal fluids, and chemical degradation and physical damage of the stent. These can also require withdrawal from the unit, raise morbidity and mortality and add up to the patient's hospital costs. Expanding knowledge of this complex interaction of eukaryotic cells/microorganisms and biomaterials can help develop new biomaterials and tissue engineering frameworks which can guide biological responses.

8.2. Non-Device-Related Chronic Biofilm Disease

In the lack of an outside organism, biofilm infection is a well-known entity. Persistent airway infection prevalent to cystic fibroid patients, chronic obstructive pulmonary disorders, TB, CWI, chronic otitis media, endocarditis, recurrent Sinusitis, Dental carditis, periodontitis, otitis media, equilibrium infection and bacterial prostatitis Virus Not linked to the introduced devices. It also suggests that biofilm microbiota may be regularly exposed to sub-minimum inhibitory antibiotic concentrations in soft tissues (e.g., the intestines and lungs). This exposure would impact bacterial physiology, including genetic and phenotypic variability of biofilm and the ability to work with antibiotics as molecules for signaling. Such results will speed up the development and spread of biofilm antibiotic-resistant bacteria.

8.2.1. Oral Cavity

The microbiome in oral form is different and is available in structurally and functionally organized biofilms as multispecies microbial communities on oral surfaces. Evidence shows how oral and pathogenic microorganisms of clinical significance engage in periodontal microbiota in many different ways, contributing to disorders impacting periodontal or periodontal safety.

Whether these opportunist pathogens play a role or are just observers in periodontal diseases they may spread to remote body sites and are likely to develop systemic infection, especially in those with immunodeficiency, because of their capacity to colonize oral mycrobiota. Periodontal illness (i.e., periodontitis and gingivithitis) is one of the most severe oral infectious diseases connected with the formation and eventual loss of the tooth of the underlying periodontal tissue, triggering immune and inflammatory host reactions (Lafaurie et al. 2017). In the supra-lingival and sub-gingival plaque in the stable oral cavity, numerous reports have shown substantial variability. Among healthy adults, bacterial supragingivals consist mainly of gram-positive coci and anaerobic-positive bars comprising streptococci and actinomyce naeslundii, while anaerobial gram-negative bars and possible anaerobic gram-positive cocci in the subgingival plaque, respectively are considerably different in bacterial composition in supragingival and subgingival plaques. Due to the anatomic proximity of periodontal biofilm to the gingival bloodstream, periodontal pocks can act as the reservoirs of and products for microbiological pathogens, as well as inflammatory mediators and immune complexes that spread to other places of the human body. Multi-synergistic and antagonistic interaction networks create microbial interrelationships and resilience of biofilms to minor environmental disturbances. Interestingly, the link between yeasts and other oral biofilm bacteria appears to be good. A forexample metagenomic study of older patients found that increased Candida load facilitated oral straptococcal disease (Del, 2018).

8.2.2. Upper Airways

Sinusitis (or rhinosinusitis) may be exacerbated or recurrent, although chronic conditions have also been identified with subacute and severe exacerbations. Since all sorts of signs have the same, discriminating between them is also medically challenging. The opinion is rising that the biofilm Growth characterizes persistent rhinosinusitis (Foreman et al., 2011).

8.2.3. Lower Airways

Biofilm infection, the form *Pseudomonas aeruginosa* is found in cystic fibrosispatients, may be due to the lower respiratory tract infection Cystic fibrosis is the inherited disease of a group of organisms (*P. aeruginosa, Stenotrophomonas maltophilia* and *Burkholderiacepacia*), characterized by the accumulation of thick, cumulative pulmonary secretions and chroniccolonization and/or infection. Interactions in mixed eukaryotic-prokaryotic biofilms (polymicrobial infections) in cystic fibrosis can lead to adverse clinical results (Leclairand Hogan, 2010). Fungal biofilms in the lung can also lead to infection already understood.

8.2.4. Gastrointestinal and Urinary Tracts

The gastrointestinal tract mucosa is heavily loaded with bacterial microbiota that is established as balanced biofilm cultures. The gastrointestinal tract itself, which is essentially a polymicrobial movie, is likely to exist and play an important role in health and illness. A polymicrobial environment that can prevent bacterial vaginosis, yeast infections and sexually transmitted diseases as well as urinary tract infections is also used in the urinary tract. A major mediator for the select inhibition of other species is the high acidity from bacterial metabolism. Biofilms were detected and shown to grow with this method on vaginal mucosa in the genitourinary tract. In several experiments for biofilm formation (Yousefi et al. 2016), isolates of the *Escherichia coli*, and Staphylococcus aureus that induce genitourinary tract (i.e., system independent infection) were shown. Often described in E. coli in diarrhea syndromes, the development of biofilm is expected to mediate the phenotype of enterology.

8.2.5. Wounds

The existence of persistent bacterial biofilm wounds has recent literature reported that they exist (Cooper et al. 2014). The presence of microbial biofilms is a significant clinical burden to non-healing wounds such as diabetes foot ulcers. Delaying in healing and potentially increasing risk of infection tend to be induced by biofilms. Biofilms can be avoided and controlled by removal and debridement of wounds after development (Percival, 2017). *S. aureus* and *P. aeruginosa* in those patients are often isolated together and a non-random association has been shown in S-wounds. *S. aureus* has fibrin receptors which can bind in the bone matrix to fibrinogenic and initiate the development of biofilm. It is easy for the pathogen to colonize skin, soft tissue and even bone (i.e., osteomyelitis), by forming a biofilm, to bind the Aureus to fibronectin, collagen, and laminin. Future studies are required in order to recognize and track colonization of biofilms at the bedside, so that care may proceed promptly. There is evidence that pathogenic species of fungi can also be involved in these infections.

8.2.6. Device-Related Biofilm Disease

In modern medicine the use of implanted medical devices has become integral. Foreign body implants nevertheless constitute a double-edged sword, as infectious problems are routinely found in addition to their excellent benefits (Fux et al., 2003.Cardiac pacemakers and prosthetic heart valves, vascular prostheses, electrical dialyzers, prosthetic joints, alloplastic orthopedic devices, intravenous catheters, indwelling urinary catheters, urethral stents, endotracheal tubes, erebrospinal fluid shunts, peritoneal dialysis catheters, biliary tract stents, breast implants, voice prostheses, intrauterine devices, abdominal drains, nephrostomy tubes, tissue fillers and contact lenses are some examples that have proved to be indispensable for patients. Nevertheless, these same devices also come with a heightened risk of biofilm-associated infection. The table shows some of the main biofilm linked human-related infections (Sharma et al., 2019).

Table 3. Bacterial species involved in biofilm associated infection and their adherent surfaces

Bacterial Species	**Surface**	**Infection/Diseases**
Mycobacterium tuberculosis	Lungs	Tuberculosis
Haemophilus influenza	Middle ear	Otitis media
Staphylococcus sp (*Staphylococcus aureus*; *Staphylococcus epidermidis*)	Middle ear, Heart valves, Bones, Surfaces/deep skin, Prostheses, Arteriovenous shunts, Sutures, Central venous catheters	Otitis media infections, Musculoskeletal Endocarditis, Chronic wounds, Nosocomial Infections
Bacterial Species	Surface	Infection/Diseases
Klebsiella pneumonia	Liver, Lungs	Pyogenic liver abscess, Urinary tract infection, Respiratory tract infection and Pneumonia.
Streptococcus mutans	Vascular grafts and Tooth surface	Endocarditis, Dental caries
Burkholderia cepacia	Lungs	Cystic fibrosis
Escherichia coli	Middle ear, Urinary catheters Prostheses, Urinary tract	Otitis media, Urinary tract infection, Bacterial prostatitis
Pseudomonas aeruginosa	Contact lenses, Lungs, Prostheses, Middle ear, Central venous catheters	Cystic fibrosis, Otitis media Nosocomial infection
Enterococcus faecalis	Central venous catheters, Tooth, Urinary catheters, Heart valves	Root canal infection Endocarditis

8.2.7. Evolutionary Pathways to Antibiotic Resistance

Our pain sensitivity and demographic composition depend on whether development takes place in well-mixed or organized settings. The evolution hypothesized in biofilms would create better genetic diversity and lead to different ways of antibiotic resistance than well-mixed environments. We tested how the biofilm lifestyles had an influence on rate, genetic processes and the pleiropic impacts of ciprofloxacin resistance in the *Acinetobacter baumannii* populations with experimental development and whole game sequencing. Dynamics of both evolution and the identities of mutations differed. Selective mutations including primary Topoisomerase medicines were experienced in Planktonic populations, whereas the biofilms-adapted populations acquired mutations in efflux pump regulators. There was an overall compromise between fitness and resistance level where the clones adapted to biofilm were less resistant than the plankton, but were more fit in the absence of drugs. But biofilm populations developed a collateral sensitivity to cephalosporins, which showed the clinical relevance of lifestyle on resistance development (Santoset al., 2019).

8.2.8. Antibiotic Tolerant and Resistant Bacterial Subpopulations in Natural Freshwater Biofilm Communities

The risk to human life is to antibiotic resistant bacteria. Sewers have recently been described as possible tanks. Thanks to the weakness of municipal irrigation resources, periodic dumping of water onto adjacent surface water is indispensable. There is little data on the impact of sporadic exposure on natural freshwater biofilms (NFBs), which ensures a basic test is required. Thus. In April, we put NFB fences in a river upstream and downstream from overflows of urban drainage. Also, there have been two screening projects. Sewage and water have been collected to gather information on the background exposure of antibiotic discoveries such as CIP, Clarithromycin (CLA) and DOX. After a 24-hour exposure with additionally dosed antibiotics, we determined the oxygen intake rate (OUR) of the NFB communities six months later. The 0.1, 1.0 and 10.0 mg/L concentrations have been selected. The mean antibiotic background levels in sewage range was 575.5 to 1289.1 ng/L, which was largely higher than those published in the literature. CIP, CLA and DOX were taken on a single basis and also in mixtures. The mean concentration was estimated in the brook between 4.6-539.0 ng/L. The first important OUR inhibition with single antibiotic doses began mainly with the 1.0 mg/L concentration. The antibiotics in a 0.1 and 1.0 mg/L combination were as active as a single dosed 10.0 mg L-1 antibiotic. The improved antibiotic toleration and the tolerance of NFB populations downstream of the CSO system was attributed to a severe impact of urban runoff. There are also hotspots of antibiotically resistant subpopulations, which are sensitive of NFB societies downstream CSO-constructions, and limits on access should be declared where there is an infection risk (Kaeseberg et al., 2018).

8.2.9. Phenazine Mediated Antibiotic Tolerance and Metabolic Heterogeneity

Bioactive metabolites and other medicines present at sites of infection may antagonize antibiotic effectiveness. A frequent cause of biofilm-based infections, *Pseudomona aeruginosas* releases metabolites called phenazines, which are used as electrons to promote the balance of cell redox. We note here that *P. aeruginosa* biofilms promote tolerance to clinically relevant antibiotics, such as ciprofloxacin, and rely on the origin of carbon for development. We incorporate a stable isotope marking with the Raman spread microscopy to show the metabolic activity of biofilm on site. Phenazine-dependent survival of ciprofloxacin is reduced in mutants without these enzymes in accordance with the functions of different respiratory complexes to facilitate the use of phenazin in biofilms. Recent research develops a chemical analysis technique for biofilm biosynthesis and points to the complex interactions between bacterial materials, the impact on biofilm's metabolism and antibiotics we use to treat infections (KT et al., 2019).

P. aeruginosa produces in its producers a redox-active pigment called phenazines, which can alter the susceptibility of antibiotics and affect gene expression, metabolism flow, andredox equilibrium (Dietrich et al., 2013). *P. aeruginosa* is a major cause of hospital-acquired infections and chronic lung infections in patients with inherited cystic fibrosis. In addition, the formation of biofilms, densely packaged communities with limited oxygen depth is an essential characteristic of *P. aeruginosa*'s infections. In anoxical environments (including the use of nitrateor, to a limited extent, arginine or pyruvate, *P. aeruginosa* has little metabolic technique for promoting redox homeostasis). Nonetheless, studied for the growth of biofilms and production of genetic films have shown that in biofilms produced under an oxic atmosphere denitrifying and pyruvate fermentation occur and that these components do not reliably express themselves in biofilm depths (Lin et al., 2018). In fact, an alternate electron acceptor is asendogenous phenazines, metabolic activity in hypoxic biofilm subregions have also been suggested to support this. According to this model, intact biofilm microelectrode measurements show a decreased extracellular pool of phenazines at depth (Jo et al. 2017), which showed that cells in the oxygen-bound biofilm bases carry out a reduction in phenazine. Metabolic mutants experiments have also included in this operation important components of the respiratory chain, P.cbB3 form terminal oxidases Cco1 and Cco2, (Jo et al., 2017). Such findings together offer indirect evidence that metabolism is qualitatively heterogeneous over biofilm profundity. In biofilms, metabolic heterogeneity may lead to antibiotic variations and hinder biofilm care. The effect on the viability of cells in biofilms that are exposed to antibiotics from *P. aeruginosal* phenazin development is discussed here. By means of combining stable isotope labeling and SRS microscopy, we develop a chemical imaging technique to monitor metabolism of biofilm on the spot. This technique is used to assess the impact on biofilm depth of metabolism by phenazinesynthesis and antibiotic treatment.

Conclusion and Future Prespectives

Biofilm is a complex microbiome structure with different bacterial colonies or single cell type in a group, adjoining the surface. These cells have a high resistance to antibiotics, as well as an extracellular polymer matrix, consisting generally of eDNA, proteins and polysaccharides. This is one of the main causes of persistence of the infection, particularly in nosocomial environments. Quorum sensing plays a key role in regulating the formation of biofilms. There are many ways for preventing pathogens, but CRISPR-CAS (genetic engineering techniques) and photodynamic therapy (PDT), in particular inducing dangerous drug-resistant, miscarriages, are recommended for use as preventive measures to reduce bacterial biofilm disease (Sharma et al., 2019).

Their ability to accept antibiotics, host protection systems and other external stress are a serious global concern for human health, which is why they contribute to persistent chronic infections. Bacterial biofilms are immobile communities that colonize and grow medical implant surfaces like sutures, catheters and dental implants, using the self-produced extracellular polymer substances and cause canonical infections that lead to inadequate treating and mental illness for their patients (Costerton et al., 2005; Hoiby et al., 2011). The biopsy includes bacteria that have bacterial secreted polymers such as exopolysaccharides (EPS), extracellular DNA, proteins and amyloidogenic proteins, and includes bacterial cells of extra-cellular matrix (ECM) (Whitchurch et al., 2002; Wingender etc.,2001). Multi-stage processes such as I the adsorption of molecules (macro and micro-molecules) on to surfaces are well controlled by the creation of biofilms; (ii) surface bacterial adhesion and release of polymer extracellular (EPS) substances; (iii) colony forming and maturation of biofilm. Compared to the planktonic community, metabolic activities of the bacterial biofilm communities have changed as e.g., increased rate of production of EPS, activation or inhibition of particular genes associated with the formation of biofilms and reduced rate of growth (Flemming et al. 2007). Biofilms offer defense not only against altered pH, osmolarities, food deficiency, mechanical and shear powers to the microorganism (Costertonet al., 1995; May et al., 2015) but also blocking access to antibiotic and host immune cell bacterial biofilm populations (Costerton et al., 1999). Biofilm matrix therefore provides additional resistance to bacteria to tolerate not only harsh conditions, but also antibiotics, leading to bad bug infections like multiple drug resistants, extensively drug resistant and completely drug resistant bacteria.

Biofilm-associated infections have been successfully treated due to a high resistance to antibiotics in the bacterial communities. Classical antibiotics chemotherapy is incapable of eradicating entirely bacterial cells located in the central biofilm region and leading to global worsening. Therefore, alternative strategies (Figure 3) and new antibiofilm agents have already been studied to overcome the resilience of bacterial biofilm community (Sharma et al., 2019).

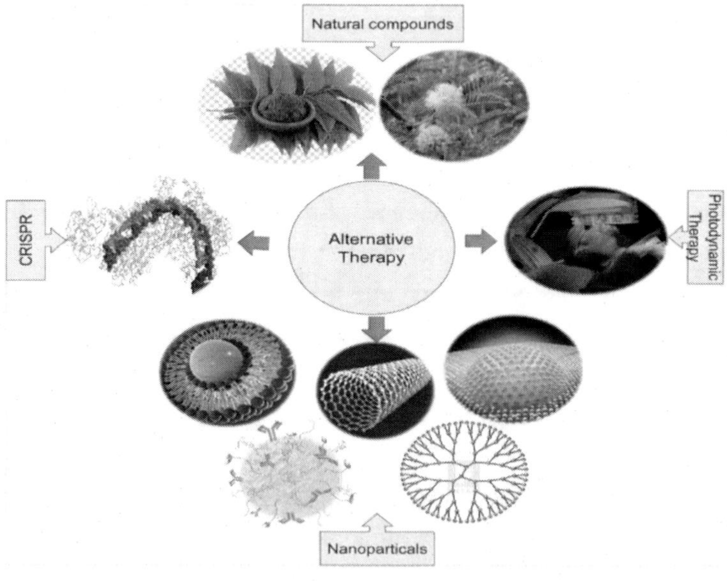

Figure 3. Diagramming of alternative approaches to antibiotic resistant biofilm communities.

About 80% of persistent and recurrent bacterial diseases within the human body have been attributed to biofilm. Bacterial cells have been more resistant to antibiotics than planktonic cells in a variety of biofilm niches, including the aquatic lakes, cliffs, deep-sea winds and hydrothermal springs. Biofilm has been found in various ecosystem niches. Infections associated with biofilms can be divided into two broadly. Biofilms can form on abiotic surfaces, particularly infections associated with medical devices and host tissue native biofilms (Burmolle et al., 2010). The original biofilm of on-medical implants, such as heart valves, catheters, contact lenses, joint prostheses, intrauterine devices and dental unit, can cause urinary tract infections and blood stream infections. These infections can only be treated by removing implants that not only increase treatment costs but also cause problems in patients (Costerton et al. 2005). Tissue-related biofilm infections, periodontitis, endocarditis, recurring urinary tract infection, chronic osteomyelitis, chronic prostatitis, chronic wounds, chronic osteomyelitis, dental cares including chronic cystic fibrosis lung infections are often chronic. Customs and other disorders are often persistent (Burmolle et al., 2010).

References

[1] Afessa B, Shorr AF, Anzueto AR, Craven DE, Schinner R, Kollef MH. Association between a silver-coated endotracheal tube and reduced mortality in patients with ventilator-associated pneumonia. *Chest.* 2010; 137(5):1015-21.

[2] Ahmed MN, Porse A, Sommer MO, Høiby N, Ciofu O. Evolution of antibiotic resistance in biofilm and planktonic *Pseudomonas aeruginosa* populations exposed to subinhibitory levels of ciprofloxacin. *Antimicrobial agents and chemotherapy.* 2018 Aug 1;62(8):e00320-18.

[3] Alkawash MA, Soothill JS, Schiller NL. Alginate lyase enhances antibiotic killing of mucoid *Pseudomonas aeruginosa* in biofilms. *Apmis.* 2006 Feb;114(2):131-8.

[4] An SQ, Murtagh J, Twomey KB, Gupta MK, O'Sullivan TP, Ingram R, Valvano MA, Tang JL. Modulation of antibiotic sensitivity and biofilm formation in *Pseudomonas aeruginosa* by interspecies signal analogues. *Nature communications.* 2019 May 27;10(1):2334.

[5] Anderl JN, Franklin MJ, Stewart PS. Role of antibiotic penetration limitation in *Klebsiella pneumoniae* biofilm resistance to ampicillin and ciprofloxacin. *Antimicrobial agents and chemotherapy.* 2000 Jul 1;44(7):1818-24.

[6] Anwar H, Costerton JW. Enhanced activity of combination of tobramycin and piperacillin for eradication of sessile biofilm cells of *Pseudomonas aeruginosa. Antimicrobial agents and chemotherapy.* 1990 Sep 1;34(9):1666-71.

[7] Babin BM, Atangcho L, van Eldijk MB, Sweredoski MJ, Moradian A, Hess S, Tolker-Nielsen T, Newman DK, Tirrell DA. Selective proteomic analysis of

antibiotic-tolerant cellular subpopulations in *Pseudomonas aeruginosa* biofilms. *MBio.* 2017 Nov 8;8(5): e01593-17.

[8] Baumann U, Stocklossa C, Greiner W, von der Schulenburg JM, von der Hardt H. Cost of care and clinical condition in paediatric cystic fibrosis patients. *Journal of Cystic Fibrosis.* 2003 Jun 1;2(2):84-90.

[9] Beloin C, Valle J, Latour-Lambert P, Faure P, Kzreminski M, Balestrino D, Haagensen JA, Molin S, Prensier G, Arbeille B, Ghigo JM. Global impact of mature biofilm lifestyle on *Escherichia coli* K-12 gene expression. *Molecular microbiology.* 2004 Feb;51(3):659-74.

[10] Bjarnsholt T, Jensen PØ, Burmølle M, Hentzer M, Haagensen JA, Hougen HP, Calum H, Madsen KG, Moser C, Molin S, Høiby N. *Pseudomonas aeruginosa* tolerance to tobramycin, hydrogen peroxide and polymorphonuclear leukocytes is quorum-sensing dependent. *Microbiology.* 2005 Feb 1;151(2):373-83.

[11] Bjarnsholt T, Jensen PØ, Fiandaca MJ, Pedersen J, Hansen CR, Andersen CB, Pressler T, Givskov M, Høiby N. *Pseudomonas aeruginosa* biofilms in the respiratory tract of cystic fibrosis patients. *Pediatric pulmonology.* 2009 Jun;44(6):547-58.

[12] Bjarnsholt T, Jensen PØ, Fiandaca MJ, Pedersen J, Hansen CR, Andersen CB, Pressler T, Givskov M, Høiby N. *Pseudomonas aeruginosa* biofilms in the respiratory tract of cystic fibrosis patients. *Pediatric pulmonology.* 2009 Jun;44(6):547-58.

[13] Bjarnsholt T, Tolker-Nielsen T, Høiby N, Givskov M. Interference of *Pseudomonas aeruginosa* signalling and biofilm formation for infection control. *Expert reviews in molecular medicine.* 2010 Apr;12.

[14] Bolitho ME, Perez LJ, Koch MJ, Ng WL, Bassler BL, Semmelhack MF. Small molecule probes of the receptor binding site in the Vibrio cholerae CAI-1 quorum sensing circuit. *Bioorganic & medicinal chemistry.* 2011 Nov 15;19(22):6906-18.

[15] Bowler PG. Antibiotic resistance and biofilm tolerance: a combined threat in the treatment of chronic infections. *Journal of wound care.* 2018 May 2;27(5):273-7.

[16] Branda SS, Vik Å, Friedman L, Kolter R. Biofilms: the matrix revisited. *Trends in microbiology.* 2005 Jan 1;13(1):20-6.

[17] Branda SS, Vik Å, Friedman L, Kolter R. Biofilms: the matrix revisited. *Trends in microbiology.* 2005 Jan 1;13(1):20-6.

[18] Brodt AM, Stovold E, Zhang L. Inhaled antibiotics for stable non-cystic fibrosis bronchiectasis: a systematic review. *European Respiratory Journal.* 2014 Aug 1;44(2):382-93.

[19] Burmølle M, Ren D, Bjarnsholt T, Sørensen SJ. Interactions in multispecies biofilms: do they actually matter?. *Trends in microbiology.* 2014 Feb 1;22(2):84-91.

[20] Burmølle M, Thomsen TR, Fazli M, Dige I, Christensen L, Homøe P, Tvede M, Nyvad B, Tolker-Nielsen T, Givskov M, Moser C. Biofilms in chronic infections–a

matter of opportunity–monospecies biofilms in multispecies infections. *FEMS Immunology & Medical Microbiology.* 2010 Aug 1;59(3):324-36.

[21] Cheema MS, Rassing JE, Marriott C. The diffusion characteristics of antibiotics in mucus glycoprotein gels. *Journal of Pharmacy and Pharmacology.* 1986 Dec;38(S12):53P-.

[22] Christensen LD, Moser C, Jensen PØ, Rasmussen TB, Christophersen L, Kjelleberg S, Kumar N, Høiby N, Givskov M, Bjarnsholt T. Impact of *Pseudomonas aeruginosa* quorum sensing on biofilm persistence in an in vivo intraperitoneal foreign-body infection model. *Microbiology.* 2007 Jul 1;153(7):2312-20.

[23] Chua SL, Sivakumar K, Rybtke M, Yuan M, Andersen JB, Nielsen TE, Givskov M, Tolker-Nielsen T, Cao B, Kjelleberg S, Yang L. C-di-GMP regulates *Pseudomonas aeruginosa* stress response to tellurite during both planktonic and biofilm modes of growth. *Scientific reports.* 2015 May 20;5:10052.

[24] Chua SL, Yam JK, Hao P, Adav SS, Salido MM, Liu Y, Givskov M, Sze SK, Tolker-Nielsen T, Yang L. Selective labelling and eradication of antibiotic-tolerant bacterial populations in *Pseudomonas aeruginosa* biofilms. *Nature communications.* 2016 Feb 19;7:10750.

[25] Chua SL, Yam JK, Hao P, Adav SS, Salido MM, Liu Y, Givskov M, Sze SK, Tolker-Nielsen T, Yang L. Selective labelling and eradication of antibiotic-tolerant bacterial populations in *Pseudomonas aeruginosa* biofilms. *Nature communications.* 2016 Feb 19;7:10750.

[26] Chugani S, Greenberg EP. An evolving perspective on the *Pseudomonas aeruginosa* orphan quorum sensing regulator QscR. *Frontiers in cellular and infection microbiology.* 2014 Oct 28;4:152.

[27] Ciofu O, Beveridge TJ, Kadurugamuwa J, Walther-Rasmussen J, Høiby N. Chromosomal β-lactamase is packaged into membrane vesicles and secreted from *Pseudomonas aeruginosa. Journal of Antimicrobial Chemotherapy.* 2000 Jan;45(1):9-13.

[28] Ciofu O, Rojo-Molinero E, Macià MD, Oliver A. Antibiotic treatment of biofilm infections. *Apmis.* 2017 Apr;125(4):304-19.

[29] Ciofu O, Tolker-Nielsen T. Tolerance and resistance of *Pseudomonas aeruginosa* biofilms to antimicrobial agents-How P. aeruginosa can escape antibiotics. *Frontiers in microbiology.* 2019;10:913.

[30] Colvin KM, Gordon VD, Murakami K, Borlee BR, Wozniak DJ, Wong GC, Parsek MR. The pel polysaccharide can serve a structural and protective role in the biofilm matrix of *Pseudomonas aeruginosa*. *PLoS pathogens.* 2011 Jan 27;7(1):e1001264.

[31] Colvin KM, Gordon VD, Murakami K, Borlee BR, Wozniak DJ, Wong GC, Parsek MR. The pel polysaccharide can serve a structural and protective role in the biofilm matrix of *Pseudomonas aeruginosa*. *PLoS pathogens.* 2011 Jan 27;7(1):e1001264.

[32] Conlon BP, Nakayasu ES, Fleck LE, LaFleur MD, Isabella VM, Coleman K, Leonard SN, Smith RD, Adkins JN, Lewis K. Activated ClpP kills persisters and eradicates a chronic biofilm infection. *Nature.* 2013 Nov;503(7476):365.

[33] Cooper RA, Bjarnsholt T, Alhede M. Biofilms in wounds: a review of present knowledge. *Journal of wound care.* 2014 Nov 2;23(11):570-82.

[34] Costerton JW, Lewandowski Z, Caldwell DE, Korber DR, Lappin-Scott HM. Microbial biofilms. *Annual review of microbiology.* 1995 Oct;49(1):711-45.

[35] Costerton JW, Montanaro L, Arciola CR. Biofilm in implant infections: its production and regulation. *The International journal of artificial organs.* 2005 Nov;28(11):1062-8.

[36] Costerton JW, Stewart PS, Greenberg EP. Bacterial biofilms: a common cause of persistent infections. *Science.* 1999 May 21;284(5418):1318-22.

[37] Costerton W, Veeh R, Shirtliff M, Pasmore M, Post C, Ehrlich G. The application of biofilm science to the study and control of chronic bacterial infections. *The Journal of clinical investigation.* 2003 Nov 15;112(10):1466-77.

[38] Curtis MM, Russell R, Moreira CG, Adebesin AM, Wang C, Williams NS, Taussig R, Stewart D, Zimmern P, Lu B, Prasad RN. QseC inhibitors as an antivirulence approach for Gram-negative pathogens. *MBio.* 2014 Dec 31;5(6):e02165-14.

[39] Darouiche RO, Dhir A, Miller AJ, Landon GC, Raad II, Musher DM. Vancomycin penetration into biofilm covering infected prostheses and effect on bacteria. *Journal of Infectious Diseases.* 1994 Sep 1;170(3):720-3.

[40] Del Pozo JL, Patel R. The challenge of treating biofilm-associated bacterial infections. *Clinical Pharmacology & Therapeutics.* 2007 Aug;82(2):204-9.

[41] Del Pozo JL, Tran NV, Petty PM, Johnson CH, Walsh MF, Bite U, Clay RP, Mandrekar JN, Piper KE, Steckelberg JM, Patel R. Pilot study of association of bacteria on breast implants with capsular contracture. *Journal of clinical microbiology.* 2009 May 1;47(5):1333-7.

[42] Del Pozo JL. Biofilm-related disease. *Expert review of anti-infective therapy.* 2018 Jan 2;16(1):51-65.

[43] Dieterich DC, Lee JJ, Link AJ, Graumann J, Tirrell DA, Schuman EM. Labeling, detection and identification of newly synthesized proteomes with bioorthogonal non-canonical amino-acid tagging. *Nature protocols.* 2007 Mar;2(3):532.

[44] Dietrich LE, Okegbe C, Price-Whelan A, Sakhtah H, Hunter RC, Newman DK. Bacterial community morphogenesis is intimately linked to the intracellular redox state. *Journal of bacteriology.* 2013 Apr 1;195(7):1371-80.

[45] Diggle SP, Cornelis P, Williams P, Cámara M. 4-Quinolone signalling in *Pseudomonas aeruginosa*: old molecules, new perspectives. *International journal of medical microbiology.* 2006 Apr 6;296(2-3):83-91.

[46] Doring G, Conway SP, Heijerman HG, Hodson ME, Hoiby N, Smyth A, Touw DJ. Antibiotic therapy against *Pseudomonas aeruginosa* in cystic fibrosis: a European consensus. *European Respiratory Journal.* 2000 Oct 1;16(4):749-67.

[47] Döring G, Hoiby N, Consensus Study Group. Early intervention and prevention of lung disease in cystic fibrosis: a European consensus. *Journal of cystic fibrosis.* 2004 Jun 1;3(2):67-91.

[48] Döring G, Pier GB. Vaccines and immunotherapy against *Pseudomonas aeruginosa. Vaccine.* 2008 Feb 20;26(8):1011-24.

[49] Dunne WM, Mason EO, Kaplan SL. Diffusion of rifampin and vancomycin through a Staphylococcus epidermidis biofilm. *Antimicrobial agents and chemotherapy.* 1993 Dec 1;37(12):2522-6.

[50] Flemming HC, Neu TR, Wozniak DJ. The EPS matrix: the "house of biofilm cells." *Journal of bacteriology.* 2007 Nov 15;189(22):7945-7.

[51] Flemming HC, Wingender J, Szewzyk U, Steinberg P, Rice SA, Kjelleberg S. Biofilms: an emergent form of bacterial life. *Nature Reviews Microbiology.* 2016 Sep;14(9):563.

[52] Foreman A, Jervis-Bardy J, Wormald PJ. Do biofilms contribute to the initiation and recalcitrance of chronic rhinosinusitis?. *The Laryngoscope.* 2011 May;121(5):1085-91.

[53] Frederiksen B, Koch C, Høiby N. Changing epidemiology of *Pseudomonas aeruginosa* infection in Danish cystic fibrosis patients (1974–1995). *Pediatric pulmonology.* 1999 Sep;28(3):159-66.

[54] Frederiksen B, Lanng S, Koch C, Hølby N. Improved survival in the Danish center-treated cystic fibrosis patients: results of aggressive treatment. *Pediatric pulmonology.* 1996 Mar;21(3):153-8.

[55] Frederiksen B, Pressler T, Hansen A, Koch C, Høiby N. Effect of aerosolized rhDNase (Pulmozyme®) on pulmonary colonization in patients with cystic fibrosis. *Acta Paediatrica.* 2006 Sep;95(9):1070-4.

[56] Fux CA, Costerton JW, Stewart PS, Stoodley P. Survival strategies of infectious biofilms. *Trends in microbiology.* 2005 Jan 1;13(1):34-40.

[57] Fux CA, Stoodley P, Hall-Stoodley L, Costerton JW. Bacterial biofilms: a diagnostic and therapeutic challenge. *Expert review of anti-infective therapy.* 2003 Dec 1;1(4):667-83.

[58] Glage S, Paret S, Winkel A, Stiesch M, Bleich A, Krauss JK, Schwabe K. A new model for biofilm formation and inflammatory tissue reaction: intraoperative infection of a cranial implant with Staphylococcus aureus in rats. *Acta neurochirurgica.* 2017 Sep 1;159(9):1747-56.

[59] Gordon CA, Hodges NA, Marriott C. Antibiotic interaction and diffusion through alginate and exopolysaccharide of cystic fibrosis-derived *Pseudomonas aeruginosa. Journal of Antimicrobial Chemotherapy.* 1988 Nov 1;22(5):667-74.

[60] Götz F. Staphylococcus and biofilms. *Molecular microbiology.* 2002 Mar;43(6):1367-78.

[61] Haagensen JA, Klausen M, Ernst RK, Miller SI, Folkesson A, Tolker-Nielsen T, Molin S. Differentiation and distribution of colistin-and sodium dodecyl sulfate-tolerant cells in *Pseudomonas aeruginosa* biofilms. *Journal of bacteriology.* 2007 Jan 1;189(1):28-37.

[62] Haagensen JA, Klausen M, Ernst RK, Miller SI, Folkesson A, Tolker-Nielsen T, Molin S. Differentiation and distribution of colistin-and sodium dodecyl sulfate-tolerant cells in *Pseudomonas aeruginosa* biofilms. *Journal of bacteriology.* 2007 Jan 1;189(1):28-37.

[63] Hall CW, Mah TF. Molecular mechanisms of biofilm-based antibiotic resistance and tolerance in pathogenic bacteria. *FEMS microbiology reviews.* 2017 Mar 25;41(3):276-301.

[64] Hansen CR, Pressler T, Høiby N. Early aggressive eradication therapy for intermittent *Pseudomonas aeruginosa* airway colonization in cystic fibrosis patients: 15 years experience. *Journal of Cystic Fibrosis.* 2008 Nov 1;7(6):523-30.

[65] Hansen MC, Palmer Jr RJ, White DC. Flowcell culture of *Porphyromonas gingivalis* biofilms under anaerobic conditions. *Journal of microbiological methods.* 2000 May 1;40(3):233-9.

[66] Harrison JJ, Turner RJ, Ceri H. Persister cells, the biofilm matrix and tolerance to metal cations in biofilm and planktonic *Pseudomonas aeruginosa. Environmental Microbiology.* 2005 Jul;7(7):981-94.

[67] Hausner M, Wuertz S. High rates of conjugation in bacterial biofilms as determined by quantitative in situ analysis. *Appl. Environ. Microbiol.* 1999 Aug 1;65(8):3710-3.

[68] Hawser SP, Douglas LJ. Biofilm formation by Candida species on the surface of catheter materials in vitro. *Infection and immunity.* 1994 Mar 1;62(3):915-21.

[69] Hengzhuang W, Wu H, Ciofu O, Song Z, Høiby N. Pharmacokinetics/pharmacodynamics of colistin and imipenem on mucoid and nonmucoid *Pseudomonas aeruginosa* biofilms. *Antimicrobial agents and chemotherapy.* 2011 Sep 1;55(9):4469-74.

[70] Høiby N, Bjarnsholt T, Givskov M, Molin S, Ciofu O. Antibiotic resistance of bacterial biofilms. *International journal of antimicrobial agents.* 2010 Apr 1;35(4):322-32.

[71] Høiby N, Ciofu O, Johansen HK, Song ZJ, Moser C, Jensen PØ, Molin S, Givskov M, Tolker-Nielsen T, Bjarnsholt T. The clinical impact of bacterial biofilms. *International journal of oral science.* 2011 Apr;3(2):55.

[72] Hoiby N, Flensborg EW, Beck B, Friis B, Jacobsen SV, Jacobsen L. *Pseudomonas aeruginosa* infection in cystic fibrosis. Diagnostic and prognostic significance of *Pseudomonas aeruginosa* precipitins determined by means of crossed

immunoelectrophoresis. *Scandinavian journal of respiratory diseases.* 1977 Apr;58(2):65-79.

[73] Høiby N, Pedersen SS. Estimated risk of cross-infection with *Pseudomonas aeruginosa* in Danish cystic fibrosis patients. *Acta Pædiatrica.* 1989 May;78(3):395-404.

[74] Hoyle BD, Alcantara JO, Costerton JW. *Pseudomonas aeruginosa* biofilm as a diffusion barrier to piperacillin. *Antimicrobial agents and chemotherapy.* 1992 Sep 1;36(9):2054-6.

[75] Islam S, Oh H, Jalal S, Karpati F, Ciofu O, Høiby N, Wretlind B. Chromosomal mechanisms of aminoglycoside resistance in *Pseudomonas aeruginosa* isolates from cystic fibrosis patients. *Clinical Microbiology and Infection.* 2009 Jan 1;15(1):60-6.

[76] Jakobsen TH, van Gennip M, Phipps RK, Shanmugham MS, Christensen LD, Alhede M, Skindersoe ME, Rasmussen TB, Friedrich K, Uthe F, Jensen PØ. Ajoene, a sulfur-rich molecule from garlic, inhibits genes controlled by quorum sensing. *Antimicrobial agents and chemotherapy.* 2012 May 1;56(5):2314-25.

[77] Jalal S, Ciofu O, Høiby N, Gotoh N, Wretlind B. Molecular mechanisms of fluoroquinolone resistance in *Pseudomonas aeruginosa* isolates from cystic fibrosis patients. *Antimicrobial agents and chemotherapy.* 2000 Mar 1;44(3):710-2.

[78] Jensen PØ, Kolpen M, Kragh KN, Kühl M. Microenvironmental characteristics and physiology of biofilms in chronic infections of CF patients are strongly affected by the host immune response. *Apmis.* 2017 Apr;125(4):276-88.

[79] Jimenez PN, Koch G, Thompson JA, Xavier KB, Cool RH, Quax WJ. The multiple signaling systems regulating virulence in *Pseudomonas aeruginosa. Microbiol. Mol. Biol. Rev.* 2012 Mar 1;76(1):46-65.

[80] Jo J, Cortez KL, Cornell WC, Price-Whelan A, Dietrich LE. An orphan cbb3-type cytochrome oxidase subunit supports *Pseudomonas aeruginosa* biofilm growth and virulence. *Elife.* 2017 Nov 21;6:e30205.

[81] Kaeseberg T, Schubert S, Oertel R, Zhang J, Berendonk TU, Krebs P. Hot spots of antibiotic tolerant and resistant bacterial subpopulations in natural freshwater biofilm communities due to inevitable urban drainage system overflows. *Environmental pollution.* 2018 Nov 1;242:164-70.

[82] Khan W, Bernier SP, Kuchma SL, Hammond JH, Hasan F, O'Toole GA. Aminoglycoside resistance of *Pseudomonas aeruginosa* biofilms modulated by extracellular polysaccharide. *International microbiology: the official journal of the Spanish Society for Microbiology.* 2010 Dec;13(4):207.

[83] Khemiri A, Jouenne T, Cosette P. Proteomics dedicated to biofilmology: what have we learned from a decade of research?. *Medical microbiology and immunology.* 2016 Feb 1;205(1):1-9.

[84] Kim HS, Lee SH, Byun Y, Park HD. 6-Gingerol reduces *Pseudomonas aeruginosa* biofilm formation and virulence via quorum sensing inhibition. *Scientific reports.* 2015 Mar 2;5:8656.

[85] Koch C, Hjelt K, Pedersen SS, Jensen ET, Jensen T, Lanng S, Valerius NH, Pedersen M, Høiby N. Retrospective clinical study of hypersensitivity reactions to aztreonam and six other β-lactam antibiotics in cystic fibrosis patients receiving multiple treatment courses. *Reviews of infectious diseases.* 1991 May 1;13 (Supplement_7):S608-11.

[86] Kumon H, Tomochika KI, Matunaga T, Ogawa M, Ohmori H. A sandwich cup method for the penetration assay of antimicrobial agents through Pseudomonas exopolysaccharides. *Microbiology and immunology.* 1994 Aug;38(8):615-9.

[87] Lafaurie GI, Sabogal MA, Castillo DM, Rincón MV, Gómez LA, Lesmes YA, Chambrone L. Microbiome and microbial biofilm profiles of peri-implantitis: a systematic review. *Journal of periodontology.* 2017 Oct 1;88(10):1066-89.

[88] Leclair LW, Hogan DA. Mixed bacterial-fungal infections in the CF respiratory tract. *Sabouraudia.* 2010 Nov;48(Supplement_1):S125-32.

[89] Levin BR, Rozen DE. Non-inherited antibiotic resistance. *Nature Reviews Microbiology.* 2006 Jul;4(7):556.

[90] Levin-Reisman I, Ronin I, Gefen O, Braniss I, Shoresh N, Balaban NQ. Antibiotic tolerance facilitates the evolution of resistance. *Science.* 2017 Feb 24;355(6327):826-30.

[91] Lewis K. Persister cells, dormancy and infectious disease. *Nature Reviews Microbiology.* 2007 Jan;5(1):48.

[92] Lewis K. Riddle of biofilm resistance. *Antimicrobial agents and chemotherapy.* 2001 Apr 1;45(4):999-1007.

[93] Li J, Nation RL, Turnidge JD, Milne RW, Coulthard K, Rayner CR, Paterson DL. Colistin: the re-emerging antibiotic for multidrug-resistant Gram-negative bacterial infections. *The Lancet infectious diseases.* 2006 Sep 1;6(9):589-601.

[94] Liao J, Sauer K. The MerR-like transcriptional regulator BrlR contributes to *Pseudomonas aeruginosa* biofilm tolerance. *Journal of bacteriology.* 2012 Sep 15;194(18):4823-36.

[95] Lin YC, Cornell WC, Jo J, Price-Whelan A, Dietrich LE. The *Pseudomonas aeruginosa* complement of lactate dehydrogenases enables use of d-and l-lactate and metabolic cross-feeding. *MBio.* 2018 Nov 7;9(5):e00961-18.

[96] Lin YC, Sekedat MD, Cornell WC, Silva GM, Okegbe C, Price-Whelan A, Vogel C, Dietrich LE. Phenazines regulate Nap-dependent denitrification in *Pseudomonas aeruginosa* biofilms. *Journal of bacteriology.* 2018 May 1;200(9):e00031-18.

[97] Lynch SV, Dixon L, Benoit MR, Brodie EL, Keyhan M, Hu P, Ackerley DF, Andersen GL, Matin A. Role of the rapA gene in controlling antibiotic resistance of

Escherichia coli biofilms. *Antimicrobial agents and chemotherapy*. 2007 Oct 1;51(10):3650-8.

[98] Mah TF, O'Toole GA. Mechanisms of biofilm resistance to antimicrobial agents. *Trends in microbiology*. 2001 Jan 1;9(1):34-9.

[99] Mah TF, Pitts B, Pellock B, Walker GC, Stewart PS, O'Toole GA. A genetic basis for *Pseudomonas aeruginosa* biofilm antibiotic resistance. *Nature*. 2003 Nov;426(6964):306.

[100] Maharjan RP, Ferenci T. A shifting mutational landscape in 6 nutritional states: Stress-induced mutagenesis as a series of distinct stress input–mutation output relationships. *PLoS biology*. 2017 Jun 8;15(6):e2001477.

[101] Marschall J, Lane MA, Beekmann SE, Polgreen PM, Babcock HM. Current management of prosthetic joint infections in adults: results of an Emerging Infections Network survey. *International journal of antimicrobial agents*. 2013 Mar 1;41(3):272-7.

[102] Marsh PD, Zaura E. Dental biofilm: ecological interactions in health and disease. *Journal of clinical periodontology*. 2017 Mar;44:S12-22.

[103] Maya-Hoyos M, Leguizamón J, Mariño-Ramírez L, Soto CY. *Sliding motility, biofilm.*

[104] McKay, Geoffrey, and Dao Nguyen. "Antibiotic resistance and tolerance in bacterial biofilms." *Handbook of antimicrobial resistance* (2014): 1-24.

[105] Moreau-Marquis S, Bomberger JM, Anderson GG, Swiatecka-Urban A, Ye S, O'Toole GA, Stanton BA. The ΔF508-CFTR mutation results in increased biofilm formation by *Pseudomonas aeruginosa* by increasing iron availability. *American Journal of Physiology-Lung Cellular and Molecular Physiology*. 2008 Jul;295(1):L25-37.

[106] Moreau-Marquis S, O'Toole GA, Stanton BA. Tobramycin and FDA-approved iron chelators eliminate *Pseudomonas aeruginosa* biofilms on cystic fibrosis cells. *American journal of respiratory cell and molecular biology*. 2009 Sep;41(3):305-13.

[107] Mulcahy H, Charron-Mazenod L, Lewenza S. Extracellular DNA chelates cations and induces antibiotic resistance in *Pseudomonas aeruginosa* biofilms. *PLoS pathogens*. 2008 Nov 21;4(11):e1000213.

[108] Neely AN, Gardner J, Durkee P, Warden GD, Greenhalgh DG, Gallagher JJ, Herndon DN, Tompkins RG, Kagan RJ. Are topical antimicrobials effective against bacteria that are highly resistant to systemic antibiotics?. *Journal of Burn Care & Research*. 2009 Jan 1;30(1):19-29.

[109] Neu HC. The crisis in antibiotic resistance. *Science*. 1992 Aug 21;257(5073):1064-73.

[110] Ng WL, Perez L, Cong J, Semmelhack MF, Bassler BL. Broad spectrum pro-quorum-sensing molecules as inhibitors of virulence in vibrios. *PLoS pathogens.* 2012 Jun 28;8(6):e1002767.

[111] Ng WL, Perez LJ, Wei Y, Kraml C, Semmelhack MF, Bassler BL. Signal production and detection specificity in Vibrio CqsA/CqsS quorum-sensing systems. *Molecular microbiology.* 2011 Mar;79(6):1407-17.

[112] Nichols WW, Dorrington SM, Slack MP, Walmsley HL. Inhibition of tobramycin diffusion by binding to alginate. *Antimicrobial agents and chemotherapy.* 1988 Apr 1;32(4):518-23.

[113] Palmer LB, Smaldone GC. Reduction of bacterial resistance with inhaled antibiotics in the intensive care unit. *American journal of respiratory and critical care medicine.* 2014 May 15;189(10):1225-33.

[114] Pamp SJ, Gjermansen M, Johansen HK, Tolker-Nielsen T. Tolerance to the antimicrobial peptide colistin in *Pseudomonas aeruginosa* biofilms is linked to metabolically active cells, and depends on the pmr and mexAB-oprM genes. *Molecular microbiology.* 2008 Apr;68(1):223-40.

[115] Park AJ, Krieger JR, Khursigara CM. Survival proteomes: the emerging proteotype of antimicrobial resistance. *FEMS microbiology reviews.* 2016 Jan 19;40(3):323-42.

[116] Parsek MR, Singh PK. Bacterial biofilms: an emerging link to disease pathogenesis. *Annual Reviews in Microbiology.* 2003 Oct;57(1):677-701.

[117] Parsek MR, Tolker-Nielsen T. Pattern formation in *Pseudomonas aeruginosa* biofilms. *Current opinion in microbiology.* 2008 Dec 1;11(6):560-6.

[118] Percival SL. Importance of biofilm formation in surgical infection. *British Journal of Surgery.* 2017 Jan;104(2):e85-94.

[119] Pérez-Martínez I, Haas D. Azithromycin inhibits expression of the GacA-dependent small RNAs RsmY and RsmZ in *Pseudomonas aeruginosa. Antimicrobial agents and chemotherapy.* 2011 Jul 1;55(7):3399-405.

[120] Potera C. *Forging a link between biofilms and disease* (1999): 283(5409):1837–39.

[121] Proesmans M, Balinska-Miskiewicz W, Dupont L, Bossuyt X, Verhaegen J, Høiby N, De Boeck K. Evaluating the "Leeds criteria" for *Pseudomonas aeruginosa* infection in a cystic fibrosis centre. *European Respiratory Journal.* 2006 May 1;27(5):937-43.

[122] Rattanaumpawan P, Lorsutthitham J, Ungprasert P, Angkasekwinai N, Thamlikitkul V. Randomized controlled trial of nebulized colistimethate sodium as adjunctive therapy of ventilator-associated pneumonia caused by Gram-negative bacteria. *Journal of antimicrobial chemotherapy.* 2010 Sep 28;65(12):2645-9.

[123] Rojo-Molinero E, Macià MD, Rubio R, Moyà B, Cabot G, López-Causapé C, Pérez JL, Cantón R, Oliver A. Sequential treatment of biofilms with aztreonam and tobramycin is a novel strategy for combating *Pseudomonas aeruginosa* chronic

respiratory infections. *Antimicrobial agents and chemotherapy.* 2016 May 1;60(5):2912-22.

[124] Romanò CL, Romanò D, Morelli I, Drago L. The concept of biofilm-related implant malfunction and "low-grade infection." In *A Modern Approach to Biofilm-Related Orthopaedic Implant Infections* 2016 (pp. 1-13). Springer, Cham.

[125] Rossi E, Falcone M, Molin S, Johansen HK. High-resolution in situ transcriptomics of *Pseudomonas aeruginosa* unveils genotype independent patho-phenotypes in cystic fibrosis lungs. *Nature communications.* 2018 Aug 27;9(1):3459.

[126] Russell CJ, Shiroishi MS, Siantz E, Wu BW, Patino CM. The use of inhaled antibiotic therapy in the treatment of ventilator-associated pneumonia and tracheobronchitis: a systematic review. *BMC pulmonary medicine.* 2016 Dec;16(1):40.

[127] Ryan RP, Dow JM. Communication with a growing family: diffusible signal factor (DSF) signaling in bacteria. *Trends in microbiology.* 2011 Mar 1;19(3):145-52.

[128] Ryan RP, Fouhy Y, Garcia BF, Watt SA, Niehaus K, Yang L, Tolker-Nielsen T, Dow JM. Interspecies signalling via the *Stenotrophomonas maltophilia* diffusible signal factor influences biofilm formation and polymyxin tolerance in *Pseudomonas aeruginosa. Molecular microbiology.* 2008 Apr;68(1):75-86.

[129] Santos-Lopez A, Marshall CW, Scribner MR, Snyder D, Cooper VS. Evolutionary pathways to antibiotic resistance are dependent upon environmental structure and bacterial lifestyle. *bioRxiv.* 2019 Jan 1:581611.

[130] Savage VJ, Chopra I, O'Neill AJ. Staphylococcus aureus biofilms promote horizontal transfer of antibiotic resistance. *Antimicrobial agents and chemotherapy.* 2013 Apr 1;57(4):1968-70.

[131] Schierholz JM, Beuth J. Implant infections: a haven for opportunistic bacteria. *Journal of Hospital Infection.* 2001 Oct 1;49(2):87-93.

[132] Schiessl KT, Hu F, Jo J, Nazia SZ, Wang B, Price-Whelan A, Min W, Dietrich LE. Phenazine production promotes antibiotic tolerance and metabolic heterogeneity in *Pseudomonas aeruginosa* biofilms. *Nature communications.* 2019 Feb 15;10(1):762.

[133] Schuster A, Haliburn C, Döring G, Goldman MH, Freedom Study Group. Safety, efficacy and convenience of colistimethate sodium dry powder for inhalation (Colobreathe DPI) in patients with cystic fibrosis: a randomised study. *Thorax.* 2013 Apr 1;68(4):344-50.

[134] Sharma D, Misba L, Khan AU. Antibiotics versus biofilm: an emerging battleground in microbial communities. *Antimicrobial Resistance & Infection Control.* 2019 Dec;8(1):76.

[135] Shigeta M, Tanaka G, Komatsuzawa H, Sugai M, Suginaka H, Usui T. Permeation of antimicrobial agents through *Pseudomonas aeruginosa* biofilms: a simple method. *Chemotherapy.* 1997;43(5):340-5.

[136] Sperandio V. Novel approaches to bacterial infection therapy by interfering with bacteria-to-bacteria signaling. *Expert review of anti-infective therapy.* 2007 Apr 1;5(2):271-6.

[137] Stass H, Delesen H, Nagelschmitz J, Staab D. Safety and pharmacokinetics of ciprofloxacin dry powder for inhalation in cystic fibrosis: a phase I, randomized, single-dose, dose-escalation study. *Journal of aerosol medicine and pulmonary drug delivery.* 2015 Apr 1;28(2):106-15.

[138] Steenackers HP, Parijs I, Foster KR, Vanderleyden J. Experimental evolution in biofilm populations. *FEMS microbiology reviews.* 2016 Feb 18;40(3):373-97.

[139] Stensballe J, Tvede M, Looms D, Lippert FK, Dahl B, Tønnesen E, Rasmussen LS. Infection risk with nitrofurazone-impregnated urinary catheters in trauma patients: a randomized trial. *Annals of internal medicine.* 2007 Sep 4;147(5):285-93.

[140] Stewart PS, Costerton JW. Antibiotic resistance of bacteria in biofilms. *The lancet.* 2001 Jul 14;358(9276):135-8.

[141] Stewart PS, Franklin MJ. Physiological heterogeneity in biofilms. *Nature Reviews Microbiology.* 2008 Mar;6(3):199.

[142] Stewart PS, Rayner J, Roe F, Rees WM. Biofilm penetration and disinfection efficacy of alkaline hypochlorite and chlorosulfamates. *Journal of applied microbiology.* 2001 Sep 12;91(3):525-32.

[143] Stewart PS. A review of experimental measurements of effective diffusive permeabilities and effective diffusion coefficients in biofilms. *Biotechnology and bioengineering.* 1998 Aug 5;59(3):261-72.

[144] Stewart PS. Antimicrobial tolerance in biofilms. *Microbiol Spectr* 3 (3): MB-0010-2014.

[145] Stewart PS. Mechanisms of antibiotic resistance in bacterial biofilms. *International Journal of Medical Microbiology.* 2002 Jan 1;292(2):107-13.

[146] Stewart PS. Theoretical aspects of antibiotic diffusion into microbial biofilms. *Antimicrobial agents and chemotherapy.* 1996 Nov 1;40(11):2517-22.

[147] Tavernier S, Crabbé A, Hacioglu M, Stuer L, Henry S, Rigole P, Dhondt I, Coenye T. Community composition determines activity of antibiotics against multispecies biofilms. *Antimicrobial agents and chemotherapy.* 2017 Sep 1;61(9):e00302-17.

[148] Trunk K, Benkert B, Quäck N, Münch R, Scheer M, Garbe J, Jänsch L, Trost M, Wehland J, Buer J, Jahn M. Anaerobic adaptation in *Pseudomonas aeruginosa*: definition of the Anr and Dnr regulons. *Environmental microbiology.* 2010 Jun;12(6):1719-33.

[149] Twomey KB, O'connell OJ, McCarthy Y, Dow JM, O'toole GA, Plant BJ, Ryan RP. Bacterial cis-2-unsaturated fatty acids found in the cystic fibrosis airway modulate virulence and persistence of *Pseudomonas aeruginosa. The ISME journal.* 2012 May;6(5):939.

[150] Van Gennip M, Moser C, Christensen LD, Bjarnsholt T, Calum H, Jensen PØ, Christophersen L, Hougen HP, Ciofu O, Molin S, Givskov M. Augmented effect of early antibiotic treatment in mice with experimental lung infections due to sequentially adapted mucoid strains of *Pseudomonas aeruginosa. Journal of antimicrobial chemotherapy*. 2009 Oct 8;64(6):1241-50.

[151] Vega NM, Allison KR, Khalil AS, Collins JJ. Signaling-mediated bacterial persister formation. *Nature chemical biology*. 2012 May;8(5):431.

[152] Vrany JD, Stewart PS, Suci PA. Comparison of recalcitrance to ciprofloxacin and levofloxacin exhibited by *Pseudomonas aeruginosa* biofilms displaying rapid-transport characteristics. *Antimicrobial Agents and Chemotherapy*. 1997 Jun 1;41(6):1352-8.

[153] Walters MC, Roe F, Bugnicourt A, Franklin MJ, Stewart PS. Contributions of antibiotic penetration, oxygen limitation, and low metabolic activity to tolerance of *Pseudomonas aeruginosa* biofilms to ciprofloxacin and tobramycin. *Antimicrobial agents and chemotherapy*. 2003 Jan 1;47(1):317-23.

[154] Whitchurch CB, Tolker-Nielsen T, Ragas PC, Mattick JS. Extracellular DNA required for bacterial biofilm formation. *Science*. 2002 Feb 22;295(5559):1487-.

[155] Williams P, Cámara M. Quorum sensing and environmental adaptation in *Pseudomonas aeruginosa*: a tale of regulatory networks and multifunctional signal molecules. *Current opinion in microbiology*. 2009 Apr 1;12(2):182-91.

[156] Williamson KS, Richards LA, Perez-Osorio AC, Pitts B, McInnerney K, Stewart PS, Franklin MJ. Heterogeneity in *Pseudomonas aeruginosa* biofilms includes expression of ribosome hibernation factors in the antibiotic-tolerant subpopulation and hypoxia-induced stress response in the metabolically active population. *Journal of bacteriology*. 2012 Apr 15;194(8):2062-73.

[157] Wingender J, Strathmann M, Rode A, Leis A, Flemming HC. [25] Isolation and biochemical characterization of extracellular polymeric substances from *Pseudomonas aeruginosa*. In *Methods in enzymology* 2001 Jan 1 (Vol. 336, pp. 302-314). Academic Press.

[158] Wozniak DJ, Wyckoff TJ, Starkey M, Keyser R, Azadi P, O'Toole GA, Parsek MR. Alginate is not a significant component of the extracellular polysaccharide matrix of PA14 and PAO1 *Pseudomonas aeruginosa* biofilms. *Proceedings of the National Academy of Sciences*. 2003 Jun 24;100(13):7907-12.

[159] Wu A, Loutherback K, Lambert G, Estévez-Salmerón L, Tlsty TD, Austin RH, Sturm JC. Cell motility and drug gradients in the emergence of resistance to chemotherapy. *Proceedings of the National Academy of Sciences*. 2013 Oct 1;110(40):16103-8.

[160] Wu X, Held K, Zheng C, Staudinger BJ, Chavez JD, Weisbrod CR, Eng JK, Singh PK, Manoil C, Bruce JE. Dynamic proteome response of *Pseudomonas aeruginosa*

to tobramycin antibiotic treatment. *Molecular & Cellular Proteomics.* 2015 Aug 1;14(8):2126-37.

[161] Yang JW, Fan LC, Lu HW, Miao XY, Mao B, Xu JF. Efficacy and safety of long-term inhaled antibiotic for patients with noncystic fibrosis bronchiectasis: a meta-analysis. *The clinical respiratory journal.* 2016 Nov;10(6):731-9.

[162] Yasuda H, Ajiki Y, Koga T, Kawada H, Yokota T. Interaction between biofilms formed by *Pseudomonas aeruginosa* and clarithromycin. *Antimicrobial agents and chemotherapy.* 1993 Sep 1;37(9):1749-55.

[163] Yousefi M, Pourmand MR, Fallah F, Hashemi A, Mashhadi R, Nazari-Alam A. Characterization of Staphylococcus aureus biofilm formation in urinary tract infection. *Iranian journal of public health.* 2016 Apr;45(4):485.

[164] Yu Q, Griffin EF, Moreau-Marquis S, Schwartzman JD, Stanton BA, O'Toole GA. In vitro evaluation of tobramycin and aztreonam versus *Pseudomonas aeruginosa* biofilms on cystic fibrosis-derived human airway epithelial cells. *Journal of antimicrobial chemotherapy.* 2012 Jul 26;67(11):2673-81.

[165] Zegans ME, Wozniak D, Griffin E, Toutain-Kidd CM, Hammond JH, Garfoot A, Lam JS. *Pseudomonas aeruginosa* exopolysaccharide Psl promotes resistance to the biofilm inhibitor polysorbate 80. *Antimicrobial agents and chemotherapy.* 2012 Aug 1;56(8):4112-22.

[166] Zogaj X, Nimtz M, Rohde M, Bokranz W, Römling U. The multicellular morphotypes of *Salmonella typhimurium* and *Escherichia coli* produce cellulose as the second component of the extracellular matrix. *Molecular microbiology.* 2001 Mar;39(6):1452-63.

In: Antimicrobial Resistance: Opportunities and Challenges ISBN: 978-1-53617-943-9
Editors: P. Kumar Prabhakar and V. Mishra

Chapter 6

METHICILLIN RESISTANT *STAPHYLOCOCCUS AUREUS* (MRSA): AN OVERVIEW

Diksha Sharma[1] and Rohit Rai[2,*]
[1]Department of Biotechnology, DAV College Jalandhar, Punjab, India
[2]Faculty of Applied Medical Sciences, Lovely Professional University, Phagwara, Punjab, India

ABSTRACT

Staphylococcus aureus with acquired methicillin resistance (MRSA) is one of the most epidemic pathogens that has extended its impact from healthcare set up to community as well as companion, live-stock and wildlife animals. MRSA is a leading cause for a wide range of infections ranging from superficial skin and soft tissue infections to invasive bloodstream infections (BSIs) leading to sepsis and death. The resistance of the organism against methicillin and other β-lactam antibiotics is associated with the expression of PBP2a (*mecA*) that has favoured MRSA to become an endemic pathogen globally. The genomics studies have revealed a cluster of genes coding for various resistance and virulence factors making this organism a successful pathogen. When it comes to treatment of the invasive MRSA infections, vancomycin is selected as the drug of first choice however many other drugs have shown potential to inhibit MRSA strains without compromising human health. In this book chapter, we provide an overview of emergence, epidemiology, genomics, diagnosis, treatment and management of MRSA infections.

Keywords: methicillin resistant *Staphylococcus aureus* (MRSA), multidrug resistance, genomics, epidemiology, diagnosis

* Corresponding Author's E-mail: rohitraisharma44@gmail.com; rohit.25274@lpu.co.in.

1. Historical Background

Staphylococcus aureus is an important gram positive bacterium that is capable of causing broad spectrum of diseases and adapting to diverse environmental forms (Grema et al. 2015). The microorganism is known to be associated with several infections ranging from superficial skin infections (furunculosis, impetigo, etc.) to invasive bloodstream infections, sepsis and death, and is a leading cause of nosocomial pneumonia (Aklilu et al. 2010; Hoban et al. 2003; Quinn 1994; Wisplinghoff et al. 2003). Nearly 50% of human adults have been reported to be either persistent or intermittent carriers of *S. aureus* (Wertheim et al. 2004). Moreover, *S. aureus* is responsible for infections of surgical wounds and prosthetic transplants. The discovery and testing of penicillin by Alexander Fleming against *S. aureus* exhibited drastic drop in rate of deaths resulting from meningitis and Staphylococcal pneumonia during Second World War in comparison with World War I which led to establishment of penicillin as a wonder drug in treating *S. aureus* infections. However, in the subsequent years, *S. aureus* started showing ability to destroy penicillin through production of penicillinase and this resistance spread quickly among various *S. aureus* strains by a plasmid bearing penicillin resistance gene (Lyon and Skurray 1987). Further advancements in research developed methicillin (modified penicillin) as a therapeutic drug against *S. aureus* in 1959 which resist the destructive action of penicillinase and exerts its effect by blocking penicillin binding proteins (PBPs) in the organism (Jevons 1961). Within time period of 2 years, first case of methicillin resistant *S. aureus* (MRSA) was reported by late Professor Patricia Jevons in UK based hospital where the organism showed expression of PBP2a (mecA gene; hallmark of MRSA). PBP2a protein is a functional analogue of PBPs that is responsible for cell wall synthesis and maintenance and is not blocked in the presence of methicillin thus allowing survival of MRSA (Jevons 1961; Lyon and Skurray 1987). The expression of PBP2a did not mean resistance to methicillin only but also to all antibiotics that have β-lactam ring structure such as cephalosporins, carbapenems and other synthetic penicillins that favoured MRSA to become an endemic pathogen globally in past decades (Panlilio et al. 1992).

2. Epidemiology of MRSA

The probability of MRSA infections is high among children, elders, soldiers, athletes, prisoners, drug abusers, those belonging to urban and underserved areas, individuals with HIV or cystic fibrosis and frequent healthcare contact (Boucher and Corey 2008; Hota et al. 2007; Turner et al. 2019). The world has witnessed rapidly increasing trend of MRSA infection between 1990s and early 2000s followed by a parallel decrease in multiple US and European countries with special reference to bloodstream and soft tissue infections (De

Kraker et al. 2013; Devriese, Van Damme, and Fameree 1972; Kallen et al. 2010; Klein et al. 2018; Landrum et al. 2012). The specific factors liable for changing trends of MRSA infections remain unfolded, however, advances in molecular epidemiology have helped in understanding the increasing complexity of MRSA population dynamics. The epidemiological typing of MRSA strains is carried out using pulse field gel electrophoresis (PFGE) and multilocus sequence typing which can distinguish strains on the basis of genes encoding staphylococcus protein A and staphylococcal chromosomal cassette (SSC) *mec* (Hallin et al. 2007; Tenover, Biddle, and Lancaster 2001; Weber 2005). MRSA epidemiological typing has resulted in identification and classification of the strains into zoonotic, humanosis and/or host specific lineages. Using these approaches, 17 strains of MRSA epidemic to human were identified in United Kingdom out of which EMRSA-15 and 16 are the most predominant (Aucken et al. 2002; Hardy et al. 2004). EMRSA-16 has been recognized as the major human MRSA pathogen in American and European countries (Holden et al. 2004). The African countries are dominated by MRSA strains ST88-IV, ST5-IV and ST239-III out of which ST5-IV and ST239-III are community associated MRSA (CA-MRSA) whereas ST88-IV is common both for community and hospital infections (Breurec et al. 2011). CC1, CC5, CC8, CC22, CC30 and CC35 are some other MRSA strains of human origin whereas CC398 lineage of MRSA is found to be dominant in pigs and other food animals (Feßler et al. 2012; Witte et al. 2007).

2.1. Healthcare Associated MRSA (HA-MRSA)

It has been observed that most of MRSA infections occur in those who have been in hospitals and/ or other healthcare set ups including nursing homes, dialysis centres etc., and are referred to as Healthcare Associated MRSA (HA-MRSA) infections. In such infections, colonization of MRSA is of utmost importance with anterior nares being the most usual site of MRSA colonization (Raymund et al. 2013; Turner et al. 2019). The other anatomical sites favourable for MRSA colonization include wounds, hands, skin, perineal region, throat, digestive and genitourinary tract (Turner et al. 2019). The major lineage of MRSA responsible for causing HA-MRSA infections in African, European and Asian countries is CC8 (ST239) (Harris et al. 2010). The predisposing factors responsible for dissemination of MRSA in healthcare settings are contact with MRSA colonized patients or contaminated objects and aerosols that may lead to serious infections and complications including dermatitis, septicemia, heart and lung diseases in immunocompromised people (Grema et al. 2015; Umaru et al. 2011).

2.2. Community-Acquired MRSA (CA-MRSA)

Community acquired MRSA (CA-MRSA) strains differ from HA-MRSA strains based on their epidemiological, clinical and microbiological characteristics however these characteristics are not exclusive to any of the two groups. CA-MRSA strains are involved in superficial infections of skin and soft tissues such as abscesses, impetigo, furuncles and cellulitis and their involvement in severe infections like sepsis, necrotising fasciitis and Waterhouse-Friderichsen syndrome is rare (Adem et al. 2005; Francis et al. 2005; Gonzalez et al. 2006; Main et al. 2005; Mcleod et al. 2006; Miller et al. 2007; Urth et al. 2005). Typically, CA-MRSA possess staphylococcal cassette chromosome *SCCmec* type IV, V or VII and shows susceptibility to most non-β-lactam antibiotics whereas HA-MRSA on the other hand contains SCC*mec* of type I, II or III and exhibits multidrug resistance (Higuchi et al. 2008; Naimi et al. 2003). Further, CA-MRSA strains also express a toxin namely phage-encoded Panton-Valentine leukocidin (PVL) known to be associated with skin infections and necrotising pneumonia responsible for necrosis of epithelial cells and lysis of human leukocytes (Boyle-Vavra and Daum 2007; Etienne 2005; Labandeira-Rey et al. 2007; Monaco et al. 2005; Tristan et al. 2007; Valentini et al. 2008). The prevalence of CA-MRSA infections in USA is higher than in Europe and the strains identified from these regions are ST80 and USA300, respectively (Valentini et al. 2008; Vandenesch et al. 2003; Witte et al. 2007).

2.3. Livestock-Associated MRSA (LA-MRSA)

Earlier, MRSA was considered to be a human infection only but later it was discovered the organism is capable of infecting animals as well. There have been reports for isolation of MRSA strains from dairy cows infected with mastitis and pigs (Devriese, Van Damme, and Fameree 1972; Stefani et al. 2012). MRSA strain CC398 has been reported as the most predominant lineage associated with livestock, however, its prevalence in Europe, USA and Asia is reported to be very low (Monecke et al. 2011; Stefani et al. 2012). Contrary to this, the occurrence of LA-MRSA strain CC398 is quite high in Belgium, Denmark and Netherlands (R Köck, Brakensiek, et al. 2009; Robin Köck, Harlizius, et al. 2009). The epidemiological studies conducted in United Kingdom indicate the transfer of livestock MRSA into hospitals through persons with frequent animal exposure and evidences for transmission from human to animals and vice versa can also be sited from the published reports (Paterson et al. 2012; Umaru et al. 2011; Voss et al. 2005).

3. Genomic Basis of MRSA Infections

The epidemiological analysis has revealed that despite great overall diversity, only few strains of MRSA dominate. The whole genome sequencing has helped in tracking the emergence and evolution of MRSA in community as well as healthcare sector. Further, phylogenetic studies on CA-MRSA strains have indicated that emergence of CA-MRSA strain USA300 is through rapid clonal expansion and not through convergent evolution (Kennedy et al. 2008). The modern genomic approaches have helped in identifying mobile genetic elements (MGEs) and core genome components as driving forces for emergence and spread of successful CA-MRSA strains. The multiple MGEs in the genome of USA300 strain have been identified as carriers of various virulence factors and drug resistance determinants (Turner et al. 2019) and the information for the same has been compiled in Table 1. Determination of methicillin resistance ways long back with first report of resistance gene *mec* been published in 1975 (Sjöström, Löfdahl, and Philipson 1975). The researchers in early 1980s noticed the production of a novel low affinity penicillin binding protein PBP2/ PBP2a (MecA) (Brown and Reynolds 1980; Hartman and Tomasz 1984; Ubukata, Yamashita, and Konno 1985; Utsui and Yokota 1985) the encoding gene (*mecA*) for which was cloned two years later (Matsuhashi et al. 1986). The *mecA* gene is carried on *Staphylococcal* chromosomal cassette (SCC)*mec* that gets integrated into *S. aureus* chromosome at open reading frame X (*orfX*) (Hiramatsu et al. 2002). Till date, nine SCC*mec* types (I to VIII and V_T) have been identified that are distinguishable by the type of *ccr* gene complex and the class of *mec* complex that they bear. The function of *ccr* gene complex is to facilitate site specific addition and deletion of SCC*mec* cassette into and out of bacterial genome (David and Daum 2010; Ito et al. 2001; Ito et al. 2004; Ito et al. 2003). The larger SCC*mec* types (Type I to III) have been found in HA-MRSA strains and reported to be transferred to the organism from a commensal staphylococcal species on multiple occasions (Ma et al. 2002). On the other hand, smaller SCC*mec* types IV and V are associated with CA-MRSA strains everywhere (Ma et al. 2002) and SCC*mec* type IV is believed to have transferred into methicillin sensitive *Staphylococcus aureus* (MSSA) at more than 20 instances (Berglund and Söderquist 2008; Daskalaki et al. 2007; Hallin et al. 2007; Robinson and Enright 2003). Type II and III SCC*mec* elements possess some junkyard regions where genes conferring non β-lactam resistance to *S. aureus* are inserted and are therefore associated with multidrug resistance phenotypes (Hiramatsu et al. 2002; Ito et al. 2004). Lack of non β-lactam antimicrobial resistance genes in SCC*mec* type IV element may account for decreased probability of occurrence of multidrug resistance in CA-MRSA strains (David and Daum 2010).

Table 1. Genes encoding drug resistance in Methicillin-resistant *Staphylococcus aureus*

Resistant character	Associated gene	Location	Product
Aminoglycoside resistance	*aacA-aphD*	Transposon	Bifunctional AAC–APH protein
	aadD	Plasmid	Aminoglycoside adenyltransferase
	ant(4')	SCC*mec*; Plasmid	*O*-nucleotidyltransferase(4')
	ant(9)	Transposon	*O*-nucleotidyltransferase(9)
Macrolide resistance	*ermA*	Transposon	rRNA methylase
	ermC	Plasmid	rRNA methylase
Tetracycline resistance	*tetK*	Transposon	Tetracycline resistance protein
	tetM	Plasmid	Tetracycline resistance protein
Penicillin resistance	*blaZ*	Transposon; Plasmid	β-lactamase
Bleomycin resistance	*bleO*	SCC*mec*; Plasmid	Bleomycin binding protein
Chloramphenicol resistance	*cat*	Plasmid	Chloramphenicol acetyltransferase
Trimethoprim resistance	*dfrA* and *dfrK*	Plasmid	Dihydrofolate reductase
Mupirocin resistance	*ileS*	Plasmid	Isoleucyl-tRNA synthetase
Methicillin resistance	*mecA*	SCC*mec*	Penicillin-binding protein 2

List compiled from (Holden et al. 2004; Kuroda et al. 2001; Malachowa and DeLeo 2010; Turner et al. 2019).

Apart from the resistance genes, a number of genes coding for putative virulence factors have been identified from CA-MRSA strains that include toxins (haemolysins), immune-evasive surface factors (protein A) and enzymes (hyaluronidase) known for promoting invasion of tissues (Turner et al. 2019). The major challenge in analysing success of the strains is variation at their multiple loci in comparison with the preceding strains within the lineage, an example of which is USA300 strain from CC8 which varies from another strain USA500 of the same lineage in terms of multiple additional virulence determinants acquired through MGEs tissues (Turner et al. 2019). Arginine catabolic mobile element (ACME) and Panton Valentine Leukocidin (PVL) are the two most studied virulence factors out of which attainment of the former is limited to USA300 and is followed by rapid spread of the strain. ACME is believed to confer tolerance for low pH environments which prove beneficial for the survival of MRSA strains on skin (Uhlemann et al. 2014). PVL mediates neutrophil lysis and shows association with necrotizing pneumonia and soft tissue infections, however, the studies have shown that PVL is not the sole determinant of MRSA infection severity (Tong et al. 2012). *Staphylococcus aureus* pathogenicity islands (SaPIs) constitute another set of accessory genes (14 – 17 kb) coding for toxic shock syndrome toxin 1 (TSST1) and enterotoxins (B and C) that are associated with toxic shock syndrome and food poisoning, respectively (Malachowa and DeLeo 2010). There are three families of genomic islands (*v*SAα, *v*SAβ and *v*SAɣ) that have been reported for carrying a variety of virulence factors like lipoproteins, leukocidins, superantigens, proteases, hyaluronidases and b-type phenol soluble modulin (PSM) genes (Lindsay and Holden 2004; Malachowa and DeLeo 2010). Bacteriocins and MGE transferrable toxins are the other toxins active against the human host that may inhibit competing or commensal bacteria (Malachowa and DeLeo 2010).

Interestingly, the chances for emergence of novel MRSA strains are very high because of the expansive genetic variations but the mankind is fortunate enough that no such MRSA strain amassing major virulence and resistance factors has emerged till date.

4. MRSA Colonization and Transmission

Nearly one third of the general population has been reported to be colonized with *S. aureus* with nares being most common site of colonization (Hassoun, Linden, and Friedman 2017; Salgado, Farr, and Calfee 2003; Turner et al. 2019). The risk factors for MRSA colonization in females are age (≥ 60 years), diabetes and poverty whereas in case of males healthcare exposure is the major risk factor.

Some studies have also identified recent hospitalization/ outpatient visit, recent antibiotic use, chronic illness, injected drug use, and contact with MRSA infected personnel as the potential risk factors (Gorwitz et al. 2008). Although most common site colonized by MRSA is anterior nares, the recent studies have shown colonization in throat as more prevalent. The other sites dwelling *S. aureus* are throat, axilla, rectum, groin and perineum (Albrecht et al. 2015; Kluytmans, Van Belkum, and Verbrugh 1997; Kumar et al. 2015). In case of nasal carriers, 20% of the population is persistently colonized with single strain and nearly 60% of the population act as intermittent carrier of the diversified strains (Williams 1963).

In case of healthcare infections, nostrils and hands of persons colonized with MRSA are highly instrumental in transmitting the strain. The release of MRSA into hospital environment is mediated in the form of aerosols, through stools and skin tissues of infected patients. The contaminated medical instruments, beddings, toiletries, furniture, clothing, etc. constitute other sources that actively transmit MRSA in hospital environment (Klotz et al. 2005). The transmission of MRSA in community is largely associated with contamination of home environment through continuous colonization of medical staff (Turner et al. 2019). In livestock settings, isolation of MRSA strains has been reported from dust particles present in herds which may be respired in by the workers and lead to spread of LA-MRSA (Schulz et al. 2012).

5. MRSA Detection

MRSA infections have outstretched from human population to pets & live-stock animals, and wildlife animals, therefore, identification of human carriers in healthcare set up and community is important for proper diagnosis and management of infections. Additionally, animals with skin and soft tissue injuries also need screening to manage the wide spread of MRSA. The appropriate sites for screening MRSA colonization in humans include nostrils whereas skin, perineum, rectal, nasal or cloacal swabs are ideal in case of animals (Khanna et al. 2008; Peacock, de Silva, and Lowy 2001). Nasal swabs in humans can identify up to 80% of the carriers, however, this number increases up to 92% with addition of throat swabs (Grundmann et al. 2006). The most suitable environmental samples involve dust samples, swabs from furniture, containers, feed, feces or bedding materials (Broens et al. 2008; Lee 2003). Milk, animal meat, and cloacal swabs of poultry animals can also be screened for detection of potent MRSA strains(Grema et al. 2015).

There are several methods for screening of MRSA strains through phenotypic as well as genotypic characterization of samples collected from infected sites. The phenotypic methods involve microbiological culture techniques followed by morphological observations, gram staining, biochemical testing (catalase and coagulase test), pigment production, etc. (Karthy, Ranjitha, and Mohankumar 2009). Determination of minimal inhibitory concentration (MIC), detection of *mec*A or $PBP2^0$ protein and screening on culture media supplemented with oxacillin are few more methods that can be used for detection of MRSA strains (Louie et al. 2001). More advanced molecular techniques such as PCR based methods can be used in combination with phenotypic methods to enhance their specificity. Although sensitivity of PCR based methods is higher as compared to traditional culture methods, higher cost along with some false positive results are the limitations associated with these methods (Blanc et al. 2011; Kelley et al. 2009; Matheson et al. 2012).

6. Treatment and Management of MRSA

The uncontrolled use of antibiotics has resulted in accumulation and dissemination of MRSA. The pathogenic strains of MRSA have acquired resistance to most of the antibiotics being used in their treatment which has in turn limited the choice of treatment. The proper treatment of MRSA infections is possible through selection of antibiotics on the basis of antimicrobial susceptibility testing. The antimicrobial therapies are generally same for CA-MRSA and HA-MRSA infections with vancomycin being first choice of drug. Antibiotics such as Trimehoprim-sulphameth-oxazole, doxycycline and clindamycin have been reported to be effective for treating CA-MRSA infections (Grema et al. 2015). Couple of new antibiotics like oritavancin, telavancin, omdacycline, tedizolid, and dalbavancin have shown promising effects results for inhibiting MRSA strains. Further, antistaphylococcal agents including daptomycin, linezolid, quinupristin-dalfopristin, and tericidal are also available for parenteral therapy (David and Daum 2010). The drugs like fosmomycin and fusidic acid are under trial for their efficiency to treat MRSA infections (Burke and Rose 2014).

The management of MRSA infections is regulated as per guidelines of several professional and government bodies. The standard guidelines suggest that "antibiotic-resistant pathogens are sensitive to routinely used hospital disinfectants, but it is essential that correct and meticulous cleaning and use of disinfectants be performed" (Muto et al. 2003). Despite specific interventions focusing identification of high-risk populations, isolation of colonized individuals, decolonization of the MRSA carriers, environmental disinfection by physico-chemical means, MRSA epidemic continues to expand (David and Daum 2010). Therefore, looking for more effective interventions has become need of the hour.

Conclusion and Future Prospective

Despite several antibiotics with their effectiveness against MRSA being approved by FDA, persistently high mortality associated with invasive MRSA infections is of great concern. The major challenge with MRSA is its ability to undergo genetic modifications and emergence as more evolved successful epidemic strains. Therefore, future efforts should be focused more on: (i) characterization of diverse forms of MRSA based on progressive genomics, epigenetics, transcriptomics, proteomics and metabolomics studies; and (ii) understanding complexity of host pathogen interactions. The advancements in these areas will strengthen the understanding of host pathogen interplay and will ultimately reduce the clinical manifestations of this persistent pathogen.

References

[1] Adem, Patricia V, Christopher P Montgomery, Aliya N Husain, Tracy K Koogler, Valerie Arangelovich, Michel Humilier, Susan Boyle-Vavra, and Robert S Daum. 2005. "*Staphylococcus aureus* sepsis and the Waterhouse–Friderichsen syndrome in children." *New England Journal of Medicine* 353 (12): 1245-1251.

[2] Aklilu, Erkihun, Z Zunita, L Hassan, and HC Chen. 2010. "Phenotypic and genotypic characterization of methicillin-resistant *Staphylococcus aureus* (MRSA) isolated from dogs and cats at University Veterinary Hospital, Universiti Putra Malaysia." *Trop Biomed* 27 (3): 483-492.

[3] Albrecht, Valerie S, Brandi M Limbago, Gregory J Moran, Anusha Krishnadasan, Rachel J Gorwitz, Linda K McDougal, and David A Talan. 2015. "*Staphylococcus aureus* colonization and strain type at various body sites among patients with a closed

abscess and uninfected controls at US emergency departments." *Journal of clinical microbiology* 53 (11): 3478-3484.

[4] Aucken, Hazel M, Mark Ganner, Stephen Murchan, Barry D Cookson, and Alan P Johnson. 2002. "A new UK strain of epidemic methicillin-resistant *Staphylococcus aureus* (EMRSA-17) resistant to multiple antibiotics." *Journal of Antimicrobial Chemotherapy* 50 (2): 171-175.

[5] Berglund, Carolina, and Bo Söderquist. 2008. "The origin of a methicillin-resistant *Staphylococcus aureus* isolate at a neonatal ward in Sweden—possible horizontal transfer of a staphylococcal cassette chromosome mec between methicillin-resistant Staphylococcus haemolyticus and *Staphylococcus aureus*." *Clinical Microbiology and Infection* 14 (11): 1048-1056.

[6] Blanc, Dominique S, Patrick Basset, Immaculée Nahimana-Tessemo, Katia Jaton, Gilbert Greub, and Giorgio Zanetti. 2011. "High proportion of wrongly identified methicillin-resistant *Staphylococcus aureus* carriers by use of a rapid commercial PCR assay due to presence of staphylococcal cassette chromosome element lacking the mecA gene." *Journal of clinical microbiology* 49 (2): 722-724.

[7] Boucher, Helen W, and G Ralph Corey. 2008. "Epidemiology of methicillin-resistant *Staphylococcus aureus*." *Clinical infectious diseases* 46 (Supplement_5): S344-S349.

[8] Boyle-Vavra, Susan, and Robert S Daum. 2007. "Community-acquired methicillin-resistant *Staphylococcus aureus*: the role of Panton–Valentine leukocidin." *Laboratory investigation* 87 (1): 3.

[9] Breurec, S, SB Zriouil, C Fall, P Boisier, S Brisse, S Djibo, J Etienne, MC Fonkoua, JD Perrier-Gros-Claude, and R Pouillot. 2011. "Epidemiology of methicillin-resistant *Staphylococcus aureus* lineages in five major African towns: emergence and spread of atypical clones." *Clinical microbiology and infection* 17 (2): 160-165.

[10] Broens, EM, EAM Graat, PJ van der Wolf, IVF van den Broek, EW Tiemersma, AW Van de Giessen, and MCM De Jong. 2008. "Prevalence study and risk factor analysis of NT-MRSA in pigs in the Netherlands." *Proceedings of the 20th International Pig Veterinary Science Congress.*

[11] Brown, Derek FJ, and Peter E Reynolds. 1980. "Intrinsic resistance to β-lactam antibiotics in *Staphylococcus aureus*." *FEBS letters* 122 (2): 275-278.

[12] Burke, Stuart L, and Warren E Rose. 2014. "New pharmacological treatments for methicillin-resistant *Staphylococcus aureus* infections." *Expert opinion on pharmacotherapy* 15 (4): 483-491.

[13] Daskalaki, Maria, Joaquin R Otero, Francisca Sanz, and Fernando Chaves. 2007. "Bacteremia due to clonally derived methicillin-resistant, gentamicin-susceptible isolates and methicillin-susceptible, gentamicin-resistant isolates of *Staphylococcus aureus*." *Journal of clinical microbiology* 45 (10): 3446-3448.

[14] David, Michael Z, and Robert S Daum. 2010. "Community-associated methicillin-resistant *Staphylococcus aureus*: epidemiology and clinical consequences of an emerging epidemic." *Clinical microbiology reviews* 23 (3): 616-687.

[15] De Kraker, MEA, V Jarlier, JCM Monen, OE Heuer, N Van De Sande, and H Grundmann. 2013. "The changing epidemiology of bacteraemias in Europe: trends from the European Antimicrobial Resistance Surveillance System." *Clinical Microbiology and Infection* 19 (9): 860-868.

[16] Devriese, LA, LR Van Damme, and L Fameree. 1972. "Methicillin (cloxacillin)-resistant *Staphylococcus aureus* strains isolated from bovine mastitis cases." *Zentralblatt für Veterinärmedizin Reihe B* 19 (7): 598-605.

[17] Etienne, Jerome. 2005. "Panton-Valentine leukocidin: a marker of severity for *Staphylococcus aureus* infection?" *Clinical Infectious Diseases* 41 (5): 591-593.

[18] Feßler, Andrea T, Richard GM Olde Riekerink, Anja Rothkamp, Kristina Kadlec, Otlis C Sampimon, Theo JGM Lam, and Stefan Schwarz. 2012. "Characterization of methicillin-resistant *Staphylococcus aureus* CC398 obtained from humans and animals on dairy farms." *Veterinary microbiology* 160 (1-2): 77-84.

[19] Francis, John S, Karen Carroll, Eric Nuermberge, and John G Bartlett. 2005. "Reply to Wargo and Eiland." *Clinical Infectious Diseases* 40 (9): 1378-1379.

[20] Gonzalez, Blanca E, Adriana M Rueda, Samuel A Shelburne, Daniel M Musher, Richard J Hamill, and Kristina G Hultén. 2006. "Community-associated strains of methicillin-resistant *Staphylococccus aureus* as the cause of healthcare-associated infection." *Infection Control & Hospital Epidemiology* 27 (10): 1051-1056.

[21] Gorwitz, Rachel J, Deanna Kruszon-Moran, Sigrid K McAllister, Geraldine McQuillan, Linda K McDougal, Gregory E Fosheim, Bette J Jensen, George Killgore, Fred C Tenover, and Matthew J Kuehnert. 2008. "Changes in the prevalence of nasal colonization with *Staphylococcus aureus* in the United States, 2001–2004." *The Journal of infectious diseases* 197 (9): 1226-1234.

[22] Grema, Hafsat Ali, Yaqub Ahmed Geidam, Galadima Bala Gadzama, James Agbo Ameh, and Abubakar Suleiman. 2015. "Methicillin resistant *Staphyloccus aureus* (MRSA): a review." *Adv Anim Vet Sci* 3 (2): 79-98.

[23] Grundmann, Hajo, Marta Aires-de-Sousa, John Boyce, and Edine Tiemersma. 2006. "Emergence and resurgence of meticillin-resistant *Staphylococcus aureus* as a public-health threat." *The Lancet* 368 (9538): 874-885.

[24] Hallin, Marie, Olivier Denis, Ariane Deplano, Ricardo De Mendonça, Rafael De Ryck, Sylvianne Rottiers, and MJ Struelens. 2007. "Genetic relatedness between methicillin-susceptible and methicillin-resistant *Staphylococcus aureus*: results of a national survey." *Journal of antimicrobial chemotherapy* 59 (3): 465-472.

[25] Hardy, KJ, PM Hawkey, F Gao, and BA Oppenheim. 2004. "Methicillin resistant *Staphylococcus aureus* in the critically ill." *British journal of anaesthesia* 92 (1): 121-130.

[26] Harris, Simon R, Edward J Feil, Matthew TG Holden, Michael A Quail, Emma K Nickerson, Narisara Chantratita, Susana Gardete, Ana Tavares, Nick Day, and Jodi A Lindsay. 2010. "Evolution of MRSA during hospital transmission and intercontinental spread." *Science* 327 (5964): 469-474.

[27] Hartman, BARRY J, and ALEXANDER Tomasz. 1984. "Low-affinity penicillin-binding protein associated with beta-lactam resistance in *Staphylococcus aureus*." *Journal of bacteriology* 158 (2): 513-516.

[28] Hassoun, Ali, Peter K Linden, and Bruce Friedman. 2017. "Incidence, prevalence, and management of MRSA bacteremia across patient populations—a review of recent developments in MRSA management and treatment." *Critical Care* 21 (1): 211.

[29] Higuchi, Wataru, Tomomi Takano, Lee-Jene Teng, and Tatsuo Yamamoto. 2008. "Structure and specific detection of staphylococcal cassette chromosome mec type VII." *Biochemical and biophysical research communications* 377 (3): 752-756.

[30] Hiramatsu, Keiichi, Yuki Katayama, Harumi Yuzawa, and Teruyo Ito. 2002. "Molecular genetics of methicillin-resistant *Staphylococcus aureus*." *International Journal of Medical Microbiology* 292 (2): 67-74.

[31] Hoban, Daryl J, Douglas J Biedenbach, Alan H Mutnick, and Ronald N Jones. 2003. "Pathogen of occurrence and susceptibility patterns associated with pneumonia in hospitalized patients in North America: results of the SENTRY Antimicrobial Surveillance Study (2000)." *Diagnostic microbiology and infectious disease* 45 (4): 279-285.

[32] Holden, Matthew TG, Edward J Feil, Jodi A Lindsay, Sharon J Peacock, Nicholas PJ Day, Mark C Enright, Tim J Foster, Catrin E Moore, Laurence Hurst, and Rebecca Atkin. 2004. "Complete genomes of two clinical *Staphylococcus aureus* strains: evidence for the rapid evolution of virulence and drug resistance." *Proceedings of the National Academy of Sciences* 101 (26): 9786-9791.

[33] Hota, Bala, Charlotte Ellenbogen, Mary K Hayden, Alla Aroutcheva, and Robert A Weinstein. 2007. "Community-associated methicillin-resistant *Staphylococcus aureus* skin and soft tissue infections at a public hospital: do public housing and incarceration amplify transmission?" *Archives of internal medicine* 167 (10): 1026-1033.

[34] Ito, Teruyo, Yuki Katayama, Kazumi Asada, Namiko Mori, Kanae Tsutsumimoto, Chuntima Tiensasitorn, and Keiichi Hiramatsu. 2001. "Structural comparison of three types of staphylococcal cassette chromosome mec integrated in the chromosome in methicillin-resistant *Staphylococcus aureus*." *Antimicrobial agents and chemotherapy* 45 (5): 1323-1336.

[35] Ito, Teruyo, Xiao Xue Ma, Fumihiko Takeuchi, Keiko Okuma, Harumi Yuzawa, and Keiichi Hiramatsu. 2004. "Novel type V staphylococcal cassette chromosome mec

driven by a novel cassette chromosome recombinase, ccrC." *Antimicrobial agents and chemotherapy* 48 (7): 2637-2651.

[36] Ito, Teruyo, Keiko Okuma, Xiao Xue Ma, Harumi Yuzawa, and Keiichi Hiramatsu. 2003. "Insights on antibiotic resistance of *Staphylococcus aureus* from its whole genome: genomic island SCC." *Drug Resistance Updates* 6 (1): 41-52.

[37] Jevons, M Patricia. 1961. "'Celbenin'-resistant staphylococci." *British medical journal* 1 (5219): 124.

[38] Kallen, Alexander J, Yi Mu, Sandra Bulens, Arthur Reingold, Susan Petit, KEN Gershman, Susan M Ray, Lee H Harrison, Ruth Lynfield, and Ghinwa Dumyati. 2010. "Health care–associated invasive MRSA infections, 2005-2008." *Jama* 304 (6): 641-647.

[39] Karthy, ES, P Ranjitha, and A Mohankumar. 2009. "Performance of CHROM agar and oxacillin resistant screening agar base media for detection of methicillin resistant *Staphylococcus aureus* (MRSA) from chronic wound." *Modern Applied Science* 3 (5): 51-55.

[40] Kelley, PG, Elizabeth A Grabsch, Benjamin P Howden, W Gao, and M Lindsay Grayson. 2009. "Comparison of the Xpert methicillin-resistant *Staphylococcus aureus* (MRSA) assay, BD GeneOhm MRSA assay, and culture for detection of nasal and cutaneous groin colonization by MRSA." *Journal of clinical microbiology* 47 (11): 3769-3772.

[41] Kennedy, Adam D, Michael Otto, Kevin R Braughton, Adeline R Whitney, Liang Chen, Barun Mathema, Jose R Mediavilla, Kelly A Byrne, Larye D Parkins, and Fred C Tenover. 2008. "Epidemic community-associated methicillin-resistant *Staphylococcus aureus*: recent clonal expansion and diversification." *Proceedings of the National Academy of Sciences* 105 (4): 1327-1332.

[42] Khanna, T, R Friendship, C Dewey, and JS Weese. 2008. "Methicillin resistant *Staphylococcus aureus* colonization in pigs and pig farmers." *Veterinary microbiology* 128 (3-4): 298-303.

[43] Klein, Eili Y, Wendi Jiang, Nestor Mojica, Katie K Tseng, Ryan McNeill, Sara E Cosgrove, and Trish M Perl. 2018. "National costs associated with methicillin-susceptible and methicillin-resistant *Staphylococcus aureus* hospitalizations in the United States, 2010–2014." *Clinical Infectious Diseases* 68 (1): 22-28.

[44] Klotz, Martina, Stefan Zimmermann, Sandra Opper, Klaus Heeg, and Reinier Mutters. 2005. "Possible risk for re-colonization with methicillin-resistant *Staphylococcus aureus* (MRSA) by faecal transmission." *International journal of hygiene and environmental health* 208 (5): 401-405.

[45] Kluytmans, Jan, Alex Van Belkum, and Henri Verbrugh. 1997. "Nasal carriage of *Staphylococcus aureus*: epidemiology, underlying mechanisms, and associated risks." *Clinical microbiology reviews* 10 (3): 505-520.

[46] Köck, R, L Brakensiek, A Mellmann, F Kipp, M Henderikx, D Harmsen, I Daniels-Haardt, C von Eiff, K Becker, and MGR Hendrix. 2009. "Cross-border comparison of the admission prevalence and clonal structure of meticillin-resistant *Staphylococcus aureus*." *Journal of Hospital Infection* 71 (4): 320-326.

[47] Köck, Robin, J Harlizius, N Bressan, R Laerberg, Lothar H Wieler, Wolfgang Witte, RH Deurenberg, A Voss, Karsten Becker, and Alexander W Friedrich. 2009. "Prevalence and molecular characteristics of methicillin-resistant *Staphylococcus aureus* (MRSA) among pigs on German farms and import of livestock-related MRSA into hospitals." *European journal of clinical microbiology & infectious diseases* 28 (11): 1375.

[48] Kumar, Neha, Michael Z David, Susan Boyle-Vavra, Julia Sieth, and Robert S Daum. 2015. "High *Staphylococcus aureus* colonization prevalence among patients with skin and soft tissue infections and controls in an urban emergency department." *Journal of clinical microbiology* 53 (3): 810-815.

[49] Kuroda, Makoto, Toshiko Ohta, Ikuo Uchiyama, Tadashi Baba, Harumi Yuzawa, Ichizo Kobayashi, Longzhu Cui, Akio Oguchi, Ken-ichi Aoki, and Yoshimi Nagai. 2001. "Whole genome sequencing of meticillin-resistant *Staphylococcus aureus*." *The Lancet* 357 (9264): 1225-1240.

[50] Labandeira-Rey, Maria, Florence Couzon, Sandrine Boisset, Eric L Brown, Michele Bes, Yvonne Benito, Elena M Barbu, Vanessa Vazquez, Magnus Höök, and Jerome Etienne. 2007. "*Staphylococcus aureus* Panton-Valentine leukocidin causes necrotizing pneumonia." *Science* 315 (5815): 1130-1133.

[51] Landrum, Michael L, Charlotte Neumann, Courtney Cook, Uzo Chukwuma, Michael W Ellis, Duane R Hospenthal, and Clinton K Murray. 2012. "Epidemiology of *Staphylococcus aureus* blood and skin and soft tissue infections in the US military health system, 2005-2010." *Jama* 308 (1): 50-59.

[52] Lee, John Hwa. 2003. "Methicillin (oxacillin)-resistant *Staphylococcus aureus* strains isolated from major food animals and their potential transmission to humans." *Appl. Environ. Microbiol.* 69 (11): 6489-6494.

[53] Lindsay, Jodi A, and Matthew TG Holden. 2004. "*Staphylococcus aureus*: superbug, super genome?" *Trends in microbiology* 12 (8): 378-385.

[54] Louie, L, A Majury, J Goodfellow, M Louie, and AE Simor. 2001. "Evaluation of a latex agglutination test (MRSA-Screen) for detection of oxacillin resistance in coagulase-negative staphylococci." *Journal of clinical microbiology* 39 (11): 4149-4151.

[55] Lyon, BRUCE R, and RA Skurray. 1987. "Antimicrobial resistance of *Staphylococcus aureus*: genetic basis." *Microbiological reviews* 51 (1): 88.

[56] Ma, Xiao Xue, Teruyo Ito, Chuntima Tiensasitorn, Mantana Jamklang, Piriyaporn Chongtrakool, Susan Boyle-Vavra, Robert S Daum, and Keiichi Hiramatsu. 2002. "Novel type of staphylococcal cassette chromosome mec identified in community-

acquired methicillin-resistant *Staphylococcus aureus* strains." *Antimicrobial agents and chemotherapy* 46 (4): 1147-1152.

[57] Main, Cheryl L, Padman Jayaratne, Allan Haley, Candy Rutherford, Fiona Smaill, and David N Fisman. 2005. "Outbreaks of infection caused by community-acquired methicillin-resistant *Staphylococcus aureus* in a Canadian correctional facility." *Canadian Journal of Infectious Diseases and Medical Microbiology* 16 (6): 343-348.

[58] Malachowa, Natalia, and Frank R DeLeo. 2010. "Mobile genetic elements of *Staphylococcus aureus*." *Cellular and molecular life sciences* 67 (18): 3057-3071.

[59] Matheson, Ann, Peter Christie, Traiani Stari, Kim Kavanagh, Ian M Gould, Robert Masterton, and Jacqui S Reilly. 2012. "Nasal swab screening for methicillin-resistant *Staphylococcus aureus*—how well does it perform? A cross-sectional study." *Infection Control & Hospital Epidemiology* 33 (8): 803-808.

[60] Matsuhashi, MICHIO, M Dong Song, FUMITOSHI Ishino, MASAAKI Wachi, MASAKI Doi, MATSUHISA Inoue, KIMIKO Ubukata, NAOKO Yamashita, and MASATOSHI Konno. 1986. "Molecular cloning of the gene of a penicillin-binding protein supposed to cause high resistance to beta-lactam antibiotics in *Staphylococcus aureus*." *Journal of Bacteriology* 167 (3): 975-980.

[61] Mcleod, JE, MI McDonald, A Ralph, and CL Stevens. 2006. "Community-acquired, methicillin resistant *Staphylococcus aureus* in Central Australia." *Communicable diseases intelligence quarterly report* 30 (4): 462.

[62] Miller, Loren G, Clifford Quan, Anthony Shay, Katayoun Mostafaie, Kiran Bharadwa, Nelly Tan, Kelli Matayoshi, Jason Cronin, Jennifer Tan, and Grace Tagudar. 2007. "A prospective investigation of outcomes after hospital discharge for endemic, community-acquired methicillin-resistant and-susceptible *Staphylococcus aureus* skin infection." *Clinical Infectious Diseases* 44 (4): 483-492.

[63] Monaco, Monica, Rosa Antonucci, Paolo Palange, Mario Venditti, and Annalisa Pantosti. 2005. "Methicillin-resistant *Staphylococcus aureus* necrotizing pneumonia."

[64] Monecke, Stefan, Geoffrey Coombs, Anna C Shore, David C Coleman, Patrick Akpaka, Michael Borg, Henry Chow, Margaret Ip, Lutz Jatzwauk, and Daniel Jonas. 2011. "A field guide to pandemic, epidemic and sporadic clones of methicillin-resistant *Staphylococcus aureus*." *PloS one* 6 (4): e17936.

[65] Muto, Carlene A, John A Jernigan, Belinda E Ostrowsky, Hervé M Richet, William R Jarvis, John M Boyce, and Barry M Farr. 2003. "SHEA guideline for preventing nosocomial transmission of multidrug-resistant strains of *Staphylococcus aureus* and enterococcus." *Infection Control & Hospital Epidemiology* 24 (5): 362-386.

[66] Naimi, Timothy S, Kathleen H LeDell, Kathryn Como-Sabetti, Stephanie M Borchardt, David J Boxrud, Jerome Etienne, Susan K Johnson, Francois Vandenesch, Scott Fridkin, and Carol O'Boyle. 2003. "Comparison of community-and health care–

associated methicillin-resistant *Staphylococcus aureus* infection." *Jama* 290 (22): 2976-2984.

[67] Panlilio, Adelisa L, David H Culver, Robert P Gaynes, Shailen Banerjee, Tonya S Henderson, James S Tolson, William J Martone, and National Nosocomial Infections Surveillance System. 1992. "Methicillin-resistant *Staphylococcus aureus* in US hospitals, 1975–1991." *Infection Control & Hospital Epidemiology* 13 (10): 582-586.

[68] Paterson, Gavin K, AR Larsen, A Robb, GE Edwards, TW Pennycott, G Foster, Dorien Mot, Katleen Hermans, Kristof Baert, and SJ Peacock. 2012. "The newly described mecA homologue, mecA LGA251, is present in methicillin-resistant *Staphylococcus aureus* isolates from a diverse range of host species." *Journal of Antimicrobial Chemotherapy* 67 (12): 2809-2813.

[69] Peacock, Sharon J, Ishani de Silva, and Franklin D Lowy. 2001. "What determines nasal carriage of *Staphylococcus aureus*?" *Trends in microbiology* 9 (12): 605-610.

[70] Quinn, PJ. 1994. "Staphylococcus species." *Clinical veterinary microbiology*: 118-126.

[71] Raymund, D, M Yi, R Belflower, D Aragon, G Dumyati, LH Harrison, FC Lessa, R Lynfield, J Nadle, and S Petit. 2013. "the Emerging Infections Program (2013)." *J. Am. Med. Assoc., Intern. Med* 173 (21): 1970-1978.

[72] Robinson, D Ashley, and Mark C Enright. 2003. "Evolutionary models of the emergence of methicillin-resistant *Staphylococcus aureus*." *Antimicrobial agents and chemotherapy* 47 (12): 3926-3934.

[73] Salgado, Cassandra D, Barry M Farr, and David P Calfee. 2003. "Community-acquired methicillin-resistant *Staphylococcus aureus*: a meta-analysis of prevalence and risk factors." *Clinical Infectious Diseases* 36 (2): 131-139.

[74] Schulz, Jochen, Anika Friese, Sylvia Klees, Bernd A Tenhagen, Alexandra Fetsch, Uwe Rösler, and Jörg Hartung. 2012. "Longitudinal study of the contamination of air and of soil surfaces in the vicinity of pig barns by livestock-associated methicillin-resistant *Staphylococcus aureus*." *Appl. Environ. Microbiol.* 78 (16): 5666-5671.

[75] Sjöström, JE, S Löfdahl, and LENNART Philipson. 1975. "Transformation reveals a chromosomal locus of the gene (s) for methicillin resistance in *Staphylococcus aureus*." *Journal of bacteriology* 123 (3): 905-915.

[76] Stefani, Stefania, Doo Ryeon Chung, Jodi A Lindsay, Alex W Friedrich, Angela M Kearns, Henrik Westh, and Fiona M MacKenzie. 2012. "Meticillin-resistant *Staphylococcus aureus* (MRSA): global epidemiology and harmonisation of typing methods." *International journal of antimicrobial agents* 39 (4): 273-282.

[77] Tenover, Fred C, James W Biddle, and Michael V Lancaster. 2001. "Increasing resistance to vancomycin and other glycopeptides in *Staphylococcus aureus*." *Emerging infectious diseases* 7 (2): 327.

[78] Tong, Amy, Steven YC Tong, Yurong Zhang, Supaporn Lamlertthon, Batu K Sharma-Kuinkel, Thomas Rude, Sun Hee Ahn, Felicia Ruffin, Lily Llorens, and

Ganesh Tamarana. 2012. "Panton-Valentine leukocidin is not the primary determinant of outcome for *Staphylococcus aureus* skin infections: evaluation from the CANVAS studies." *PloS one* 7 (5): e37212.

[79] Tristan, Anne, Michele Bes, Helene Meugnier, Gerard Lina, Bülent Bozdogan, Patrice Courvalin, Marie-Elisabeth Reverdy, Mark C Enright, François Vandenesch, and Jerome Etienne. 2007. "Global distribution of Panton-Valentine leukocidin–positive methicillin-resistant *Staphylococcus aureus*, 2006." *Emerging infectious diseases* 13 (4): 594.

[80] Turner, Nicholas A, Batu K Sharma-Kuinkel, Stacey A Maskarinec, Emily M Eichenberger, Pratik P Shah, Manuela Carugati, Thomas L Holland, and Vance G Fowler. 2019. "Methicillin-resistant *Staphylococcus aureus*: an overview of basic and clinical research." *Nature Reviews Microbiology*: 1.

[81] Ubukata, KIMIKO, NAOKO Yamashita, and MASATOSHI Konno. 1985. "Occurrence of a beta-lactam-inducible penicillin-binding protein in methicillin-resistant staphylococci." *Antimicrobial agents and chemotherapy* 27 (5): 851-857.

[82] Uhlemann, Anne-Catrin, Janina Dordel, Justin R Knox, Kathy E Raven, Julian Parkhill, Matthew TG Holden, Sharon J Peacock, and Franklin D Lowy. 2014. "Molecular tracing of the emergence, diversification, and transmission of *S. aureus* sequence type 8 in a New York community." *Proceedings of the National Academy of Sciences* 111 (18): 6738-6743.

[83] Umaru, GA, J Kabir, NB Adamu, and YA Umar. 2011. "A review of emerging methicillin-resistant *Staphylococcus aureus* (MRSA): A growing threat to veterinarians." *Nigerian Veterinary Journal* 32 (3).

[84] Urth, Tinna, Gitte Juul, Robert Skov, and Henrik C Schønheyder. 2005. "Spread of a methicillin-resistant *Staphylococcus aureus* ST80-IV clone in a Danish community." *Infection Control & Hospital Epidemiology* 26 (2): 144-149.

[85] Utsui, YUKIO, and TAKESHI Yokota. 1985. "Role of an altered penicillin-binding protein in methicillin-and cephem-resistant *Staphylococcus aureus*." *Antimicrobial agents and chemotherapy* 28 (3): 397-403.

[86] Valentini, Piero, Gabriella Parisi, Monica Monaco, Francesca Crea, Teresa Spanu, Orazio Ranno, Mirella Tronci, and Annalisa Pantosti. 2008. "An uncommon presentation for a severe invasive infection due to methicillin-resistant *Staphylococcus aureus* clone USA300 in Italy: a case report." *Annals of clinical microbiology and antimicrobials* 7 (1): 11.

[87] Vandenesch, François, Timothy Naimi, Mark C Enright, Gerard Lina, Graeme R Nimmo, Helen Heffernan, Nadia Liassine, Michèle Bes, Timothy Greenland, and Marie-Elisabeth Reverdy. 2003. "Community-acquired methicillin-resistant *Staphylococcus aureus* carrying Panton-Valentine leukocidin genes: worldwide emergence." *Emerging infectious diseases* 9 (8): 978.

[88] Voss, Andreas, Frans Loeffen, Judith Bakker, Corne Klaassen, and Mireille Wulf. 2005. "Methicillin-resistant *Staphylococcus aureus* in pig farming." *Emerging infectious diseases* 11 (12): 1965.

[89] Weber, J Todd. 2005. "Community-associated methicillin-resistant *Staphylococcus aureus*." *Clinical Infectious Diseases* 41 (Supplement_4): S269-S272.

[90] Wertheim, HFL, MC Vos, HAM Boelens, Andreas Voss, CMJE Vandenbroucke-Grauls, MHM Meester, JAJW Kluytmans, PHJ Van Keulen, and HA Verbrugh. 2004. "Low prevalence of methicillin-resistant *Staphylococcus aureus* (MRSA) at hospital admission in the Netherlands: the value of search and destroy and restrictive antibiotic use." *Journal of Hospital Infection* 56 (4): 321-325.

[91] Williams, REO. 1963. "Healthy carriage of *Staphylococcus aureus*: its prevalence and importance." *Bacteriological reviews* 27 (1): 56.

[92] Wisplinghoff, Hilmar, Adriana E Rosato, Mark C Enright, Michael Noto, William Craig, and Gordon L Archer. 2003. "Related clones containing SCCmec type IV predominate among clinically significant Staphylococcus epidermidis isolates." *Antimicrobial agents and chemotherapy* 47 (11): 3574-3579.

[93] Witte, Wolfgang, Birgit Strommenger, Christian Stanek, and Christiane Cuny. 2007. "Methicillin-resistant *Staphylococcus aureus* ST398 in humans and animals, Central Europe." *Emerging infectious diseases* 13 (2): 255.

In: Antimicrobial Resistance: Opportunities and Challenges ISBN: 978-1-53617-943-9
Editors: P. Kumar Prabhakar and V. Mishra

Chapter 7

MECHANISM OF DRUG RESISTANCE IN *MYCOBACTERIUM TUBERCULOSIS*

Jiju Janardhanan[1], Sarika Sharma[2] and Sandeep Sharma[1,*]
[1]Department of Medical Laboratory Sciences,
Lovely Professional University, Phagwara, Punjab, India
[2]Department of Life Sciences, Arni University. Kathgarh, Indora,
Kangra. H. P., India

ABSTRACT

Tuberculosis (TB) is a disease caused by *Mycobacterium tuberculosis* and is the only a bacterial disease considered to be as major cause of morbidity and mortality around the world. Around 33% of the world's population is infected with TB asymptotically. The emergence of drug resistance in *Mycobacterium tuberculosis* poses a serious threat to global health agencies. It was reported that the TB cases in India itself are two to three times higher than in last few years. Different kinds of mechanism were adopted by the bacteria to become multidrug resistant including target site mutation, drug efflux, alternation in the target site, and inactivation of drug by enzymes. High communicability of the bacteria, long anti-TB drug therapy among densely populated areas is the biggest threat in TB control program especially in malnourished, immunocompromised and other health conditions of individuals in developing countries. There is great need for new and novel drugs with a different mode of action to encounter the emergence of drug resistance in *Mycobacterium tuberculosis*. The current chapter will focus on different mechanisms adopted by bacteria to develop the arsenal of multidrug resistant Mycobacteria.

* Corresponding Author's E-mail: sandeep4380@gmail.com.

1. INTRODUCTION

Tuberculosis is one of the ancient diseases of mankind. Tuberculosis is caused by *Mycobacterium* family which is known as *Mycobacterium tuberculosis*. Mycobactericeae family includes two pathogenic species like *M. tuberculosis* and *M. leprae* causing tuberculosis and leprosy respectively. There are other non-mycobacterium species that cause some opportunistic disease. *Mycobacterium tuberculosis* was first isolated by a German doctor named Robert Koch in 1882. The resistance of *M. tuberculosis* towards immune mechanisms is due to a special feature in its morphology. The cell wall of the organism consists of waxy material known as mycolic acid, which imparts acid-fast nature; hence it is called Acid-fast bacillus. The organisms primarily affect the pulmonary area because it spread through inhalation. According to WHO when a person coughs 3000 droplets are coughed out into the environment. (Cole & Cook, 1998,). When the organism enters the lungs first they are phagocytized by alveolar macrophages but these macrophages cannot kill the organism, it will remain viable and began to multiply within the host. During this time specific immune mechanisms, specifically, T cells will get activated and thereby release various cytokines like interferon-gamma, interleukins, and tumor necrosis factors. Tuberculosis will also affect other parts of the body like kidney, bones, gastrointestinal tract and brain. Common symptoms of tuberculosis include low grade fever, night sweats, productive cough for more than three weeks, chest pain, hemoptysis, unintentional weight loss and loss of appetite

According to WHO in 2018 ten million people are affecting tuberculosis all over the world, (https://www.who.int/news-room/fact-sheets/detail/tuberculosis) so it still remains a public threat to human beings. In recent years tuberculosis has reported with HIV patients because of their low immunity. It was also reported that tuberculosis is the reason for the high mortality in AIDS-HIV patients (Narain et al. 1992). The major reasons for the death of patients with tuberculosis are the delay in diagnosis and emergence of antibiotic resistance.

Currently, available diagnoses of tuberculosis have classified into two types, they are direct detection methods and indirect detection methods. Direct detection methods include staining (Acid-Fast staining), culture includes Lowenstein-Jensen medium (LJ medium), Middle brooks media, etc. The growth of the organism is very slow, *in-vitro* generation time is 18- 20 hours. Colonies will appear in 2 weeks and sometimes take up to 8 weeks. Solid media has been classified into agar-based media and egg-based media. Agar based medias are Middlebrook 7H10, Middlebrook 7H10 selective, Middlebrook 7H11 selective and Middlebrook biplate (7H10/7H11S agar). Egg-based medias are LJ- media with pyruvic acid, LJ with iron, Petragnini media and Dorset media. Other medias include Tarshi's media, Loffler's serum slope and Pawlowsky media. In solid media *M. tuberculosis* forms dry, rough, raised, irregular colonies with a wrinkled surface. They are creamy white, becoming yellowish or buff-colored on further incubation.

Commonly used liquid media for *M. tuberculosis* are: Dubo's media, Middlebrook media, Proskauer and beck's media, Sula's and Sauton's media, BacTec 12B media, and Middle- brook 7H9 broth. Liquid media reduces the time for isolation of acid-fast bacilli to 10 days. In liquid cultures, *M. tuberculosis* grows as twisted rope-like colonies termed serpentine cords. These media were also used for sensitivity testing, chemical analysis and preparation of antigens for vaccines. (Panicker and Ananthanarayanan 2006)

Biochemical reactions like NANCAN, radiological diagnoses like Chest X-ray and CT scan of lungs and nucleic acid amplification were also used for early diagnosis. Indirect diagnosis is based on immunological reactions which include ELISA, and hypersensitivity test which is known as the Mantoux test otherwise called as a skin test. Nowadays the most important diagnostic method used to identify the mycobacteria and its antibiotic resistance are Cartridge-based NAAT (CB-NAAT), Line probe assay (LPA), and Loop-mediated amplification (LAMP) Oommen et al. (2019). These methods are mostly used worldwide because of the emerging drug resistance in tuberculosis.

Based on the pathogenesis and clinical conditions TB has been classified into two types. Primary TB and post-primary TB. In primary TB initial infection is produced in the lungs and leads to the multiplication of bacilli in the alveolar macrophages. It produces an initial lesion or Ghon focus usually in the lower lobe or lower part of upper lobe. Ghon focus together with enlarged hylar lymph node forms the primary complex. Bacillus is transported by lymphatic and hematogenous dissemination into many organs and tissues including other parts of lungs (Greenwood et al. 1997).

The antigen of *M. tuberculosis* are processed by APC's in 10 days and the T- cells are activated, activated T- cell release cytokines, interferon-γ together with calcitriol, activated macrophages and cause them to form a compact cluster or granuloma around the site of infection (activated macrophages- epithelioid cells). The centre of granuloma contains a mixture of necrotic tissue and dead macrophages. Due to its cheese like appearance and consistency, it is called as caseation. To limit the primary infection the lesions become quiescent and surrounding fibroblasts produce dense scar tissue which may become calcified. After an interval of months, years or decades reactivation of tubercle bacillus or exogenous reinfection will lead to post-primary tuberculosis. Endogenous reactivation may occur spontaneously or after intercurrent illness and other conditions that lower the host immune response. Post-primary TB mainly seen in the upper lobe of lungs and the same processes of granuloma formation occur similar to primary TB. Dissemination of bacillus in lymph node and other organs are unusual.

Prophylaxis of tuberculosis is mainly done by vaccination i.e., BCG and treatment for TB is by antibiotics, commonly used antibiotics for tuberculosis have classified into two types First-line drugs and Second-line drugs. First-line drugs are isoniazid (INH), rifampin (RIF), ethambutol (EMB), and pyrazinamide (PZA). Rifabutin (RBT) and Rifapentine (RPT) may also be considered as first-line drugs under certain circumstances. The second-line drugs are Cycloserine, Capreomycin, ρ-Aminosalicylic acid, Levofloxacin, Moxifloxacin, Gatifloxacin, Amikacin/Kanamycin, Ethionamide, linezolid.

In the year 1950, Renee Dubos predicted that the bacteria will eventually lead to developing antibiotic resistance through random mutations and natural selection at this time itself tuberculosis was getting resistant to streptomycin which was used to treat TB. Over to next two decades, tuberculosis was getting resistant to most of the drugs and it was coined as Multi Drug-resistant tuberculosis (MDR) and Extensively drug-resistant tuberculosis (XDR) so to avoid the drug resistance World Health Organization (WHO) put forward a treatment regimen which is known as DOTS direct observation therapy short course. Even though new treatment therapy was put forward by WHO the drug resistance was getting more severe around the world.

Multidrug-resistant tuberculosis means resistance towards the first-line drugs like Isoniazid and Rifampicin and can no longer kill the bacteria by these two drugs and it will become difficult to treat normal tuberculosis.

Extensively drug resistance XDR is less common drug resistance seen in tuberculosis it is mainly seen in HIV patients with tuberculosis. In XDR tuberculosis they are resistant to first-line drugs as well second-line drugs like Fluoroquinolone, and at least one of the other three injectable anti-TB drugs: Amikacin, kanamycin, or capreomycin. As a result, XDR TB needs up to 2 years of extensive drug treatment and is the most challenging to treat.

2. Cultural Characteristics

M. tuberculosis is a facultative, aerobic intracellular pathogen. It is slow-growing, non-motile, non-spore-forming, aerobic and weakly gram-positive that can survive and multiply inside macrophages and other mammalian cells. It has a unique cell wall, due to the dominant presence of mycolic acids that make up more than 50% of its dry weight. One of the characteristic features of bacteria included in *M. tuberculosis* is the ability to retain stain after decolourization with acids. Hence, these organisms have been named as Acid Fast Bacilli (AFB) upon Ziehl-Neelsen staining.

The bacilli grow slowly with a generation time of 18-20 hrs *in vitro*. The Colonies appear in about three weeks and may sometimes take up to eight weeks. The optimum temperature for the growth of *M. tuberculosis* is 37°C with the optimum pH 6.4-7.0 and growth does not occur below 25°C or above 40°C. Several solid and liquid media have been defined for the cultivation of *M. tuberculosis* bacilli. The solid media most widely used for routine culture is Lowenstein Jensen (L-J) media. Middle brook 7H9 and Middle brook 7H10 is for drug susceptibility testing (Recommended by the International Union Against Tuberculosis). On solid media, *M. tuberculosis* forms opaque, dry colonies with a wrinkled or nodular surface. Colonies are creamy white, becoming yellowish or buff coloured on further incubation, while in liquid media *M. tuberculosis* grows in the bottom and form serpentine cords or clusters.

3. Factors that Lead to Drug Resistance

Drug resistance in tuberculosis is also due to misuse and mismanagement of drugs that are used for the treatment some of the reasons include:

- Patients do not complete the full course of the treatment
- Health care providers prescribe wrong treatment
- Proper availability of drug
- Drugs of poor quality
- Develop of tuberculosis again after a previous TB infection
- Peoples who are coming for the work from TB resistant areas
- Contact for long time with TB resistant patients
- Co-infections

3.1. Mechanisms of Antibiotic Resistance

Mainly two types of drug resistance are seen in case of bacteria they are intrinsic drug resistance and acquired drug resistance. In case of intrinsic resistance, it will occurs in a natural process that may be chromosomally encoded. The acquired drug resistance occurred through mutations in existing DNA or by acquisition of new genes that means extra chromosomal methods. Intrinsic resistance may have occurred through a lack of target sites, innate efflux pump mechanisms and drug inactivation. Acquired resistance is mainly acquired through mutations that means change in DNA structure of gene. Plasmids extrachromosomal genetic material which carries Gene's resistant R- genes called as R plasmids. These R genes can readily be transferred from one gene to other plasmids. (JL Martinez et al. 2014)

The shift of R- Genes from one bacterium to another bacteria will take place by conjugation, transduction and transformation. Other mechanisms include moving of R-genes between plasmids within the bacterium by means of transposons, and integrons.

Transposons are a sequence of DNA that can move around in different positions within a genome of single cell. Transposons will cointegrate with acceptor plasmid and replicate during cointegration. After cointegration the two plasmids will separate, and each contains the R- gene carrying the transposons (Babakhani and oloomi 2018)

Integrons are large mobile DNA which can lead to spread multidrug resistance. Each integron is fill up with multiple gene cassettes which include resistance that attached to a small recognition site and encodes several bacterial functions including resistance and virulence (Gillings 2014).

Other biochemical mechanisms of antibiotic resistance are,

- Prevention of drug accumulation within the bacterium
  In this case, bacteria will prepare new porin channels in the bacterial cell wall and lead to decrease permeability due to porin loss and bacteria will not allow antibiotics to enter into the cell.
  Another is efflux pump mechanisms this means the removal of all toxic material from the cell wall of the bacteria including antibiotics efflux mechanisms are the main reasons for antibiotic resistance in most of the bacteria (Poole 2007).
- Modification of target site
  Modification is mainly achieved by ribosomal point mutations, alteration in DNA gyrase enzymes, modified penicillin-binding proteins (e.g., In case of strep. Pneumonia), mutation in DNA dependent RNA polymerase in case of *Mycobacterium tuberculosis*.
- By production of enzymes
  Bacteria will construct new enzymes that can inactivate antibiotics which are normally used for example production of beta-lactamase, chloramphenicol acetyltransferase and adenyl transferases (Egorov et al. 2018).
- Quorum sensing
  Quorum sensing is the regulation of gene expression in response to fluctuations in cell population density. In this bacteria produce and release chemical signals molecules called auto inducers that increase concentration as a function of cell density it also helps to coordinate gene expressions for virulence, conjugation, apoptosis, mobility and resistance (Rutherford and Bassler 2012).

3.2. Drug Resistance in Mycobacterium Tuberculosis

In *Mycobacterium tuberculosis* resistance is mainly by voluntarily mutation in chromosomal genes producing the selection of resistant strains during drug therapy.

Drug resistance in tuberculosis is classified into

- Mono resistance tuberculosis
- Polydrug resistance tuberculosis
- Multidrug-resistant tuberculosis(MDR-TB)
- Extensively drug-resistant tuberculosis (XDR-TB)

Other types include

- Primary drug resistance - no previous treatment has taken but lead to resistance
- Acquired drug resistance–acquired through previously treatment or during the treatment because of the inappropriate treatment procedure.
- New drug resistance- no prior treatment has taken but leads to resistance
- Previously treated drug resistance and transmitted drug resistance (Schaaf and Zumla 2009)

3.3. First-Line Drug Resistance in Tuberculosis

3.3.1. Isoniazid (INH)

One of the main drugs for tuberculosis since its recognition of its clinical activity against *Mycobacterium tuberculosis* in 1952 (Robitzek and Selikoff, 1952). INH is a prodrug that requires activation by the catalase-peroxidase enzyme encoded by *katG* genes (DA Mitchison, 1979). This drug mainly inhibits cellular pathways including synthesis of nucleic acid, phospholipids, NAD metabolisms. Primarily this drug inhibits mycolic acid synthesis by inhibiting NADH depend on enoyl ACP reductase which encoded by *inhA* gene (Rawat et al. 2003). Resistance to this drug is due to mutations in several genes such as *katG*, *inhA*, *ahpC*, *kasA* and NDH

Two main mechanisms are

- Mutation in *katG* and mutation in *inhA* in the promoter region
- Point mutation in *katG* gene

Mechanism is the substitution of threonine for serine at residue 315 (S315T) is the major reason for resistance. The mutation results in reduced affinity of the enzymes NADH without effecting its enoyl reductase activity. Mutation reduces the activity of NADH dehydrogenase and defective NADH dehydrogenase leads to an increased ration of NADH/NAD interfering with *katG* mediated peroxidation of INH or by displacing the INH/NAD abduct from the *inhA* active site. Mutation in the promoter of *ahp*C gene found that increase in the expressions of *ahpC* seems to be more mutation or the loss of catalase-peroxidase activity. (Vilcheze et al. 2019).

3.3.2. Rifampicin

Rifampicin introduced in 1972 one of the most effective anti TB drug. This drug is given together with isoniazid leads the basis of the multidrug treatment regimen for tuberculosis. The mode of action of rifampicin acts on the B subunit of RNA polymerase and inhibits the elongation of mRNA (Blanchard, Jhon 1996). The resistance towards this drug is due to mutation in the *rpoB* gene that encodes the β subunit of RNA polymerase. Mutation takes place in a hot spot region of 81 bp of *rpoB* gene, this zone is also known as rifampicin resistance determining region. Also, mutation in codons 516, 526, 513 are seen with rifampicin. Rifampicin resistant strains will show high level of resistance to other drugs particularly to isoniazid (Munir et al. 2019)

3.3.3. Pyrazinamide

Pyrazinamide is a structural analogue of nicotinamide. It is a type of prodrug converted to its active form pyrazinoic acid by the enzyme pyrazinamidase/nicotinamidase (PZase). PZase is encoded by the gene *pncA*. (Konno et al. 1967), Scorpio, A.; Zhang, Y 1996). The resistance is due to mutation in *pnc*A gene alteration occur in 561 bp region of the open reading frame or in an 82 bp region. Another mutation taking place in *rpsA* gene which encodes for ribosomal proteins this leads to over expressions of *rpsA* gene that may induce resistance for pyrazinamide (carter et al. 2019).

3.3.4. Streptomycin

Streptomycin is an aminoglycoside antibiotic and it was the first antibiotic used for the treatment of *Mycobacterium tuberculosis*. Streptomycin mainly inhibits the initiation of translation by binding to the 16s rRNA this is encoded by the genes *rpsL* and *rrs* (Finken et al. 1993). Resistance acquired through mutation in *rpsL* and *rrs* which produces alterations in STM binding site this leads to substitution in codon 43 that is lysine to arginine. In *rrs* gene mutation occurs around the nucleotides 530 and 915. (Haling et al. 2017)

3.3.5. Ethambutol

This drug was first used in 1966 mainly used to treat *Mycobacterium avium complex*, ethambutol is a bacteriostatic drug which inhibits the enzyme arabinosyl transferase and interferes with biosynthesis of cell wall arabinogalactan (Takayama. K; Kilbur J.o 1989). The gene *embCAB* code for arabinosyl transferase (Mikusova et al. 1995). Resistance in case of ethambutol is due to mutation in the gene *embB* at position emb306 codon. (Sun Lin et al. 2019)

3.4. Second-Line Drug Resistance in Tuberculosis

3.4.1. Fluoroquinolones

Fluoroquinolones are the currently used second-line drug in the treatment of MDR-TB. Mainly ciprofloxacin and ofloxacin are synthetic derivatives of nalidixic acid (Goss et al. 1965). The mode of action of fluoroquinolones is that it will inhibit the DNA gyrase and topoisomerase IV enzymes required for bacterial viability. These proteins are coded by the genes *gyrA*, *gyrB*, *parC* and *pare* (Fabrega et al. 2009). The resistance shown in fluoroquinolones is by a chromosomal mutation in concerned genes. Recurrent mutations are seen at positions 90 and 94 of *gyrA*. Also mutation at Ala-74, Gly-88 and ser-91 are observed. Another important reason for resistance to ciprofloxacin and sparfloxacin is the presence of *MfpA* proteins. The existence of natural polymorphism at position 95 in *gyrA* gene is also a cause for resistance in *M. tuberculosis* (Che yang et al. 2017)

3.4.2. Kanamycin, Capreomycin, Amikacin, and Viomycin

These drugs are used in the treatment of MDR-TB. The mode of action of these drugs will inhibit protein translation at the level of 16S rRNA. Resistance towards these drugs are due to mutations with an A1401G in *rss* gene coding for 16S rRNA (Angkanang et al. 2014). Resistance towards kanamycin is by mutation promoter region of *eis* gene. Mutation in *tlyA* gene leads to resistance against capreomycin and viomycin (Sophia. B. et al. 2012).

3.4.3. Ethionamide

Ethionamide is imitative of isonicotinic acid which is similar to isoniazid. It is also a prodrug activated by monooxygenase enzyme which encodes by *ethA* gene (Carette et al. 2011). The action of this drug by interfering with mycolic acid synthesis. Ethionamide resistance is due to mutation *etaA* gene and *ethR* gene and mutation in *inhA* gene (Dookie et al. 2018)

3.4.4. Para-Aminosalicylic Acid

Para- aminosalicylic acid was one of the first drug used in combination with isoniazid and streptomycin. It is used for the treatment of MDR –TB. This drug is an analogue of para- aminobenzoic acid and it will inhibit the enzyme dihydropteroate synthase, needed for folate synthesis (Zheng et al. 2013). Resistance against para- aminosalicylic acid acquired through transposon mutagenesis in the *thyA* gene also missense mutation in *folC* gene encoding for dihydrofolate synthase (Lou et al. 2019).

3.4.5. Cycloserine

It is a correlative of D- alanine and that by obstructing the activity of D- alanine ligase activity and prevent peptidoglycan synthesis (Zhang, Ying 2005). The resistance towards cycloserine is due to mutation in *alrA* and *cycA* gene (Caccres et al. 1997), (Evegelopoulos et al. 2019).

3.4.6. Macrolides

Macrolides are common drugs used for the treatment of tuberculosis. The resistance in the case of macrolides because of the expressions of *erm 37* genes which encodes an enzyme that methylates a specific site in the 23S rRNA and prevents antibiotic binding (Paartt et al. 2015), (Bongulto et al. 2019)

3.4.7. Linezolid

Linezolid was the first compound belonging to oxazolidinones and approved by FDA on April 2000 used to treat drug-resistant tuberculosis (Singh et al. 2019). This drug was limited in use because of its side effects. The mode of action of this drug is to bind to the ribosomal 50s subunit and thus inhibit early step in protein synthesis (Zhang, Ying 2005). Resistance towards linezolid in *Mycobacterium tuberculosis* is very rare but it has found to contain mutation at G2061T and G2576 in the 23s rRNA gene. Recent studies showed that mutation is also seen in T460C in *rplC* gene which encodes for ribosomal L3 protein (Zhang et al. 2016).

3.4.8. New Drugs, Targets, and New Drug Resistance

To avoid the emergence of drug resistance World Health Organization (WHO) with Food and Drug Administration (FDA) find out newer drugs and give approval for the treatment. Newer drugs were found out to minimize drug resistance and to treat drug-resistant tuberculosis. New drugs include

- Bedaquiline
- Delamind
- PA - 824 (Pretomanid)
- SQ- 109
- Benzothiazinones

The recent studies shown that emergence of drug resistance towards newer drugs.

3.4.9. Bedaquiline

New antibiotic belonging to the class to the diary quinolines and it is the most widely used drug against MDR-TB (Low, M pipeline report 2017). The action of bedaquiline is by preventing the ATP synthase of *Mycobacterium tuberculosis* (Huitric et al. 2007). The only mutation associated with bedaquiline was at *atpE* gene which coded for the c part of the f0 subunit of the ATP synthase. The most important mutation is in the *atp*E gene found in bedaquiline resistant mutant is A63P (Polsfuss et al. 2019).

3.4.10. Delamind

It is a derivative of nitro-dihydro-imidazooxazole. This is also a prodrug activated by *Mycobacterium tuberculosis*. The mode of action of this drug is by inhibiting the synthesis of mycolic acid especially methoxy and keto-mycolic acids (Carlos Palomino et al. 2013), (Matsumoto et al. 2006). But experimentally a mutation was found in *Rv*3547 gene and do not metabolize the drug (Polsfuss et al. 2019).

3.4.11. PA- 824 (Pretomanid)

PA- 824 is a bicyclic derivative of nitroimidazole that affects the synthesis of protein and cell wall lipids and acts as a nitric oxide donor leads to respiratory poisoning (Stover et al. 2000). The mechanism of resistance to PA -824 has been shown to be most frequently associated with loss of specific glucose- 6- phosphate dehydrogenase (FGD1) or the deazaflavin cofactor 420 (Bapista et al. 2018).

3.4.12. SQ- 109

SQ – 109 is a synthetic analogue of ethambutol. The action is by interfering with assembly of mycolic acids into bacterial cell wall core resulting in the accumulation of trehalose momomycolate a precursor of the trehalose dimycolate (Protopopova et al. 2005). The resistance in SQ-109 is due to mutation in *mmpL3* gene which is the target site of this drug (Tahlan et al. 2012), (Vasava et al. 2019).

3.4.13. Benzothiazinones

A new class of mycobacterial agent, the mode of action is inhibition of cell wall synthesis that is effect on drpE1 subunit of the enzyme decaprenyl phosphoryl– β-D- ribose 2- epimerase and inhibits this enzyme which leads to preventing cell wall synthesis by lacking the formation of decaprenyl phopshoryl arabinose. No resistance has been found yet but experimentally over expressions of nitroreductase *NfnB* has been seen that leads to inactivation of drug (Richter et al. 2018). These are different types of drugs used for the treatment of resistant *Mycobacterium tuberculosis*. The drug resistance in tuberculosis remains largely a man-made phenomenon that leads the bacteria resistant to most of the TB drugs.

3.5. DE- Novo Evolution of Drug Resistance

This includes three important factors that lead to de novo evolution of drug resistance are population size of bacteria relates to number of binary fission, the mutation rate defines by the replication fidelity of bacterial DNA polymerase, mutational target size which depends on the number of potential sites which may be mutated to confer the drug resistance; an important factor involved in determining the rate of drug-resistant evolution (Sebastian et al., 2017).

Table 1.

Drug	Target	Target gene involved	Resistance mechanism	Associated MIC (µg/ml)	References
Rifampicin	inhibition of RNA synthesis	*rpoB*	drug target alteration	0.05-1	Telenti et al. (1993)
Isoniazid	inhibition of cell wall mycolic acid synthesis	*katG*	Abrogated prodrug activation	0.02-0.2	Heym et al. (1995
					Banerjee et al. (1994); Morlock et al.(2003) Hazbon et al. (2006)
		inhA	Drug target alteration		
		katA			
Ethambutol	Inhibition of cell wall arabinogalactan biosynthesis	*emb* *ubiA*	Drug target alteration	1-5	Sreevatsan et al. (1997)
Pyrazinamide	Reduction of membrane energy by inhibition of trans translation and inhibition of panthothenate and coenzyme A synthesis	*pncA* *rpsA* *panD*	Abrogated prodrug activation	10-100	Konno, Feldmann, and McDermott (1967); Scorpio and Zhang (1996)
Ethionamide	Inhibition of cell wall mycolic acid synthesis	*eth* *msh* *ndh* *inhA*	Drug target alteration	2.5-25	Banerjee et al. (1994) Morlock et al. (2003)
		inhA promoters	Drug target over expression		
Streptomycin	Inhibition of protein synthesis	*rrs* *rps* *eis* *tlyA*	Drug target alteration	2-8	Meier et al. (1994); Maus, Plikaytis and Shinnick (2005) Nair et al. (1993)
Fluoroquinolones	Inhibition of DNA synthesis	*gyrA* *gyrB*	Drug target alteration	0.5-2.5	Takiff et al. (1994); Xu et al. (1996); Malik et al. (2012)
Amikacin, Kanamycin, capreomycin	Inhibition of protein synthesis	*rrs* *eis* *tlyA*	Drug target alteration Over expression of drugs inactivating	2-4	Alangaden et al. (1998); Maus, Plikaytis and Shinnick (2005)
			enzymes Abrogation of drug target methylation		Kambli et al. (2016)
Bedaquiline	Inhibition of mycobacterial ATPase syn nthase	*atpE* *rvo678* *pepQ*	Drug target alteration	0.06-1	Huitric et al. (2010) Hartkoorn, Uplekar and Cole (2014); Bloemberg et al. (2015)
Para amino salicylic acid	Inhibition of folic acid and thymine nucleotide metabolism	*thyA* *folC* *ribP*	Drug target bypassing Abrogation of prodrug activation	1-8	Zhao et al. (2014); Minato et al. (2015)

Drug	Target	Target gene involved	Resistance mechanism	Associated MIC (µg/ml)	References
Linezolid	Inhibition of protein synthesis	*rplC* *rrl*	Drug target alteration	0.25-0.5	Beckert et al. (2012); Hillemann, Rusch-Gerdes and Richter (2008)
Delamind/Pretomanid (PA-824)	Specific and selective inhibition of mycolic acid biosynthesis essential for cell wall formation	*fgd1* *fb1C* *fb1A* *fb1B* *ddn*	Abrogation of prodrug activation	0.006-0.24 (D) 0.015-0.25 (P)	Manjunatha et al. (2006); Bloemberg et al. (2015); Haver et al. (2015)
Clofazamine	Inhibits mycobacterial growth and binds preferentially to mycobacterial DNA. It may also bind to bacterial potassium transporters thereby inhibits the normal function	*rvo678* *rv1979* *rv2535* *ndh* *pepQ* *mmpR*	Over expression of efflux pump mmpl5	0.1-1.2	Hartkoorn, Uplekar and Cole (2014); Bloemberg et al. (2015)

(Source: Sebastian et al., 2017)

3.6. Current Scenario in Tuberculosis

Tuberculosis is a global issue and have a high mortality rate. So, the prevention of tuberculosis needs special attention both and national as well as at International level. Based on the facts that WHO has put forward lots of TB control programs associated with different countries around the world.

In India, TB control program is known as Revised National TB Control Program (RNTCP). According to RNTCP, TB incidence in India is twenty-seven lakhs. In 2018, there is 16% increase when compared to 2017 and the highest so far. Out of the total notifications, 25% of cases were reported from private sectors. Among the Indian population, the main age group infected with tuberculosis is 15-69 years. A total of 132,711 paediatric TB cases were reported during 2018. Another condition associated with TB is the co-infection with HIV; it was nearly 15000 cases reported by RNTCP.

To identify the drug resistance in TB, RNTCP adopted different molecular methods like CBNAAT and line probe assays (LPA). In 2018, twenty-four lakh CBNAAT tests were performed. Out of these 66000 rifampicin-resistant strains were detected and 23000 patients with second-line drug resistance. Out of this 23000, 3794 XTR- TB patients were detected; and 13000 cases were resistant to isoniazid.

In 2018, 200876 numbers of first-line LPA test was done out of this 6% of among them were MDR TB and in second-line LPA test 6.3% cases were detected as XDR TB. For identifying the active cases in 2018, RNTCP was screened 18.93 crore population and found out 47307 additional TB cases. During January 2018- to September 2018 approximately 41250 drug-resistant TB patients were diagnosed.

In India out of total cases, 17% cases were reported from Uttar Pradesh (around 4.2 lakh cases); 187 cases per lakh population and 10% cases in Maharashtra. India contributes 27% of global burden with estimated 27.5 lakhs TB patients as per global TB report 2018.

According to the global TB report 2019, published by WHO globally an estimated 10.0 million people were infected with tuberculosis. In every year 500 new cases per 100,000 of population were reported. 1.2 million TB death was reported in 2018 and it is comparatively less than in 2017. About 251000 death was reported among HIV positive patients with TB.

Geographically, most TB cases were in the WHO regions of Southeast Asia (44%), Africa (24%), and Western Pacific (18%). Eight countries accounted for two-thirds of the global total, i.e., India (27%), China (9%), Indonesia (8%), Philippines (6%), Pakistan (6%), Nigeria (4%), Bangladesh (4%) and South Africa (3%).

In 2018, half a million new cases of rifampicin-resistant TB were reported, 78% was multidrug-resistant TB. Globally 3.4% of new TB cases and 18% previously treated cases had multidrug-resistant tuberculosis. A global total of 186,772 cases of MDR- TB were reported in 2018.

To overcome the drug resistance and to prevent TB in recommendations of WHO, RNTCP made a structure between state government and the central government. At national level, RNTCP is managed by central TB division (CTD) comes under the Ministry of health and family welfare. Under CTD four national institutes are established and also framed national level committees. Altogether 14 committees are made to improve the facilities for TB control programs, they include,

- National level coordination committee
- National technical expert group
- National technical expert group on treatment
- National TB comorbidity coordination committee
- National technical working group on TB comorbidity
- Technical working group on latent TB infection management in India
- National technical expert group on paediatric TB
- Technical expert committee on TB in women including gender issues
- National task force for medical colleges
- National operation on research committee
- National technical working group on private sector management
- Interministerial coordination committee for TB elimination
- National TB forum
- National ACSM committee

National committees will deal with state-level committees and at state level there will be state TB cells and they coordinate with district-level program committees and sub-district level committees. State-level committee will report to national-level committee.

To avoid the drug resistance WHO has put forward a new diagnostic strategy and new drug to achieve drug resistance. In India, RNTCP has started six national level TB laboratories and adopted molecular methods in detection of MDR- TB and XDR- TB.

For the treatment of TB, RNTCP provided different strategies like

- Treatment for drug-sensitive TB
- Policy decision on discontinuation of regimen for discontinuation of previously treated TB patients.
- Programmatic management of drug-resistant TB
- Introduction of shorter MDR- TB regimen
- Countrywide expansion of bedaquiline containing regimen.

RNTCP in combination with WHO, they are establishing the importance of newer drugs like bedaquiline and delamind to treat MDR- TB and XDR- TB and to achieve the prevention of tuberculosis within 2022. In the year 2030 WHO planning to eliminate tuberculosis from all over the world with participation of all countries. (WHO 2019, RNTCP 2019)

3.6.1. Challenges and Future Strategies in TB Control

Emergence of drug resistance in Tuberculosis is a global challenge and difficult to treat especially in immunocompromised patients. The new drug discovery is unable to discover new molecules to encounter drug resistance. In the last five to seven-decade only bedaquiline is the molecule introduced in TB therapy to treat MDR-TB and is FDA approved for its use. Moreover, long term therapy, more toxicity of drugs, less effectiveness, and more expensive drugs are the biggest hindrance in TB control. At society level increase in population, much crowded and densely populated areas are always be a threat to spread TB. People were not aware of the TB therapy and misuse the drug by self-medicating themselves.

Current TB research focuses on three main aspects: early diagnostics, effective and short treatment, and the new vaccine, especially for immunocompromised patients. Combination therapy with new drug candidates like Bedaquiline could be a better option for shortening TB treatment. A large number of molecules have been in clinical trials for many years. Few of them had been approved by FDA by 2013 (Keith D Green and Garneau-Tsodikova 2013).

REFERENCES

Alangaden, George J; Barry N. Kreiswirth, Arlette Aouad, Minoo Khetarpal, Felicitas R. Igno, Soraya L. Moghazeh, Elias K. Manavathu, Stephen A. Lerner. "Mechanism of resistance to amikacin and kanamycin in Mycobacterium tuberculosis." *Antimicrobial agents and chemotherapy*, 42, no. 5, (1998), 1295-1297.

Almeida Da Silva, Pedro Eduardo, Juan Carlos Palomino. "Molecular basis and mechanisms of drug resistance in *Mycobacterium tuberculosis*: classical and new drugs." *Journal of antimicrobial chemotherapy*, 66, no. 7, (2011), 1417-1430.

Banerjee, Asesh, Eugenie Dubnau, Annaik Quemard, V. Balasubramanian, Kyung Sun Um, Theresa Wilson, Des Collins, Geoffrey de Lisle, William R. Jacobs. "inhA, a gene encoding a target for isoniazid and ethionamide in Mycobacterium tuberculosis." *Science*, 263, no. 5144, (1994), 227-230.

Baptista, Rafael, David M. Fazakerley, Manfred Beckmann, Les Baillie, Luis AJ Mur. "Untargeted metabolomics reveals a new mode of action of pretomanid (PA-824)." *Scientific reports*, 8, no. 1, (2018), 5084.

Beckert, Patrick, Doris Hillemann, Thomas A. Kohl, Jörn Kalinowski, Elvira Richter, Stefan Niemann, Silke Feuerriegel. "rplC T460C identified as a dominant mutation in linezolid-resistant Mycobacterium tuberculosis strains." *Antimicrobial agents and chemotherapy*, 56, no. 5, (2012), 2743-2745.

Blanchard, John S. "Molecular mechanisms of drug resistance in Mycobacterium tuberculosis." *Annual review of biochemistry*, 65, no. 1, (1996), 215-239.

Bloemberg, Guido V; Sebastien Gagneux, Erik C. Boettger. "Acquired resistance to bedaquiline and delamanid in therapy for tuberculosis." *The New England journal of medicine*, 373, no. 20, (2015), 1986.

Bongulto, Kenneth Aquino, Concepcion F. Ang, Esperanza C. Cabrera. "Antibiograms and Molecular Characterization of Drug Resistance of Mycobacterium abscessus complex from Patients with Multidrug-Resistant Pulmonary Tuberculosis (MDR TB) Infection." *bioRxiv*, (2019), 864447.

Caceres, Nancy E; Beth Harris, N; James F. Wellehan, Zhengyu Feng, Vivek Kapur, Raul G. Barletta. "Overexpression of the D-alanine racemase gene confers resistance to D-cycloserine in Mycobacterium smegmatis." *Journal of bacteriology*, 179, no. 16, (1997), 5046-5055.

Carette, Xavier, Nicolas Blondiaux, Eve Willery, Sylviane Hoos, Nathalie Lecat-Guillet, Zoé Lens, Alexandre Wohlkönig; et al. "Structural activation of the transcriptional repressor EthR from Mycobacterium tuberculosis by single amino acid change mimicking natural and synthetic ligands." *Nucleic acids research*, 40, no. 7, (2011), 3018-3030.

Carlos Palomino, Juan, Anandi Martin. "Tuberculosis clinical trial update and the current anti-tuberculosis drug portfolio." *Current medicinal chemistry*, 20, no. 30, (2013), 3785-3796.

Che, Yang, Qifa Song, Tianchi Yang, Guohua Ping, Mei Yu. "Fluoroquinolone resistance in multidrug-resistant Mycobacterium tuberculosis independent of fluoroquinolone use." *European Respiratory Journal*, 50, no. 6, (2017), 1701633.

Cole, Eugene C; Carl E. Cook. "Characterization of infectious aerosols in health care facilities: an aid to effective engineering controls and preventive strategies." *American journal of infection control*, 26, no. 4, (1998), 453-464.

Dookie, Navisha, Santhuri Rambaran, Nesri Padayatchi, Sharana Mahomed, Kogieleum Naidoo. "Evolution of drug resistance in Mycobacterium tuberculosis: a review on the molecular determinants of resistance and implications for personalized care." *Journal of Antimicrobial Chemotherapy*, 73, no. 5, (2018), 1138-1151.

Egorov, AM; Ulyashova, MM; Yu Rubtsova, M. "Bacterial Enzymes and Antibiotic Resistance." *Acta naturae*, 10, no. 4, (2018), 33.

Evangelopoulos, Dimitrios, Gareth A. Prosser, Angela Rodgers, Belinda M. Dagg, Bhagwati Khatri, Mei Mei Ho, Maximiliano G. Gutierrez, Teresa Cortes, Luiz Pedro S. de Carvalho. "Comparative fitness analysis of D-cycloserine resistant mutants reveals both fitness-neutral and high-fitness cost genotypes." *Nature communications*, 10, no. 1, (2019), 1-11.

Fàbrega, Anna, Sergi Madurga, Ernest Giralt, Jordi Vila. "Mechanism of action of and resistance to quinolones." *Microbial biotechnology*, 2, no. 1, (2009), 40-61.

Finken, Marion, Philip Kirschner, Albrecht Meier, Annette Wrede, Erik C. Böttger. "Molecular basis of streptomycin resistance in Mycobacterium tuberculosis: alterations of the ribosomal protein S12 gene and point mutations within a functional 16S ribosomal RNA pseudoknot." *Molecular microbiology*, 9, no. 6, (1993), 1239-1246.

Georghiou, Sophia B; Marisa Magana, Richard S. Garfein, Donald G. Catanzaro, Antonino Catanzaro, Timothy C. Rodwell. "Evaluation of genetic mutations associated with Mycobacterium tuberculosis resistance to amikacin, kanamycin and capreomycin: a systematic review." *PloS one*, 7, no. 3, (2012), e33275.

Gillings, Michael R. "Integrons: past, present, and future." *Microbiol. Mol. Biol. Rev.*, 78, no. 2, (2014), 257-277.

Goss, William A; William H. Deitz, Thomas M. Cook. "Mechanism of action of nalidixic acid on *Escherichia coli* II. Inhibition of deoxyribonucleic acid synthesis." *Journal of Bacteriology*, 89, no. 4, (1965), 1068-1074.

Greenwood, David, Richard CB Slack, John F. Peutherer, Michael R. Barer. "Medical Microbiology: A Guide to Microbial Infections: Pathogenesis." *Immunity, Laboratory Diagnosis and Control, 15th Edition, Churchill Livingstne, Edinburgh, United Kingdom*, (1997), 690.

Gygli, Sebastian M; Sonia Borrell, Andrej Trauner, Sebastien Gagneux. "Antimicrobial resistance in Mycobacterium tuberculosis: mechanistic and evolutionary perspectives." *FEMS microbiology reviews*, 41, no. 3, (2017), 354-373.

Hartkoorn, Ruben C; Swapna Uplekar, Stewart T. Cole. "Cross-resistance between clofazimine and bedaquiline through upregulation of MmpL5 in Mycobacterium tuberculosis." *Antimicrobial agents and chemotherapy*, 58, no. 5, (2014), 2979-2981.

Haver, Hana L., Adeline Chua, Pramila Ghode, Suresh B. Lakshminarayana, Amit Singhal, Barun Mathema, René Wintjens, Pablo Bifani. "Mutations in genes for the F420 biosynthetic pathway and a nitroreductase enzyme are the primary resistance determinants in spontaneous *in vitro*-selected PA-824-resistant mutants of Mycobacterium tuberculosis." *Antimicrobial agents and chemotherapy*, 59, no. 9, (2015), 5316-5323.

Hazbón, Manzour Hernando, Michael Brimacombe, Miriam Bobadilla Del Valle, Magali Cavatore, Marta Inírida Guerrero, Mandira Varma-Basil, Helen Billman-Jacobe et al. "Population genetics study of isoniazid resistance mutations and evolution of multidrug-resistant Mycobacterium tuberculosis." *Antimicrobial agents and chemotherapy*, 50, no. 8, (2006), 2640-2649.

Heym, Beate, Pedro M. Alzari, Nadine Honore, Stewart T. Cole. "Missense mutations in the catalase-peroxidase gene, katG, are associated with isoniazid resistance in Mycobacterium tuberculosis." *Molecular microbiology*, 15, no. 2, (1995), 235-245.

Hillemann, Doris, Sabine Rüsch-Gerdes, Elvira Richter. "*In vitro*-selected linezolid-resistant Mycobacterium tuberculosis mutants." *Antimicrobial agents and chemotherapy*, 52, no. 2, (2008), 800-801.

Hlaing, Yin Moe, Pongsri Tongtawe, Pramuan Tapchaisri, Jeeraphong Thanongsaksrikul, Unchana Thawornwan, Buppa Archanachan, Potjanee Srimanote. "Mutations in streptomycin resistance genes and their relationship to streptomycin resistance and lineage of Mycobacterium tuberculosis Thai isolates." *Tuberculosis and respiratory diseases*, 80, no. 2, (2017), 159-168.

Huitric, E; Verhasselt, P; Anil Koul, Andries, K; Hoffner, S; Dan, I. Andersson. "Rates and mechanisms of resistance development in Mycobacterium tuberculosis to a novel diarylquinoline ATP synthase inhibitor." *Antimicrobial agents and chemotherapy*, 54, no. 3, (2010), 1022-1028.

Huitric, Emma, Peter Verhasselt, Koen Andries, Sven E. Hoffner. "*In vitro* antimycobacterial spectrum of a diarylquinoline ATP synthase inhibitor." *Antimicrobial agents and chemotherapy*, 51, no. 11, (2007), 4202-4204.

India TB report 2019 – central TB division.

Jennings, Benjamin C; Kristin J. Labby, Keith D. Green, Sylvie Garneau-Tsodikova. "Redesign of substrate specificity and identification of the aminoglycoside binding residues of Eis from Mycobacterium tuberculosis." *Biochemistry*, 52, no. 30, (2013), 5125-5132.

Kambli, Priti, Kanchan Ajbani, Chaitali Nikam, Meeta Sadani, Anjali Shetty, Zarir Udwadia, Sophia B. Georghiou, Timothy C. Rodwell, Antonino Catanzaro, Camilla Rodrigues. "Correlating rrs and eis promoter mutations in clinical isolates of Mycobacterium tuberculosis with phenotypic susceptibility levels to the second-line injectables." *International journal of mycobacteriology*, 5, no. 1, (2016), 1-6.

Konno, K; Feldmann, FM; McDermott, W. Pyrazinamide susceptibility and amidase activity of tubercle bacilli. *Am. Rev. Respir. Dis.*, 1967, 95, 461–469.

Low, Marcus. "The Tuberculosis Treatment Pipeline: A Breakthrough Year for the Treatment of XDR-TB." *Pipeline Report*, (2017), 129.

Luo, Ming, Kun Li, Huizheng Zhang, Xiaofeng Yan, Jing Gu, Zhen Zhang, Yu Chen, Jungang Li, Jing Wang, Yaokai Chen. "Molecular characterization of para-aminosalicylic acid resistant Mycobacterium tuberculosis clinical isolates in southwestern China." *Infection and drug resistance*, 12, (2019), 2269.

Mabhula, Amanda, Vinayak Singh. "Drug-resistance in Mycobacterium tuberculosis: where we stand." *Med Chem Comm*, (2019).

Malik, Seidu, Melisa Willby, David Sikes, Oleg V. Tsodikov, James E. Posey. "New insights into fluoroquinolone resistance in Mycobacterium tuberculosis: functional genetic analysis of gyrA and gyrB mutations." *PloS one*, 7, no. 6, (2012), e39754.

Manjunatha, Ujjini H., Helena Boshoff, Cynthia S. Dowd, Liang Zhang, Thomas J. Albert, Jason E. Norton, Lacy Daniels, Thomas Dick, Siew Siew Pang, Clifton E. Barry.

"Identification of a nitroimidazo-oxazine-specific protein involved in PA-824 resistance in Mycobacterium tuberculosis." *Proceedings of the National Academy of Sciences*, 103, no. 2, (2006), 431-436.

Martínez, José Luis, Fernando Baquero. "Emergence and spread of antibiotic resistance: setting a parameter space." *Upsala Journal of Medical Sciences*, 119, no. 2, (2014), 68-77.

Matsumoto, Makoto, Hiroyuki Hashizume, Tatsuo Tomishige, Masanori Kawasaki, Hidetsugu Tsubouchi, Hirofumi Sasaki, Yoshihiko Shimokawa, Makoto Komatsu. "OPC-67683, a nitro-dihydro-imidazooxazole derivative with promising action against tuberculosis *in vitro* and in mice." *PLoS medicine*, 3, no. 11, (2006), e466.

Maus, Courtney E; Bonnie, B. Plikaytis; Thomas, M. Shinnick. "Molecular analysis of cross-resistance to capreomycin, kanamycin, amikacin, and viomycin in Mycobacterium tuberculosis." *Antimicrobial agents and chemotherapy*, 49, no. 8, (2005), 3192-3197.

Meier, Albrecht, Philip Kirschner, Franz-Christoph Bange, Ulrich Vogel, Böttger, EC. "Genetic alterations in streptomycin-resistant Mycobacterium tuberculosis: mapping of mutations conferring resistance." *Antimicrobial agents and chemotherapy*, 38, no. 2, (1994), 228-233.

Mikusova, Katarina, Richard A. Slayden, Gurdyal S. Besra, Patrick J. Brennan. "Biogenesis of the mycobacterial cell wall and the site of action of ethambutol." *Antimicrobial agents and chemotherapy*, 39, no. 11, (1995), 2484-2489.

Mitchison, Denis A. "Basic mechanisms of chemotherapy." *Chest*, 76, no. 6, (1979), 771-780.

Minato, Yusuke, Joshua M. Thiede, Shannon Lynn Kordus, Edward J. McKlveen, Breanna J. Turman, Anthony D. Baughn. "Mycobacterium tuberculosis folate metabolism and the mechanistic basis for para-aminosalicylic acid susceptibility and resistance." *Antimicrobial agents and chemotherapy*, 59, no. 9, (2015), 5097-5106.

Morlock, Glenn P; Beverly Metchock, David Sikes, Jack T. Crawford, Robert C. Cooksey. "ethA, inhA, and katG loci of ethionamide-resistant clinical Mycobacterium tuberculosis isolates." *Antimicrobial agents and chemotherapy*, 47, no. 12, (2003), 3799-3805.

Munir, Asma, Narender Kumar, Suresh Babu Ramalingam, Sembulingam Tamilzhalagan, Siva Kumar Shanmugam, Alangudi Natarajan Palaniappan, Dina Nair; et al. "Identification and characterization of genetic determinants of Isoniazid and Rifampicin resistance in Mycobacterium tuberculosis in Southern India." *Scientific reports*, 9, no. 1, (2019), 1-13.

Nair, Jaygopal, David A. Rouse, Gill-Han Bai, and Sheldon L. Morris. "The rpsL gene and streptomycin resistance in single and multiple drug-resistant strains of Mycobacterium tuberculosis." *Molecular microbiology*, 10, no. 3, (1993), 521-527.

Narain, Jai P; Mario C. Raviglione, Arata Kochi. "HIV-associated tuberculosis in developing countries: epidemiology and strategies for prevention." *Tubercle and lung disease*, 73, no. 6, (1992), 311-321.

Oommen, Seema, Nandita Banaji. "Laboratory diagnosis of tuberculosis: Advances in technology and drug susceptibility testing." *Indian journal of medical microbiology*, 35, no. 3, (2017), 323.

Palomino, Juan Carlos, Anandi Martin. "Drug resistance mechanisms in *Mycobacterium tuberculosis*." *Antibiotics*, 3, no. 3, (2014), 317-340.

Polsfuss, Silke, Sabine Hofmann-Thiel, Matthias Merker, David Krieger, Stefan Niemann, Holger Rüssmann, Nicolas Schönfeld, Harald Hoffmann, Katharina Kranzer. "Emergence of Low-level Delamanid and Bedaquiline Resistance During Extremely Drug-resistant Tuberculosis Treatment." *Clinical Infectious Diseases*, (2019).

Paniker. CKJ. *Ananthnarayan and Paniker's Textbook of Microbiology*. (2006).

Poole, Keith. "Efflux pumps as antimicrobial resistance mechanisms." *Annals of medicine*, 39, no. 3, (2007), 162-176.

Protopopova, Marina, Colleen Hanrahan, Boris Nikonenko, Rowena Samala, Ping Chen, Jackie Gearhart, Leo Einck, Carol A. Nacy. "Identification of a new antitubercular drug candidate, SQ109, from a combinatorial library of 1, 2-ethylenediamines." *Journal of antimicrobial chemotherapy*, 56, no. 5, (2005), 968-974.

Rawat, Richa, Adrian Whitty, Peter J. Tonge. "The isoniazid-NAD adduct is a slow, tight-binding inhibitor of InhA, the Mycobacterium tuberculosis enoyl reductase: adduct affinity and drug resistance." *Proceedings of the National Academy of Sciences*, 100, no. 24, (2003), 13881-13886.

Richter, Adrian, Ines Rudolph, Ute Möllmann, Kerstin Voigt, Chun-wa Chung, Onkar MP Singh, Michael Rees; et al. "Novel insight into the reaction of nitro, nitroso and hydroxylamino benzothiazinones and of benzoxacinones with Mycobacterium tuberculosis DprE1." *Scientific reports*, 8, no. 1, (2018), 13473.

Rutherford, Steven T; Bonnie L. Bassler. "Bacterial quorum sensing: its role in virulence and possibilities for its control." *Cold Spring Harbor perspectives in medicine*, 2, no. 11, (2012), a012427.

Schaaf, H. Simon, Alimuddin Zumla. *Tuberculosis E-Book: A Comprehensive Clinical Reference*. Elsevier Health Sciences, 2009.

Scorpio, A; Zhang, Y. Mutations in *pncA*, a gene encoding pyrazinamidase/ nicotinamidase, cause resistance to the antituberculous drug pyrazinamide in tubercle bacillus. *Nat. Med.*, 1996, 2, 662–667.

Selikoff, Irving J; Edward H. Robitzek, George G. Ornstein. "Treatment of pulmonary tuberculosis with hydrazide derivatives of isonicotinic acid." *Journal of the American Medical Association*, 150, no. 10, (1952), 973-980.

Singh, Bhagteshwar, Derek Cocker, Hannah Ryan, Derek J. Sloan. "Linezolid for drug-resistant pulmonary tuberculosis." *Cochrane Database of Systematic Reviews*, 3, (2019).

Sowajassatakul, Angkanang, Therdsak Prammananan, Angkana Chaiprasert, Saranya Phunpruch. "Molecular characterization of amikacin, kanamycin and capreomycin resistance in M/XDR-TB strains isolated in Thailand." *BMC microbiology*, 14, no. 1, (2014), 165.

Sreevatsan, Srinand, Kathryn E. Stockbauer, Pan, XI; Barry N. Kreiswirth, Soraya L. Moghazeh, William R. Jacobs, Amalio Telenti, James M. Musser. "Ethambutol resistance in Mycobacterium tuberculosis: critical role of embB mutations." *Antimicrobial agents and chemotherapy*, 41, no. 8, (1997), 1677-1681.

Stover, C. Kendall, Paul Warrener, Donald R. VanDevanter, David R. Sherman, Taraq M. Arain, Michael H. Langhorne, Scott W. Anderson; et al. "A small-molecule nitroimidazopyran drug candidate for the treatment of tuberculosis." *Nature*, 405, no. 6789, (2000), 962.

Sun, Lin, Liqun Zhang, Ting Wang, Weiwei Jiao, Qinjing Li, Qingqin Yin, Jieqiong Li; et al. "Mutations of Mycobacterium tuberculosis induced by anti-tuberculosis treatment result in metabolism changes and elevation of ethambutol resistance." *Infection, Genetics and Evolution*, 72, (2019), 151-158.

Tahlan, Kapil, Regina Wilson, David B. Kastrinsky, Kriti Arora, Vinod Nair, Elizabeth Fischer, S. Whitney Barnes; et al. "SQ109 targets MmpL3, a membrane transporter of trehalose monomycolate involved in mycolic acid donation to the cell wall core of Mycobacterium tuberculosis." *Antimicrobial agents and chemotherapy*, 56, no. 4, (2012), 1797-1809.

Takayama, K; Kilburn, JO. Inhibition of synthesis of arabinogalactan by ethambutol in *Mycobacterium smegmatis. Antimicrob. Agents Chemother.*, 1989, 33, 1493–1499.

Telenti, Amalio, Paul Imboden, Francine Marchesi, Matter, L; Schopfer, K; Bodmer, T; Lowrie, D; Colston, MJ; Cole, S. "Detection of rifampicin-resistance mutations in Mycobacterium tuberculosis." *The Lancet*, 341, no. 8846, (1993), 647-651.

Van der Paardt, Anne-Fleur, Bob Wilffert, Onno W. Akkerman, Wiel CM de Lange, Dick van Soolingen, Bhanu Sinha, Tjip S. van der Werf, Jos GW Kosterink, Jan-Willem C. Alffenaar. "Evaluation of macrolides for possible use against multidrug-resistant Mycobacterium tuberculosis." *European Respiratory Journal*, 46, no. 2, (2015), 444-455.

Vasava, Mahesh S; Sneha G. Nair, Sanjay K. Rathwa, Dhaval B. Patel, Hitesh D. Patel. "Development of new drug-regimens against multidrug-resistant tuberculosis." *Indian Journal of Tuberculosis*, 66, no. 1, (2019), 12-19.

Vilchèze, Catherine, William R. Jacobs Jr. "The isoniazid paradigm of killing, resistance, and persistence in *Mycobacterium tuberculosis*." *Journal of molecular biology*, (2019).

World Health Organization. *Global tuberculosis report 2019*. World Health Organization, 2019.

World Health Organization (WHO). "Infant and young child feeding." *Fact sheet. Saatavilla:* https://www. who. int/en/newsroom/ fact-sheets/detail/infant-and-young-child-feeding. *Luettu*, 26, (2018), 2019.

Xu, Chen, Barry N. Kreiswirth, Srinand Sreevatsan, James M. Musser, Karl Drlica. "Fluoroquinolone resistance associated with specific gyrase mutations in clinical isolates of multidrug-resistant Mycobacterium tuberculosis." *Journal of Infectious Diseases*, 174, no. 5, (1996), 1127-1130.

Zhang, Shuo, Jiazhen Chen, Peng Cui, Wanliang Shi, Xiaohong Shi, Hongxia Niu, Denise Chan, Wing Wai Yew, Wenhong Zhang, Ying Zhang. "Mycobacterium tuberculosis mutations associated with reduced susceptibility to linezolid." *Antimicrobial agents and chemotherapy*, 60, no. 4, (2016), 2542-2544.

Zhang, Ying. "The magic bullets and tuberculosis drug targets." *Annu. Rev. Pharmacol. Toxicol.*, 45, (2005), 529-564

Zhao, Fei, Xu-De Wang, Luke N. Erber, Ming Luo, Ai-zhen Guo, Shan-shan Yang, Jing Gu; et al. "Binding pocket alterations in dihydrofolate synthase confer resistance to para-aminosalicylic acid in clinical isolates of *Mycobacterium tuberculosis*." *Antimicrobial agents and chemotherapy*, 58, no. 3, (2014), 1479-1487.

Zheng, Jun, Eric J. Rubin, Pablo Bifani, Vanessa Mathys, Vivian Lim, Melvin Au, Jichan Jang; et al. "para-Aminosalicylic acid is a prodrug targeting dihydrofolate reductase in *Mycobacterium tuberculosis*." *Journal of Biological Chemistry*, 288, no. 32, (2013), 23447-23456.

In: Antimicrobial Resistance: Opportunities and Challenges ISBN: 978-1-53617-943-9
Editors: P. Kumar Prabhakar and V. Mishra

Chapter 8

VIRAL DRUG RESISTANCE

Namita Giri[1] and Vijay Mishra[2,*]
[1]School of Pharmacy, Ferris State University, Big Rapids, Michigan, US
[2]School of Pharmaceutical Sciences,
Lovely Professional University, Phagwara Punjab

ABSTRACT

There are various diseases caused by viruses ranging from common cold to life threatening serious disorders like AIDS. These infections can range from an epidemic such as a seasonal flu to a global pandemic such as SARS, Measles, Swine flu, Ebola and most recently COVID 19. Although, they lack the major structural and functional components of their own, they have mastered the skill to find the host and use their machinery to survive, replicate and infect other healthy cells. Their borderline existence, ease of transmission, variable incubation period and dormancy period are quite some characteristics that make very hard to manage and treat viral infections. Drugs that have the potential to intervene any step in a viral infection process including viral particle attachment to releasing new virions are known as antiviral agents or antiviral drugs. Antiviral agents are used either to completely eradicate virus from the host or manage the viral infection by maintaining the viral load to minimum level. Constant mutations in viral genetic material and exposure to suboptimal dose of antiviral agents are primary reasons for the emergence of antiviral resistance, making it a challenging situation to deal with. These constant mutations can bring changes in the virus making it less susceptible to the antiviral drug. This reduced susceptibility to the antiviral drugs can be a prime indication of resistance development. The risk of resistance is usually higher in immunocompromised patients, hence antiviral resistance testing has now become a common and useful tool for clinical management of the complex and life-threatening viral infections. Both genotype and phenotype testing are performed using sequencing methods, which help to develop a specific treatment plan with reduced chance of resistance development. In the present chapter we have discussed the major types of viruses, their structural features, replication

* Corresponding Author's E-mail: vijaymishra2@gmail.com.

mechanism, common antiviral drugs that are used for treatment and the specific mutations that are responsible for development of resistance. Although a great progress was made in last 25 years for developing new antiviral drugs, there is a constant need of new agents due to the resistance emergence. It is critical that adherence in patients is monitored and specific treatment plan is developed based on antiviral resistance testing to reduce the emergence of resistant strains of the viruses.

Keywords: virus, antivirus, resistance, viral replication

INTRODUCTION

Viruses are the borderline organisms, which are not considered to be "alive" and cannot be "killed". This borderline existence and their essential need and dependence for biochemical parts, molecules, and energy and reproduction on the host cell sways the viruses away from being defined as living organisms and hence makes them hard to kill. Viruses have the smart system in place that hijacks the host cell machinery to replicate and infect other non-infected cells. During the process of replication and further infection to other healthy cells, the viruses can also kill the host cells, and that's when they become more dangerous (Cangelosi and Meschke, 2014). This dependence on host cell machinery puts the viruses under the parasite class, which cannot survive independently without the living cells. Viruses are responsible for many diseases. Figure 1 presents a schematic for the common infections caused by virus affecting various physiological systems.

Drugs that used for treating viral infections depend on the type of specific virus species and strain causing the particular infection, making an antiviral therapy different for each virus. The development of an ideal antiviral drug agaisnt a specific virus is the one that targets exclusively a viral-specific function with no or minimal damage to the host cell. An ideal or a perfect antiviral drug has not been developed so far as most of the viruses are an obligate parasite to the host cells necessitates the host cell mechanisms for replication (Pillay and Zambon, 1998). Reduced susceptibility to a drug in a laboratory culture system expressed as Inhibitory concentration (IC) IC_{50} or IC_{90} referred as resistance. IC_{50} or IC_{90} - inhibitory concentration is defined as the concentration required to inhibit the cell growth by 50% and 90% respectively. However, the susceptibility measure of IC_{50} and IC_{90} value is not used for the estimation of the antiviral drug concentration required to inhibit the viral replication in animal models or in patient population. Drug resistance in anti-viral agents is an utmost concern due to the continuous viral replication, mutations and prolonged drug exposure leading to the development and emergence of resistant strains (Strasfeld and Chou, 2010). The success of antiviral therapy is under a threat due to an increasing prevalence of resistant strains (Hughes and Anderson, 2015). Drug resistance may lead to multiple serious consequences ranging from toxicity related to the use of second-line antivirals agents to severe disease or maybe even death from progressive viral infection in absence of an alternative effective treatment (Strasfeld and Chou, 2010). Resistance in antivirals is mediated through the modification of the target sites, as almost all the antiviral agents target crucial parts of the virus or steps in the viral replication mechanism (Pillay 1998). Drug resistance in antiviral drugs is conferred through the mutations in targeted molecular targets of an antiviral therapy. This is also strong and convincing evidence that antiviral action of the drugs is based on specifically inhibiting the virus/viral mechanisms, not the host cells (Shafer and Chou, 2015). A wide variety of anti-viral drugs are available to treat viral infections.

VIRAL DRUG RESISTANCE IN TREATMENT AND MANAGEMENT OF HUMAN IMMUNODEFICIENCY VIRUS (HIV) INFECTION

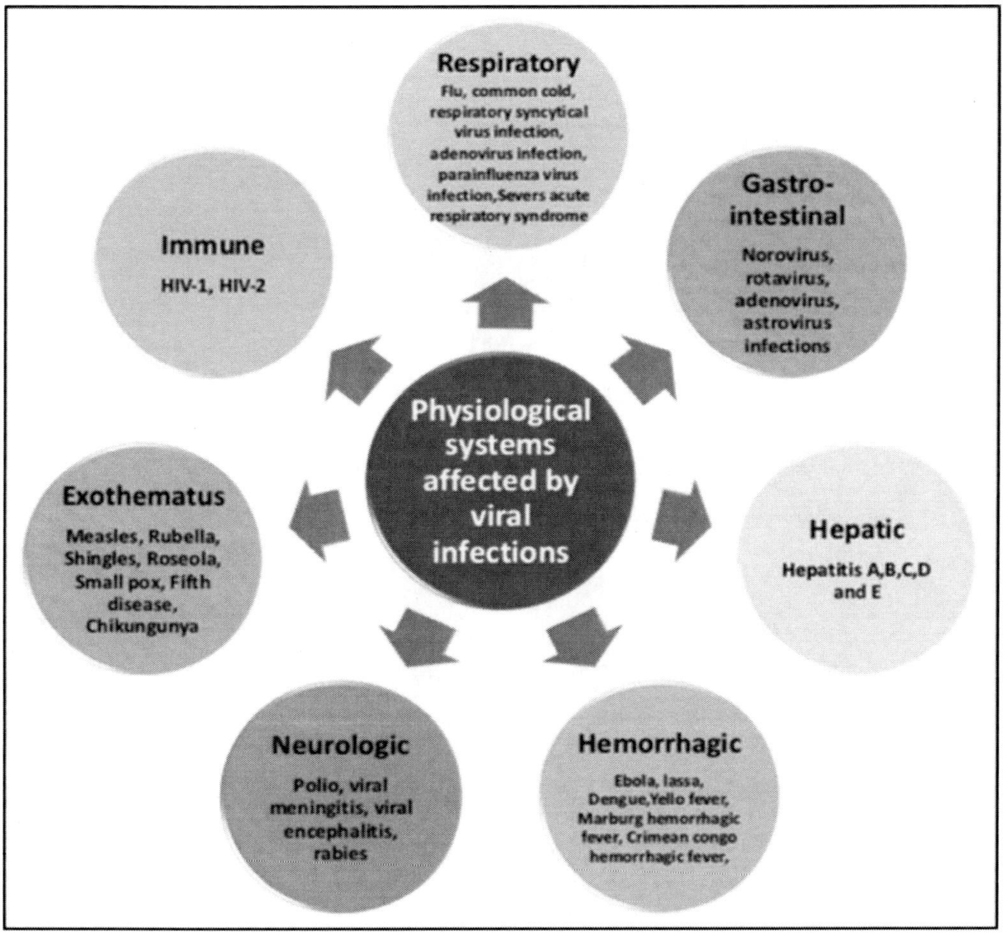

Figure 1. Schematic for viral diseases affecting different physiological systems and specific infections.

HIV virus is the most challenging virus to deal with due to its strong self-sufficient machinery and rapid rate of mutations. HIV uses ribonucleic acid (RNA) as its genetic material and hence falls under retrovirus class. Figure 2 shows the structural features of the HIV virus.

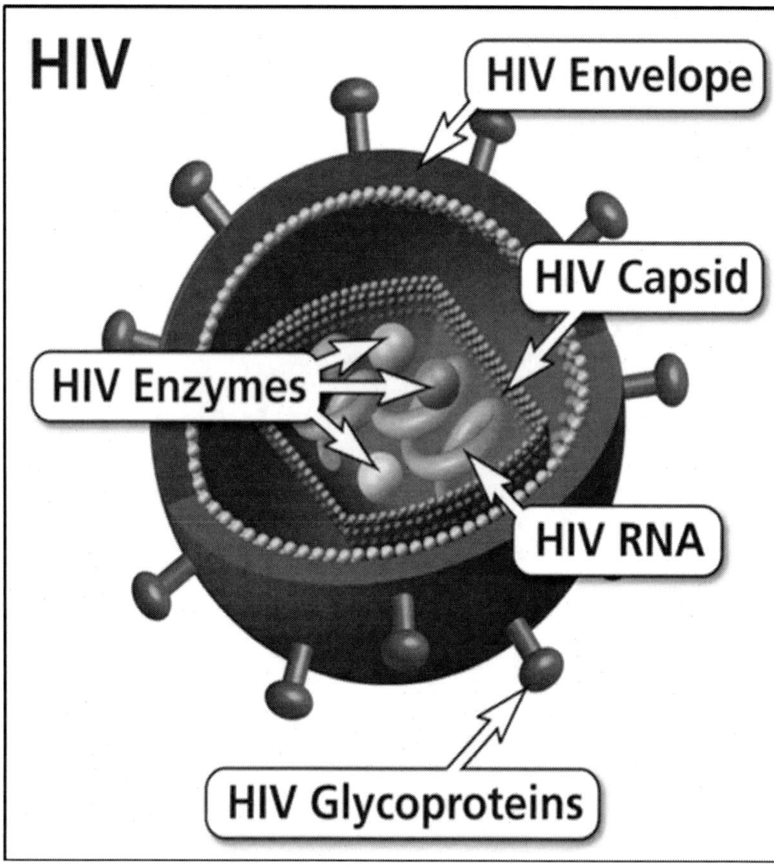

Figure 2. Human immunodeficiency virus (Human immunodeficiency virus (HIV) 2019).

Once the host is exposed to the virus, it begins to multiply and during the multiplication process, it undergoes mutation. These mutations developed while the host/patient undergoing antiretroviral therapy may lead to the development or emergence of drug-resistant strains of HIV. Along with the mutations, poor adherence to the regimen is also one of the key reasons for resistance development. Adherence can be defined as an extent to which the patient follows a prescribed regimen (Young et al., 1986; Daniels et al., 1994).

Antiviral agents that are used to treat retroviral infections focusses on individual viral replication steps. So far, a total of 22 antiretroviral agents has been approved by the United States food and drug administration (FDA). Table 1 presents FDA approved antiretroviral agents, Class, Mechanism of action, Mechanism of resistance, and Specific mutations responsible for the resistance.

viral infections

Antiviral agents, their mechanism of action (MOA) and resistance (MOR) for the treatment and management of HIV					
S. No	**Drug class**	**Generic name**	**Anti-viral mechanism of action**	**Mechanism of Resistance**	**Specific mutations responsible for resistance (Shafer and Chou, 2015)**
1.	**Nucleoside Reverse Transcriptase Inhibitors (NRTIs)**	Abacavir Emtricitabine Lamivudine Tenofovir disoproxil fumarate Zidovudine	NRTIs block reverse transcriptase, an enzyme HIV needs to make copies itself	Mutations that allow reverse transcriptase discrimination between NRTIs and naturally occurring nucleosides triphosphates	M184V/I confer resistance to lamivudine and emtricitabine. K65R confers resistance to didanosine, abacavir, tenofovir, stavudine, lamivudine, and emtricitabine and increases susceptibility to zidovudine; L74V/I confer resistance to didanosine and abacavir; Q151M confers intermediate resistance to tenofovir, lamivudine, and emtricitabine and high-level resistance to remaining NRTIs. Usually occurs in combination with two or more of the following mutations: A62V, V751, F77L, F116Y; T69 deletions occur in combination with K65R or Q151M.
				Mutations that promotes ATP dependent hydrolytic removal of chain-terminating nucleoside monophosphates (TAMs)	M41L, D67N, K70R, L210W, T215F/Y, and K219Q/E develop in isolates from persons receiving zidovudine and stavudine but confer clinically significant cross-resistance to each of the other NRTIs except possibly lamivudine and emtricitabine; T69 insertion most commonly T69SSS occur with TAMs and in this context causes intermediate lamivudine and emtricitabine resistance and high level of resistance to the remaining NRTIs; D67 deletions occur in combination with T69G and TAMs with similar effect to T69 insertions.
2.	**Non-Nucleoside Reverse transcriptase inhibitor (NNRTIs)**	Doravirine Efavirenz Etravirine Nevirapine Rilpivirine	NNRTIs bind to and later alter reverse transcriptase an enzyme HIV needs to make copies of itself	Mutations in the NNRTI binding pocket RT that interfere inhibitor binding	Efavirenz: L100I, K103N/S, K101P, V106A/M, Y181C/I/V, Y188C/H/L, G 1890 A/SE, F227C, M230L; Etravirine: L100I, K101P, V106A/M, Y181C/I/V, G190E, F227C, M230L; Rilpivirine: L100I, K101P, E138A/Q/G/K, Y181C/I/V, Y188L, G190E, F227C, M230L; Accessory: A98G, K101H, V1081, V179D/E/F, P225H, F227L, K238T;
3	Protease Inhibitors (PIs)	Atazanavir Darunavir Fosamprenavir Ritonavir Saquinavir Tipranavir	PIs block HIV protease, an enzyme that HIV needs to make copies	Protease mutations in or near the substrate cleft that interfere with inhibitor binding or compensate for the decreased replication associated with substrate cleft mutations	Substrate cleft mutations: L231, D30N, V321, 147V/A, G48V/M, 150V/L, V82A/F/S/T/M/C, 184 V/A/C; Flap mutations: M46I/L, F53L, I54V/M/T/L/A/S; Mutations at other conserved sites: L10F/R, V11I, K20T, L24I/F, L33F, K43T, Q58E, A71I/L, G73S/C/T/A, T74P, L76V, N88D/S, L89V, L90M; Common polymeric accessory mutations: L10I/V, K20R/I/M, M361, L63P, A71V/T
4	Fusion Inhibitors	Enfuvirtide	Fusion inhibitors block HIV from entering the CD4 cells of the immune system	Mutations in the first heated repeat region (HR1) of the gp41 transmembrane protein	Mutations in a highly conserved region between gp41 residues. G36D/E/V/S, 137V, V38E/A/M/G, Q48H, N42T, N43D/K/S, L44M, L45M. Alone G36D/E, V 38E/A, Q40H, and N43D reduces enfuvirtide susceptibility more than 10-fold. Two mutations are usually required to reduce susceptibility by more than 100-fold.
5	CCR5 Antagonists	Maraviroc	CCR5 antagonists block CCR5 receptor on the surface of certain immune cells that HIV needs to enter the cells	Expansion of preexisting CXCR4 tropic variants not detected at the start of therapy *In vitro* and rarely *in vivo* resistance is caused by gp120 env mutations that facilitate gp120 binding to an inhibitor bound CCR5 molecule	CXCR4-tropic variants: Positively charged residues at positions 11 and 25 of the V3 loop of gp120 and several other combinations of mutations primarily within the V 3 loop are associated with CXCR4 tropism. No consistent pattern of gp120 mutations is associated with virus binding to an inhibitor bound CCR5 receptor.
6	Integrase Inhibitors	Dolutegravir Raltegravir Elvitegravir	Integrase inhibitors block HIV integrase, an enzyme HIV needs to make copies of itself	Mutations in residues surrounding the enzymes active site	Raltegravir: 148H/R/K±G140S/A/C ±E138A/K; N155H±E92Q; Y143C/H/R Elvitergravir: E92Q±N155H; Q148H/R/K, ±G140S/A/C ±E138A/K, T66A/I/K; S147G Dolutegrevir: Q148H/R/K, +G140S/A/C± E138A/K Common accessory mutations: H51Y, L74M, T97A, V151Y, G163R/K, R263K; Rare potent mutations: F121Y, P145S, Q146P Mutations selected by *in vitro* models by dolutegravir G118R, R263K

Table 1. (Continued)

Antiviral agents, their mechanism of action (MOA) and resistance (MOR) for the treatment and management of HIV					
S. No	**Drug class**	**Generic name**	**Anti-viral mechanism of action**	**Mechanism of Resistance**	**Specific mutations responsible for resistance (Shafer and Chou, 2015)**
7	Post Attachment Inhibitors	Ibalizumab-uiyk	Post attachment inhibitors block CD4 receptors on the surface of certain immune cells that HIV needs to enter the cells	Information not available	Information not available
8.	Pharmacokinetic Enhancers	Cobicistat	Increase the effectiveness of HIV medicine included in the HIV regimen	Information not available	Information not available
Antiviral agents, their mechanism of action (MOA) and resistance (MOR) for the treatment and management of Herpes Virus (HSV) infection					
1		Acyclovir Valacyclovir (Prodrug of acyclovir) Penciclovir Famiciclovir (Prodrug of Penciclovir)	Converts to triphosphate form which competitively inhibits viral DNA polymerase	Thymidine Kinase and/or DNA polymerase mutations	TK frameshift (commonly at strings of multiple G or C bases) or substitution mutations causing a TK deficient phenotype and cross-resistance among the four drugs. Less commonly TK mutations conferring altered substrate specificity or DNA polymerase mutations clustered in conserved regions (e.g., II and III)
2		Cidofovir	converts to an active metabolite in diphosphate form which selectively inhibits viral DNA polymerase	DNA polymerase Mutations	Exonuclease and V regions with cross-resistance to ganciclovir. Low-grade cross-resistance from some region III foscarnet-associated mutations. Mutations in exonuclease and conserved regions II III and V.
3		Ganciclovir	Converts to triphosphate form which preferentially inhibits viral DNA polymerase	UL97 kinase mutations and/or DNA polymerase Mutations	Usually UL97 mutations at residues 460, 520, and 590-607 with M460V/I, H520Q, C592G, A594V, L595S, and C603W most common. DNA polymerase mutations are less common mainly at exonuclease and V region e.g., at residues 408, 412, 22, 545, 987) and usually add to preexisting UL97 mutations.
4		Foscarnate	Selective inhibition at the pyrophosphate binding site on virus-specific DNA Polymerases at concentrations that do not affect cellular DNA Polymerases	DNA polymerase Mutations	Mutations in polymerase catalytic domains involved in pyrophosphate mutations binding and nucleotide recognition (clustered in and between regions II, VI, and III, e.g., at residues 700, 715, 756, 781, 802, 809) some may also decrease ganciclovir ± cidofovir susceptibility. Mutations clustering in regions II-VI and II (probable phosphate-binding region)
Antiviral agents, their mechanism of action (MOA) and resistance (MOR) for the treatment and management of Hepatitis-B virus (HBV) infection					
1	Interferon	Interferon α,β		Unknown	Unknown
2	Nucleoside/nucleotide analogs (NRTIs)	lamivudine, Telbivudine, Emtricitabine		Mutations that allow RT to discriminate between NRTIs and naturally occurring nucleoside triphosphates. Whether some of these mutations also cause primer unblocking is not known.	M204V/I ± Ll8OM, Vl73L, L8OI: lamivudine, emtricitabine, and telbivudine; partial cross resistance to entecavir M204VI + Ll8OM ± 1169T, T184G, S2021 M250V: entecavir A181V/T: adefovir, lamivudine; partial crossresistance to tenofovir N236T: adefovir, partial cross-resistance to tenofovir
3	Deoxyguanosine analog	Entecavir3			
4	Acyclic nucleotides	Adefovir Tenofovir			

Antiviral agents, their mechanism of action (MOA) and resistance (MOR) for the treatment and management of Hepatitis-C virus (HCV) infection					
S. No	**Drug class**	**Generic name**	**Anti-viral mechanism of action**	**Mechanism of Resistance**	**Specific mutations responsible for resistance (Shafer and Chou, 2015)**
1	Interferon			Unknown	Genotype 1 isolates respond poorly than genotype 2 or 3 but the molecular basis for this is unknown.
2		Ribavirine	Guanosine analog that stops viral RNA synthesis and viral mRNA capping	Interferes with dNTP metabolism, inhibits HCV RNA polymerase and/ or increases viral mutagenesis	Uncertain
3	Protease Inhibitors (PIs)	Telpavir Boceprevir Simeprevir Faldaprevir Vaniprevir Asunaprevir Danoprenavir	Blocks the enzyme protease and thereby prevent maturing of the virus thereby cannot infect other healthy cells	Mutations in the NS3 protease substrate-binding cleft particularly the P2 and P3 pockets	Telaprevir: V36A/M/L/C, T54A/S, R155K/T, A156S/V/T Boceprevir: V36A/M/L/C, T54A/S, R155K/T, A156S/V/T, D170A Simeprevir: Q80K, R155K/T, A156S/V/T, D168A/V
4	Nucleoside Inhibitors (NIs)	Sofosbuvir Mericitabine	Inhibits Hepatitis C NS5B protein Selective cytidine nucleoside analogue and non-cytotoxic Hepatitis C polymerase inhibitor	Mutations near the active site that inhibit nucleoside analog incorporation	Sofosbuvir: S282T+compensatory Ns5B mutations are selected in-vitro and associated with approx. 10-fold reduction in susceptibility. S282T was detected transiently following sofosbuvir relapse in one genotype 2 viruses. L159F is a nonpolymorphic mutation detected in six sofosbuvir relapse subjects with genotype 3a viruses, it is associated with a 2-fold reduction in sofosbuvir susceptibility in genome 1a and 1b replicons. Mericitabine: S282T + compensatory NSSB mutations are selected *in vitro* and associated with -10-fold reduced susceptibility. L159F+ L320F, which are associated with reduced Mericitabine and sofosbuvir susceptibility and several inconsistent patterns of NS5B mutations have been observed in patients receiving Mericitabine.
5	NS5A Inhibitors	Daclatasvir Ledipasvir	Stops HCV viral RNA replication and protein translation by inhibiting HCV protein NS5A	Unknown	Daclatasvir and ledipasvir M28T, Q30E/H/R, L31M/F/V, P32L, Y93C/H/N. The phenotypic effect of these mutations is greater than the genotype 1a than the 1b virus. Q30 and Y93H mutations each occur in about 10% of DAA naive patients with genotype 3 virus.
	Non-nucleoside inhibitors (NNIs)			NS5B mutations that prevent binding to the four NNI binding domains	Thumb 1 domain: P495S/A/L, P496S/A, V499A Thumb 2 domain: L419V/M, M423T, I482L/V/T, V494A/I Palm 1 domain: H95RS, C316Y, M414T, C451R, G554D, G558R Pam 2 domain: L314, C316N/Y, S365 T, M414T, Y144H/C
Antiviral agents, their mechanism of action (MOA) and resistance (MOR) for the treatment and management of Influenza infection					
1	M2 Channel blockers	Amantadine Rimantadine		M2 trans-membrane mutation that blocks the hydrogen ion transfer	Mutations at residues 27, 30, 31, and 34, particularly S31N Influenza Band most circulating lineages of A(H3N2), A(H1Nl) pdm09, and the pathogenic avian influenza strain A(H5Nl) and A(H7N9) are resistant as a result of S31N.
2	Neuraminidase Inhibitors	Oseltamivir Zunamivir		NAI mutations in the receptor-binding site associated with decreased inhibitor binding	A(HlNl) and A(H1Nl) pdm09: oseltamivir: H274Y > Yl55H, I222M/R/ K, S246N/G, S294S; zanamivir: Ql36K, Yl55H, 1222R An (H3N2): oseltamivir: Ell9V/l, R292K, N294S; zanamivir: Ell91, R292K A (H5Nl): oseltamivir: (H274Y A (H7N9): oseltamivir: R292K; zanamivir: R292K B: oseltamivir: Rl52K, R371K > Dl98N/E, I222T; zanamivir: Rl52K > Dl98N/E, 1222T, R371K

The genetic variability in HIV occurs as a result of (i) Reverse transcriptase (RT) enzyme errors (ii) High replication rate of the virus (iii) Recombination of different sequences from two separate viruses in the same cell, and (iv) Proviral variant accumulation during the course of the viral infection. Drug-resistant strains selection is based on (i) viral replication extent under sub-optimal suppression therapy, (ii) convenience of acquisition of a specific mutation, and (iii) effect of drug resistance mutation on virus replication and drug susceptibility (Coffin, 1995). Highly active antiretroviral therapy (HAART) that utilizes a combination of three or more drugs of different drug class has played a vital role in the clinical management of HIV positive patients through re-establishing the immune function and consequently decreasing the morbidity and mortality rates (Assemie et al., 2019). HIV-1 suppression by HAART can persist indefinitely once a complete suppression is achieved, provided the therapy is not interrupted. The challenges arise as a result of antiretroviral therapy (ART) unable to inhibit proviral deoxyribonucleic acid (DNA), as complete viral eradication cannot be achieved. Resistance to the antiviral drugs can be either acquired or transmitted. When the virus replication occurs in the presence of an incomplete suppressive therapy, it's called an acquired resistance. Whereas, a transmitted resistance can be attributed to the resistance resulted from interrupted therapy or poor adherence (Coffin, 1995). A reduced virological response should be expected and must be addressed in patients with acquired or transmitted resistance before starting a new antiviral regimen.

There are five Nucleoside/Nucleotide reverse transcriptase inhibitors (NRTIs) that are routinely used clinically namely, Abacavir, Emtricitabine, Lamivudine, Tenofovir, and Zidovudine. HAART recommends the use of NRTIs as two-drug combinations along with either Non-nucleoside reverse transcriptase inhibitors (NNRTIs), Protease Inhibitors (PIs), Integrase inhibitors (INIs) or Chemokine receptors (CCR5) inhibitors. At least one cytosine analogs, either Lamivudine or Emtricitabine must be included in first-line and salvage therapy (Panel on Antiretroviral guidelines, 2013). NRTIs needs an intracellular phosphorylation for its convertion to its active form, which makes it difficult to assess its *in-vitro* activity. This is due to (i) different rates of phosphorylation in activated lymphocytes that are used for susceptibility testing and (ii) different types of infected cells *in-vivo*. This could also be an underlying reason for the difference in resistance levels in NRTIs (Meyer et al., 2000; Whitcomb et al., 2003).

There are two different biochemical mechanisms that attributes to the resistance among NRTIs, the first mechanism is mediated by mutations reducing reverse transcriptase (RT) affinity for an NRTI preventing its addition to the growing viral DNA chain. However, the second mechanism is mediated by mutations targetting primer unblocking, which are selected by thymidine analog inhibitors are also called as thymidine analog mutations or TAMs. A greater cross-resistance is observed in Didanosine, Abacavir and Tenofovir among a certain subset of TAMs (Lanier et al., 2004; Miller et al., 2004; Molina et al., 2005).

A single mutation of M184V can present more than 200-fold reduction in virals susceptibility towards Lamivudine and Emtricitabine, as they have a low genetic barrier to the resistance. It is also found to be the most common mutation in the patient population that develops a virologic failure to the first-line regimen. Antiviral activity of Lamivudine and Emtricitabine still manages to show activity as these mutations decrease the HIV-1 fitness, as well as increases its susceptibility to Zidovudine (Parikh et al., 2005; Melikian, 2012). Multidrug resistance (MDR) is also observed by specific mutations such as T69SSS and Q151M. T69SSS mutation inserts double amino acid at HIV-1 RT at position 69 whereas Q151M causes an immediate resistance to Tenofovir, Lamivudine and Emtricitabine and a higher resistance to the remaining drugs from the class (Shafer et al., 1994; Shirasaka et al., 1995).

NNRTIs binds to the hydrophobic pocket and inhibit HIV-1-RT (Singh et al., 2010). The pocket is not very well conserved than the active site and hence, HIV-2 and HIV-1 group O are resistant to the most of the NNRTIs (Yang et al. 1996). A greater inter-isolate variability is observed among HIV-1 group M viruses susceptibility towards NNRTIs than NRTIs (Parkin et al. 2004).

Monotherapy with NNRTIs generates a rapid resistance and indicates that this resistance is caused by the selection of pre-existing low abundance populations of mutant viruses (Shafer and Chou, 2015). The low genetic barrier to resistance is observed in Nevirapine and Efavirenz, as just a single mutation in binding pockets may cause a high-level resistance. Whereas two or more mutations are required to cause a high-level resistance in Etravirine, which has a high-level genetic barrier against resistance (Vingerhoets et al., 2004). This high genetic barrier resistance is attributed to its flexibility to re-arrangement that allows the adoption of multiple binding modes within NNRTIs binding pockets. Drug with similar structure, Rilpavirine has a lower genetic barrier as compared to etravirine due to its low dose of administration (Das et al., 2004).

Protease Inhibitors (PIs) are the class of drugs that are approved by USFDA. Ritonavir, a P4503-A pathway inhibitor is the one used most commonly and administered at subtherapeutic doses only as a booster drug, which helps to boosts the intracellular concentration of other PIs. Nelfinavir and Indinavir have a very poor pharmacokinetic profile and toxicity issues hence, the monotherapy is rarely used nowadays. Currently, Ritonavir boosted Atazanavir and Darunavir is the most commonly used regimen. This also exhibits a higher genetic barrier to resistance; as multiple mutations are required before an altered or sub-optimal antiviral activity is observed (King et al., 2007; De Meyer et al., 2008). Monotherapy with Lopinavir and Darunavir has shown activity at reducing the HIV-1-RNA level below detection limits within one year of therapy and hence considered as the most simple but highly effective regimen (Bierman et al., 2009).

The most common mechanism for antiviral drug regimen failure is the expansion of pre-existing C-X-C chemokine receptor 4 (CXCR4) tropic viruses that are already resistant to C-C motif chemokine receptor (CCR5) inhibitors (Westby et al., 2006). Whereas in *in-*

vitro studies, a different mechanism of resistance was observed for CCR5 inhibitors- it's the gp120 mutations enabling HIV-1 binding CCR5/CCR5 inhibitor complex causing a plateau at maximum percent inhibition at maximal CCR5 inhibitor concentration (Westby et al., 2007). However, this pattern is rarely observed in patients.

Enfuvirtide is the only drug approved under fusion inhibitor class (FIs), it's synthetic peptide with a high activity, which inhibits the fusion through a two-step process, namely binding to heptad repeat region (HR1) and subsequently preventing its bundling with HR2. A single or double mutation at positions 36 and 45 of gp41 HR1 is observed in Enfuvirtide resistant isolates. Mutation position defines the extent of alteration in susceptibility i.e., single mutation causes 10-fold decrease whereas, a double mutation causes a 100-fold decrease in susceptibility of Enfuvirtide. Although Enfuvirtide is a potent candidate, its genetic barrier to resistance is low and a rapid virological rebound may emerge if Enfuvirtide is not administered with enough other active inhibitors (Lu et al., 2006).

Integrase strand transfer inhibitors (INIs) binds to catalytic core domains (CCD) active site and chelates the divalent metal ions required for an efficient enzymatic function. Raltegravir, Elvitegravir and Dolutegravir are the approved drugs under integrase inhibitors. The mutations that significantly reduce INI susceptibility are the ones with highly conserved residues in the INI binding site and residues 140 to 149 which forms a flexible loop required for catalysis following DNA-IN binding (Chen et al., 2008). Raltegravir resistance is attributed to an overlapping genetic pathway, (i) Q148H/R/K ± G140S/A/C ± E138A/K, (ii) N155H ± E92Q and (iii) Y143C/R ± T97A (77-79). Whereas two mutational pathways by T66A/I/K and S147G are responsible for Eltegravir resistance. A high genetic barrier to the resistance is observed for Dolutegrevir (Hare et al., 2010; Johnson et al., 2006).

Viral Drug Resistance in the Treatment and Management of Herpes Virus Infection

Infections caused by herpes simplex virus (HSV), Varicella zoster virus (VZV) and Cytomegalovirus are the most common among the eight species of the herpes virus class.

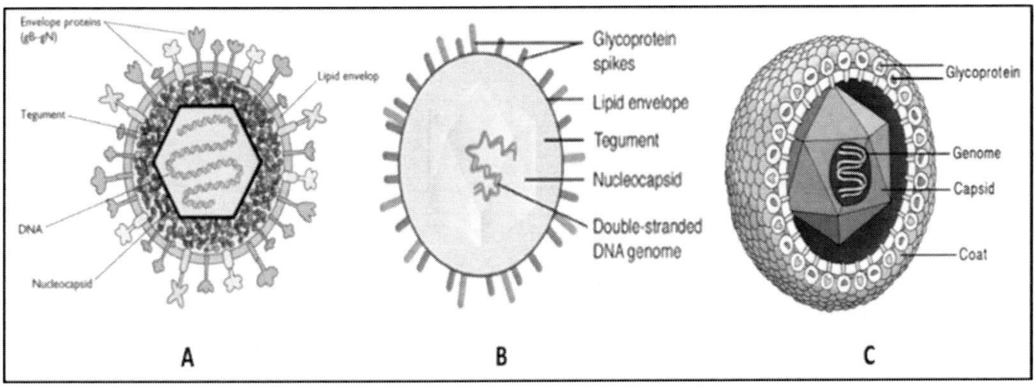

Figure 3. Structure of (A) Herpes Simplex virus, (B) Varicella-Zoster Virus and (C) Cytomegalovirus. Adopted from https://www.ck12.org/book/CK-12-Biology-Advanced-Concepts/section/11.22/, https://www.onlinebiologynotes.com/herpes-simplex-virushsv-structure-genome-mode-transmission-pathogenesis-infection-laboratory-diagnosis-treatment/, https://microbenotes.com/varicella-zoster-virus/.

Antiviral drugs for Herpes infection targets viral DNA polymerase. Initial therapy for Herpes infection management consists of nucleoside\ analogs that are selectively phosphorylated by viral enzymes. Prodrugs were used to improve the oral bioavailability of Acyclovir (ACV), Penciclovir (PCV) and Ganciclovir (GCV) in the form of Valacyclovir, Famciclovir and Valganciclovir, respectively. A pyrophosphate analogue, Foscarnate (Fos) and a cytosine analogue, Cidofovir can also acts as a supplement and are usually used in second- and third-line therapy in response to the first-line agent's failure to treat the infection (Shafer and Chou, 2015). HSV thymidine kinases (TK) phosphorylates Acyclovir and Penciclovir generates a monophosphate derivative, cellular enzymes then convert this monophosphate to a triphosphate derivative which finally terminates the DNA replication and forms a dead-end complex that eventually inhibits the viral DNA polymerase (Reardon and Spector, 1989). The viral thymidine kinase is not essential for viral replication, its mutation may decrease the extent of phosphorylation of nucleoside analogs, which is believed to be the underlying mechanism of Acyclovir and Penciclovir resistance (Earnshaw et al., 1992; Piret and Boivin, 2011). Cross-resistance is not observed in drugs with a mechanism of action independent of viral meditated phosphorylation such as Foscarnate and Cidofovir (CDV). Drug-resistant mutants are referred to as Thymidine Kinase negatives- resulted from frameshift or point mutations that deletes functional domain responsible for phosphorylation of natural nucleosides or antiviral agents (Piret and Boivin, 2011; Andrei and Snoeck, 2013; Burrel et al., 2013). Thymidine kinase mutations specific to PCV and ACV may differ from each other and hence there is a possibility of cross-resistance exixtance.

Antiviral resistance in HSV due to the DNA polymerase (pol) mutations is rarely observed. DNA polymerase is critical for gene replication, the mutations involved are usually a single amino acid change in the functional domain and are more selective. However, the mutations conferring to Acyclovir or Foscarnate resistance can be of several types including conserved pol regions or clustering of mutation in a specific region II, II and IV. It is reported that there are some mutations that confer resistance to both acyclovir as well as Foscarnate (Andrei and Snoeck, 2013).

Oral prodrugs such as Valacyclovir and Famciclovir are used as a first-line treatment for Varicella Zoster virus (VZV) infections. Patient response is based on the early initiation of potent therapy than the antiviral susceptibility. Risk of resistance is higher in immunocompromised individuals due to multiple, prolonged and recurrent infections (Nikkels et al., 1999; Van der Beek et. al., 2013). VZV and HSV share a similar mechanism of resistance, TK gene mutations (stop, frameshift and substitution) resulting in reduced TK availability or altered specificity to the substrate (Andrei et al., 2012; Morfin et al., 1999). Genotypic resistance assays are available for the detection of such mutations (Sauerbrei et al., 2011). FoS and CDV are the alternative drugs available for ACV resistant VZV (Hatchette et al., 2008; Bryan et al., 2008).

Thymidine kinase is not present in cytomegalovirus (CMV), however, the UL97 gene is presently responsible for another kinase, which is critical for the production of infectious virus in the *in-vitro* model and serves as potential drug targets for antiviral agents (Prichard, 2009). The standard choice of drug for CMV treatment is Ganciclovir, which is converted to its monophosphate form by UL97, and eventually to a triphosphate form, an active metabolite (Sullivan et al., 1992). Prolonged viral replication along with an extended antiviral drug exposure is thought to be a potential risk factor for the development of resistant strains. A majority of GCV resistant CMV strains have UL97 mutations that impairs GCV phosphorylation (Lurain and Chou, 2010). There are 7 most common mutations that are known to increase the resistance of GCV by a magnitude of 7-10-fold (Chou, 2010). Resistance to FoS and CDV is neither observed not expected through UL97 resistance. Prolong exposure to GCV, the mutation in UL54 DNA polymerase along with preexisting UL97 mutations contributes to the increased level of the drugs (Chou et al., 2005; Chou et al., 2005; Smith et al., 1997). GCV resistant pol mutations are also expected to exhibit cross-resistance among other drugs. FoS resistance mutations presented a 3-5-fold increase in effective concentration (EC) EC_{50} values (Lorain and Chou, 2010).

Viral Drug Resistance in Treatment and Management of Hepatitis Virus Infection

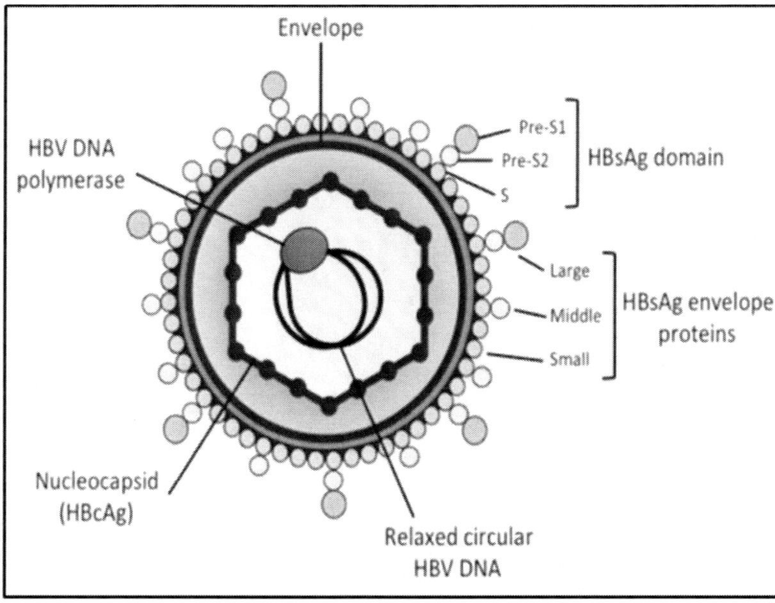

Figure 4. Structure of Hepatitis B virus (Adopted from http://www.antimicrobe.org).

Hepatitis B Virus (HBV) is approximately 42 nanometers in size, partially double-stranded DNA virus that belongs to the Hepadnaviridae family, which has an outer lipid envelope containing viral glycoproteins and aninner nucleocapsid. The glycoprotein in the lipid envelops of the virus, constitutes the hepatitis surface antigen (HBsAg) whereas, the nucleocapsid consists of hepatitis-B core antigen (HBcAg) encloses a copy of double-stranded circular HBV-DNA and the HBV DNA polymerase (Liang, 2009). Once the virus infects the host cell, the genome places itself inside and further converts to its covalently closed circular DNA (cccDNA) form which serves as the template for messenger Ribonucleic acid (mRNA) and genomic RNA transcription. Reverse transcription of this genomic RNA generates viral DNA, resulting in the viral core that can bud into the endoplasmic reticulum and may export from the cell or can return back to the nucleus and converts back agaun into cccDNA. Antiviral therapy cannot eliminate cccDNA, but cell turnover can, hence eradication of the infection is not observed frequently (Werle-Lapostolle et al., 2004; Wong et al., 2013). Replication for HBV is high, around 10^{11} virions per day whereas in the absence of an antiviral therapy the HBV DNA can be as high as 10^{8}-10^{10} copies/mL. HBV polymerase has reverse transcriptase activity and shares functional and structural similarities to the HIV virus, the error rates are similar to that of other retroviruses (Nowak et al., 1996; Hollinger, 2007). Five Nucleosides analogs (Lamivudine, Telbivudine, Emtricitabine and two forms of Interferon (alpha and PEGylated alpha interferon) are available as the treatment options for the management of

HBV infections. Emtricitabine shares structural similarities as well as antiviral activity with lamivudine. They both are used to treat HBV infections as it is co-formulated with Tenofovir for HIV-1. Tenofovir and Entecavir are recommended as the first-choice of drugs in an initial HBV therapy as they both are very potent and drug resistance is rarely observed in either drug when used as monotherapies (European Association, 2012; Tenney et al., 2009; Kitrinos et al., 2014). High resistance is observed in Lamivudine monotherapy, hence is not recommended anymore (Lai et al., 2003).

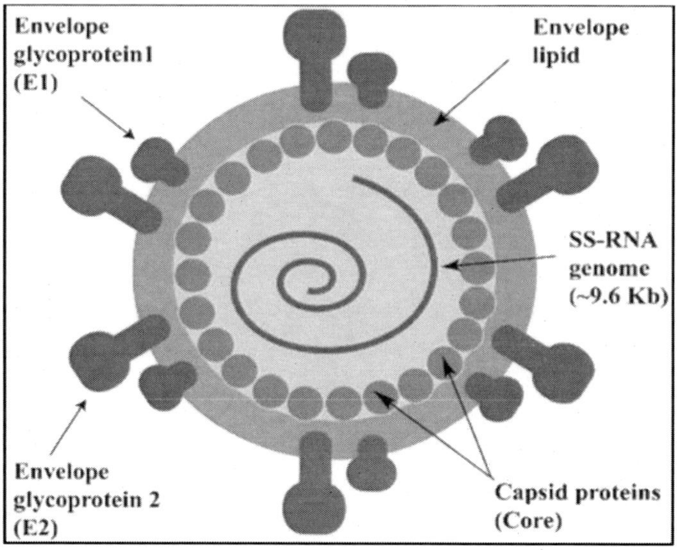

Figure 5. Hepatitis C structure (Sharma, 2010).

HCV is the prototype member of the *Hepacivirus* genus which is classified into further seven genotypes differing in their nucleotide sequence. Its structure has an inner core encapsulating a single-strand viral RNA which is approximately 9.6 kb in size. HCV RNA replication takes place in the cytoplasm of hepatocytes as the HCV genome cannot enter the nucleus (Sharma, 2010). The viral-RNA genome has a single open reading frame flanked by 5' and 3' non translated RNA segments. The cis-acting replication elements (CREs) are located in both the 5' and 3' non translated RNA segments (NTRs) and in the NS5B coding sequence. The 5′ UTR (+) is ~341 nucleotides in length and contains an internal ribosomal entry site (IRES). The HCV (IRES) is folded into four stem-loop motifs referred to as I, II, III and IV. The IRES is required for the cap-independent translation of the viral RNA, which is carried out by the host cell ribosome. The domain IIId of the IRES constitutes the key anchoring site for the 40S subunit. The IRES domains III-IV have also been shown to be an activator of protein kinase R (PKR) nucleus (Sharma, 2010). FDA approved directly acting agents such as PEGylated Interferon, protease inhibitors (Telaprevir, Boceprevir) for treating the genotype-1 HCV. Seventy percent of HCV genotype-1 is cured by the combination of interferon, protease inhibitors and Ribavirin although, this combination was also associated with toxicity and a high incidence of resistance towards protease inhibitors. A new protease inhibitor Simeprevir is approved to be used in combination with PEGylated Interferon and Ribavirin (Food and Drug Administration, 2013a). Later, Sofosbuvir a nucleotide analog is also approved that could be used in combination with ribavirin alone or with Interferon plus ribavirin for the treatment of other genotypes of HCV (Reardon, 1989; Earnshaw et al., 1992; Piret and Bovin, 2011; Andrei and Snoeck, 2013; Food and Drug Administration, 2013b). A higher cure rate was observed with several interferon-free combinations as compared to the currently approved Telaprevir and Boceprevir based combinations (Gane et al., 2013; Zeuzem et al., 2013; Lawitz et al., 2014; Poordad et al., 2013). NIs have shown activity and genetic barriers against all HCV genotypes, whereas PIs, NS5A inhibitors and NNIs have different activity against different genotypes and subtypes along with with a lower genetic barrier to resistance (McCown et al., 2008). Development and discovery of new drugs acting against HCV, there is a dire need of acheivement of multiple milestones such as developing an *in-vitro* model for HCV replication and cell-based assays, determination of inhibitory concentration, identification drug-resistant strains, and assessment of susceptibility.

VIRAL DRUG RESISTANCE IN THE TREATMENT AND MANAGEMENT OF INFLUENZA VIRUS INFECTION

The Center for disease control (CDC) describes 4 types of Influenza viruses, type A, B, C and D. Among these, types A and B are the ones that cause seasonal epidemics every year. Influenza virus C causes a mild respiratory illness whereas, subtype D is not known to cause any illness in humans and affects only cattle. Based on the surface proteins known as Hemagglutinin (HA) and Neuraminidase (NA) present, the Influenza type A virus can thus be classified into 18 hemagglutinin and 11 neuraminidase subtypes. These subtypes can be further divided into strains. However, Influenza virus B is not divided into any subtypes but can be subclassified into lineages and strains.

The genetic material of Influenza virus, the RNA is located inside the protein shell lining the viral lipid membrane. Hemagglutinin, Neuraminidase, and a channel protein are all embedded in the membrane. There are two classes of antiviral agents available for treatment or management of influenza viral infection based on their specificity towards the specific surface proteins, namely M2 channel blockers and Neuraminidase inhibitors. Amantadine and Rimantadine belongs to M2 channel blocker whereas, Oseltamivir and Zanamivir belongs to Neuraminidase inhibitors. M2 channel blockers are currently inactive against Influenza Type A virus and intrinsically inactive against type B also (Burnham et al., 2013). However, Neuraminidase inhibitors show activity against most A and B virus types (Harter et al., 2010).

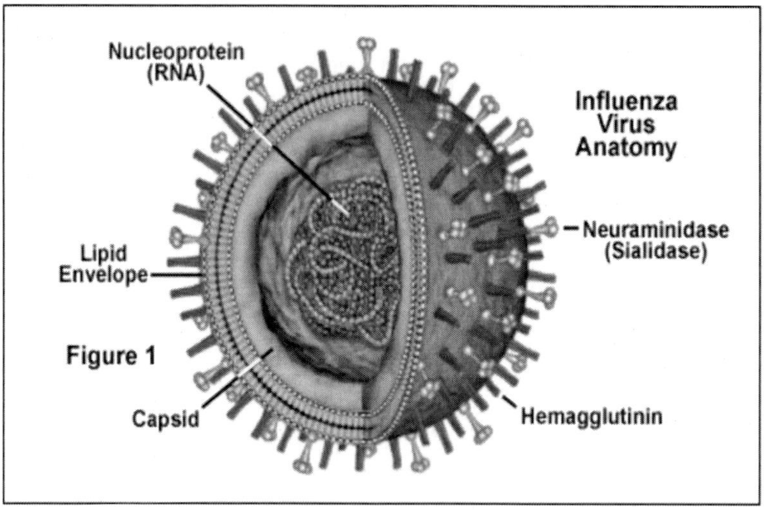

Figure 6. Structure of the Influenza virus (Adopted from https://micro.magnet.fsu.edu/cells/viruses/influenzavirus.html).

M2 channel is a tetrameric, proton selective, pH activated channel that plays a critical role in the uncoating of the virus (Cady et al., 2009). Amantadine and Rimantadine interrupt the hydrogen ion penetration through the M2 channels and prevent ribonucleotide-complex transport into the nucleus (Pielak et al., 2009). A single amino acid substitution in the M2 membrane domain may result in cross-resistance in both the drugs. Some mutations tend to inhibit M2 channel binding whereas, other mutations may allow the binding but may not affect other channel functions at all (Wang et al., 2011).

The initiation of viral infection requires the binding of hemagglutinin (HA) to sialic (N acetylneuraminic) acid with cellular receptors. Alpha-ketosidic bond that links the terminal sialic residue to adjacent oligosaccharide moietiy are cleaved by neuraminidase and further prevents the formation of HA and cell surface sialic acids that allows newly formed virion to be released from the infected cells (Smith et al., 2002). A balance between the two activities, HA binding to cellular receptors to initiate infection and destruction of cellular receptors by NA to release the new virus, is critical for an efficient infection (Wagner et al., 2002). Replication fitness of NAIs resistance mutations is modulated by HA (Okomo-Adhiaambo, 2013).

Sialic acid analogs Oseltamivir and Zanamivir are competitive inhibitors of HA and are active against influenzas' human strains as well as avian strains but have less activity against type B than typeA (Burnham et al., 2013; Von-Itzstein, 2007). Among Oseltamivir and Zanamivir, a less occurrence of resistance is observed for Zanamivir because of its close structural resemblance with the natural sialic acid (Nguyen et al., 2012). Whereas,

Oseltamivir is a carboxylic acid analog of sialic acid and has a bulky side chain that aids in a conformational change in NA to allow the binding (Collins et al., 2009).

It's the type and subtype of the virus that is responsible for mutations conferring resistance to NAI inhibitors. Due to the genetic lineage within the virus subtype may differ in susceptibility levels to an anti-influenza agent and also to the predisposition to the specific neuraminidase resistance mutations. Hence, the genotypic and phenotypic screenings for the identification of any change in intrinsic susceptibility of circulating DNA variants and their spectrum of mutations selected in patients on the NAIs regimen (Shafer and Chou, 2015) are preferred.

Susceptibility assays, Genotype assays, and enzymatic assays provide an excellent tool to keep a close watch on resistance emergence, characterization of novel variants and overall surveillance (Okomo-Adhiambo et al., 2013; Zambon, 2013).

Drug resistance in antiviral agents presents a burden on global health, as it causes high costs of treatment, deaths and patients morbidity that could be avoided by taking measures to avoid the development of resistance. Complete adherence to the treatment regimen, use of right combinations and an advance genotypic and phenotypic testing to address already resistant strains are some of the measures that could help us in addressing the growing issue of resistance in all drug classes. Besides this, there is an urgent need to develop new drugs to support effective therapy but also to develop a better understanding of the underlying mechanisms and forces that drives resistance development.

REFERENCES

[1] Andrei, G. & Snoek, R. (2013). Herpes simplex virus drug-resistance: new mutations and insights. *Curr Opin Infect Dis*, 26, 551-560.

[2] Andrei, G., Topalis, D., Fiten, P., McGuigan, C., Balzarini, J., Opdenakker, G. & Snoeck, R. (2012). *In vitro*-selected drug-resistant varicella-zoster virus mutants in the thymidine kinase and DNA polymerase genes yield novel phenotype-genotype associations and highlight differences between antiherpesvirus drugs. *J Virol*, *86*, 2641-2652.

[3] Assemie, M. A., Alene, M., Ketema, D. B. & Mulatu, S. (2019). Treatment failure and associated factors among first line patients on highly active antiretroviral therapy in Ethiopia: A systematic review and meta-analysis. *Glob Health Res Policy*, ***4***, 32.

[4] Blanco, J. L., Varghese, V., Rhee, S. Y., Gatell, J. M. & Shafer, R. W. (2011). HIV-I integrase inhibitor resistance and its clinical implications. *J Infect Dis*, *203*, 1204-1214.

[5] Bryan, C. J., Prichard, M. N., Daily, S., Jefferson, G., Hartline, C., Cassady, K. A., Hilliard, L. & Shimamura, M. (2008). Acyclovir-resistant chronic verrucous vaccine strain varicella in a patient with neuroblastoma. *Pediatr Infect Dis J*, *27*, 946-948.

[6] Burnham, A. J., Baranovich, T. & Govorkova, E. A. (2013). Neu-raminidase inhibitors for influenza B virus infection: efficacy and resistance. *Antiviral Res*, *100*, 520-534.

[7] Burrel, S., Aime, C., Hermet, L., Ait-Arkoub, Z., Agut, H. & Boutolleau, D. (2013). Surveillance of herpes simplex virus resistance to antivirals: A 4-year survey. *Antiviral Res*, *100*, 365-372.

[8] Cady, S. D., Luo, W., Hu, F. & Hong, M. (2009). Structure and function of the influenza A M2 proton channel. *Biochemistry*, *48*, 7356-7634.

[9] Cangelosi, G. A. & Meschke, J. S. (2014). Dead or Alive: molecular Assessment of Microbial Viability. *Appli Environ Microbiol.*, *80* (19), 5884-5891.

[10] Chen, X., Tsiang, M., Yu, F., Hung, M., Jones, G. S., Zeynalzade, gan A., Qi, X., Jin, H., Kim, C. U., Swaminathan, S. & Chen, J. M. (2008). Modeling, analysis, and validation of a novel HIV integrase structure provide insights into the binding modes of potent integrase inhibitors. *J Mol Biol*, *380*, 504-519.

[11] Chou, S., Marousek, G., Guentzel, D., Follansbee, S. E., Poscher, M. E., Lalezari, J. P., Miner, R. C. & Drew, W. L. (1997). Evolution of mutations conferring multidrug_resistance during prophylaxis and therapy for cycomegaloyirus disease.

[12] Chou, S., Van Wechel, L. C., Lichy, H. M. & Marousek, G. I. (2005). Phenotyping of cytomegalovirus drug resistance mutations by using recombinant viruses incorporating a reporter gene. *Antimicrnb Agents Chemother*, *49*, 2710-2715.

[13] Chou, S. (2010). Recombinant phenotyping of cytomegalovirus UL97 kinase sequence variants for ganciclovir resistance. *Antimicrob Agents Chemother*, *54*, 2371-2378.

[14] Coffin, J. M. (1995). HIV population dynamics *in vivo*, implications for genetic variation, pathogenesis and therapy. *Science*, *267*, 483-489.

[15] Collins, P. J., Haire, L. F., Lin, Y. P., Liu, J., Russell, R. J., Walker, P. A., Martin, S. R., Daniels, R. S., Gregory, V., Skehel, J. J., Gamblin, S. J. & Hay, A. J. (2009). Structural basis for oseltamivir resistance of influenza viruses. *Vaccine*, *27*, 6317-6323.

[16] Daniels, D., Rene, A., Daniels, V. (1994). Race: an explanation of patient compliance – fact or fiction? *J Natl Med Assoc, 86*, 20–25.

[17] Das, K., Clark, A. D., Jr. T., ewi, P. J., Heeres, J., De Jonge, M. R., Koymans, L. M., Vinkers, H. M., Daeyaert, F., Ludovici, D. W., Kukla, M. J., De Corte, B., Kavash, R. W., Ho, C. Y., Ye, H., Lichtenstien, M. A., Andries, K., Pauwels, R., De Bethune, M. p., Boyer, P. L., Clark, P., Hughs, S. H., Janssen, P. A. & Arnold, E. (2004). Roles of conformational and positional adaptability in structure-based design of TMC125-R165335 (etravirine) and related non-nucleoside reverse transcriptase inhibitors that are highly potent and effective against wild-type and drug-resistant HIV-1 variant. *J Med Chem*, *47*, 2550-2560.

[18] Earnshaw, D. L., Bacon, T. H., Darlison, S. J., Edmonds, K., Perkins, R. M. & Vere Hodge, R. A. (1992). Mode of antiviral action of penciclovir in MRC -5 cells infected with herpes simplex virus type 1 (HSV-1), HSV-2, and varicella-zoster virus, *Antimicrob Agents Chemother*, *36*, 2747-2757.

[19] European Association for the study of teh liver. (2012). EASL clinical practice guidelines: management of chronic hepatiis B virus infection. *J Hepatol*, *57*, 167-185.

[20] Flisiak, R., Jaroszewicz, J., Flisiak, I. & Lapinski, T. (2012). Update on alisporivir in t1eatment of viral hepatitis C. *Expert Opin Investig Drugs*, *21*, 375-382.

[21] Food and Drug Administration. (2013a). *Olysio (sime-previr), prescribing information.* http://www.accessdata.fda.gov/drugsatfdadocs/label/2013/ 205123s00 1lbl.pdf.

[22] Food and Drug Administration. (2013b). *Sovaldi (sofus-buvir): prescribing information.* https://www.gilead.com/-/media/Files/pdfs/medicines/liverdisease/ sovaldi/sovaldi_pi.pdf.

[23] Fransen, S., Gupta, S., Frantzell, A., Petropoulos, C. J. & Huang, V. W. (2012). Substitutions at amino acid positions 143, 148, and 155 of HIV-1 integrase define distinct genetic barriers co raltegravir resistance *in vivo*. *J Virol*, *86*, 7249-7255.

[24] Frentz, D., Boucher, C. A. & van de Vijver, D. A. (2012). Tem-poral changes in the epidemiology of transmission of drug-resistant HIV-1 across the world. *AIDS Rev*, *14*, 17-27.

[25] Gane, E. J., Stedman, C. A., Hyland, R. H., Ding, X., Svarovskaia, E., Symonds, W. T., Hindes, R. G. & Berrey, M. M. (2013). Nucleotide polymerase inhibitor sofosbuvir plus ribavirin for hepatitis C. *N Engl Med*, *368*, 34-44.

[26] Gonzalez, S. A. Hepatitis B Virus, Antimicrobe accessed on nov 15, 2019 adopted from http://www.antimicrobe.org/v22.asp .

[27] Hare, S., Gupta, S. S., Valkov, E., Engelman, A. & Cherepanov, P. (2010). Retroviral intasome assembly and inhibition of DNA strand transfer. *Nature*, *464*, 232-236.

[28] Harter, G., Zimmermann, O., Maier, L., Schubert, A., Mertens, T., Kern, P. & Wohrle, J. (2010). Intravenous zanamivir for patients with pneumonitis due to pandemic (HlNl) 2009 influenza virus. *Clin Infect Dis*, *50*, 1249-1251.

[29] Hatchette, T., Tipples, G. A., Peters, G., Alsuwaidi, A., Zhou, J. & Mailman, T. L. Foscarnet salvage therapy for acyclovir-resistant varicella zoster: report of a novel thymidine kinase mutation and review of the literature. *Pediatr Infect Dis J*, *27*, 75-77.

[30] Hollinger, F. B. (2007). Hepatitis B virus genetic diversity and its impact on diagnostic assays. *Viral Hepat*, *14*, (Suppl 1), 11-15.

[31] Hughes, D. & Anderson, D. I. (2015). Evolutionary consequences of drug resistance: shared principles across diverse targets and organisms. *Nat Rev Genet*, *16*, 459-471.

[32] *Human immunodeficiency virus (HIV).* (2019). Retrieved from https://aidsinfo.nih.gov/understanding-hiv-aids/glossary/325/human-immunodeficiency-virus.

[33] Hurt, C. B., Sebastian, J., Hicks, C. B. & Eron, J. J. (2014). Resistance to HIV integrase strand transfer inhibitors among clinical specimens in the United States, 2009-2012. *Clin Infect Dis*, *58*, 423-431.

[34] Janssen, H. L., Reesink, H. W., Lawitz, E. J., Zeuzem, S., Rodriguez-Torres, M., Patel, K., van der Meer, A. J., Patick, A. K., Chen, A., Zhou, Y., Persson, R., King, B. D., Kauppinen, S., Levin, A. A. & Hodges, M. R. (2013). Treatment of HCV infection by targeting microRNA. *N Engl Med*, *368*, 1685-1694.

[35] Johnson strand, A. A., transfer. Santos, W., Nature Pais, G. C., Marchan, C., Amin, R., Burke, T. R., Jr. Verdine, G. & Pommier, Y. (2006). Integration requires a specific interaction of the donor DNA terminal 5'-cytosine with glutamine 148 of the HIV-1 integrase flexible loop. *J Biol Chem*, *281*, 461-467.

[36] Kitrinos, K. M., Corsa, A., Liu, Y., Flaherty, Snow Lampart A., Marcellin, P., Borroto-Esoda, K. & Miller, M. D. (2014). No detectable resistance to tenofovir disoproxil fumarate after 6 years of therapy in patients with chronic hepatitis B, *Hepatology*, *59*, 434-442.

[37] Lai, C. L., Dienstag, J., Schiff, E., Leung, N. W., Atkins, M., Hunt, C., Brown, N., Woessner, M., Boehme, R. & Condreay, L. (2003). Prevalence and clinical correlates of YMDD variants during lamivudine therapy for patients with chronic hepatitis B. *Clin Infect Dic*, *36*, 687-696.

[38] Lanier, E. R., Ait-Khaled, M., Scott, J., Stone, C., Melby, T., Sturge, G., St Clair, M., Steel, H., Hetherington, S., Pearce, G., Spreen, W. & Lafon, S. (2004). Antiviral efficacy of abacavir in antiretroviral therapy-experienced adults harbouring HJV-1 with specific patterns of resistance to nucleoside reverse transcriptase inhibitors. *Amivir Ther*, *9*, 37-45.

[39] Lawitz, E., Poordad, F. F., Pang, P. S., Hyland, R. H., Ding, X., Mo, H., Symonds, W. T., McHutchison, J. G. & Membreno, F. E. (2014). Sofosbuvir and ledipasvir fixed-dose combination with and without ribavirin in treatment-naive and previously treated patients with genotype l hepatitis C virus infection (LONESTAR): an open-label, randomised, phase 2 trial. *Lancet*, *383*, 515-523.

[40] Liang, T. J. (2009). Hepatitis B: the virus and disease. *Hepatol*, *49*(5 Suppl), S13-21.

[41] Lu, J., Deeks, S. G., Hoh, R., Beatty, G., Kuritzkes, B. A., Martin, J. N. & Kuritzkes, D. R. (2006). Rapid emergence of enfuvirtide resistance in HIV-1-infected patients: results of a clonal analysis. *J Acquir Immune Defic Syndr*, *43*, 60-64.

[42] Lurain, N. S. & Chou, S. (2010). Antiviral drug resistance of human cytomegalovirus. *Clin Microbial Rev*, *23*, 689-712.

[43] McCown, M. F., Rajyaguru, S., Le Pogam, S., Ali, S., Jiang, R., Kang, H., Symons, J., Cammack, N. & Najera, I. (2008). The hepatitis C virus replicon presents a higher

barrier to resistance to nucleoside analogs than to nonnucleoside polymerase or protease inhibitors. *Antimicrob Agents Che-mother*, *52*, 1604-1612.

[44] Melikian, G. L., Rhee, S. Y., Taylor, J., Fessel, W. J., Kaufman, D., Towner, W., Troia-Cancio, P. V., Zolopa, A., Robbins, G. K., Kagan, R., lsraelski, D. & Shafer, R. W. (2012). *Standardized comparison of the relative impacts of HIV-1 reverse transcriptase (RT) mutations on nucleoside RT inhibitor susceptibility Antimicrob Agents Chemother*, *56*, 2305-2313

[45] Meyer, P. R., Matsuura, S. E., Schinazi, R. F., So, A. G. & Scott, W. A. (2000). Differential removal of thymidine nucleotide analogues from blocked DNA chains by human immunode-ficiency virus reverse transcriptase in the presence of physi-ological concentrations of 2'-deoxynucleoside triphosphates. *Antimicrob Agents Chemother*, *44*, 3465-3472.

[46] Miller, M. D., Margot, N., Lu, B., Zhong, L., Chen, S. S., Cheng, A. & Wulfsohn, M., (2004). Genotypic and phenotypic prehibitors of the magnitude of response to tenofovir disoprovil fumarate treatment in antiretroviral-experienced patient *J Infect Dis*, *191*, 840-847.

[47] Molina, J. M., Marcelin, A. G., Pavie, J., Heripret, L., De Boever, C. M., Troccaz, M., Leleu, G. & Calvez, V. (2005). Didanosine in HIV-I-infected patients experiencing failure of antiretroviral therapy: a randomized placebo-controlled trial *J Infect Dis*, *191*, 840-847.

[48] Morfin, F., Thouvenot, D., De Turenne-Tessier, M., Lina, B., Aymard, M. & Ooka, T. (1999). Phenotypic and genetic characterization of thymidine Kinase from clinical strains of varicella-zoster virus resistant to acyclovir. *Antimicrob Agents Chemother*, *43*, 2412-2416.

[49] Nguyen, H. T., Fry, A. M. & Gubareva, L. V. (2012). Neuraminidase inhibitor resistance in influenza viruses and laboratory testing methods. *Antivir Ther*, *17*, 159-173.

[50] Nikkels, A. F., Snoeck, R., Rentier, B. & Pierard, G. E. (1999). Chronic verrucous varicella zoster virus skin lesions: clinical, historical, molecular and therapeutic aspects. *Clin Exp Dermatol*, *24*, 346-353.

[51] Okomo-Adhiambo, M., Sleeman, K., Lysen, C., Nguyen, H. T., Xu, X., Li, Y., Klimov, A. l. & Gubareva, L. V. (2013). Neuraminidase inhibitor susceptibility surveillance of influenza viruses circulating worldwide during the 2011 Southern Hemisphere season. *Influenza Ocher Respir Viruses*, *7*, 645-658.

[52] Panel on Antiretroviral Guidelines for Adults and Adolescents. (2013). *Guidelines for the use of antiretroviral agents in HIV-I-infected adults and adolescents.* Department of Health and Human Services. http: 1/aidsinfo.nih.gov/contentfiles/ lvguidelines/ AdultandAdolescentGL.pdf.

[53] Parikh, U. M., Koontz, D. L., Chu, C. K., Schinazi, R. F. & Mellors, J. W. (2005). *In vitro activity of structurally diverse nucleoside analogs against human*

immunodeficiency virus type 1 with the K65R mutation in reverse transcriptase Antimicrobe Agents Chemother, *49*, 1139-1144.

[54] Parkin, N. T., Hellmann, N. S., Whitcomb, J. M., Kiss, L., Chappey, C. & Petropoulos, C. J. (2004). Natural variation in drug susceptibility in wild-type human immunodeficiency virus type 1. *Antimicrob Agents Chemother*, *48*, 437-443.

[55] Pielak, R. M., Schnell, J. R. & Chou, J. J. (2009). Mechanism of rug inhibition and drug resistance of influenza A M2 channel. *Proc Natl Acad Sci US A*, *106*, 7379-7384.

[56] Pillay, D. & Zambon, M. (1998). Antiviral drug resistance. *BMJ.*, *5*, 317(7159), 660–662.

[57] Pillay, D. (1998). Emergence and control of resistance to antiviral drugs in resistance in herpes virus, hepatitis B virus and HIV. *Commun Dis Public Health*, *1*(1), 5-13.

[58] Piret, J. & Boilin, G. (2011). Resistance of herpes simplex viruses to nucleoside analogues: mechanisms, prevalence, and management. *Antimicrob Agents Chemother*, *55*, 495-472.

[59] Poordad, F., Lawitz, E., Kowdley, K. V., Cohen, D. E., Podsa-decki, T., Siggelkow, S., Heckaman, M., Larsen, L., Menon, R., Koev, Tripathi R., Pilot-Matias, T. & Bernstein, B. (2013). Eploratory study of oral combination antiviral therapy for hepatitis C. *N Engl J Med*, *368*, 45-53

[60] Prichard, M. N. (2009). *Function of human cytomegalo virusUL97 kinase in viral infection and its inhibition by maribavir Rev Med Viral*, *19*, 215-229.

[61] Reardon, J. E. & Spector, T. (1989). Herpes simplex virus type 1 DNA polymerase. Mechanism of inhibition by acyclovir triphosphate. *J Biol Chem*, *264*, 7405-7411.

[62] Sauerbrei, A., Taut, J., Zell, R. & Wutzler, P. (2011). Resistance testing of clinical varicella-zoster virus strains. *Antiviral Res*, *90*, 242-247.

[63] Shafer, R. W., Kozal, M. J., Winters, M. A., Iverson, A. K., Katzenstien, D. A., Ragni, M. V., Meyer, W. A., Gupta, P., Rasheed Coombs, R. & Merigan, T. C. (1994). Combination therapy with zidovudine and didanosine selects for drug-resistant human immunodeficiency virus type 1 strains with unique patterns of pol gene mutations. *J Infect Dis*, *169*, 722-729.

[64] Shafer, R. W. & Chou, S. (2015). Mechanisms of resistance to antiviral agents. JH Jorgensen, KC Carroll, G Funke, MA Pfaller, ML Landry, SS Richer, DW Warnock. *Manual of clinical microbiology*., Vol 2. Washington DC. ASM Press., 1895-1912.

[65] Sharma, S. D. (2010). Hepatitis C virus: Molecular biology and current therapeutic options. *Indian J Med Res*, *131*.

[66] Shirasaka, T., Kavlick, M. F., Ueno, T., Gao, W. Y., Kojima, E., Alcaide, M. L., Chokekijchai, S., Roy, B. M., Arnold, E., Yarchoan, R. & Mitsuya, H. (1995). Emergence of human immunodeficiency virus type 1 variants with resistance to multiple dideoxynucleosides in patients receiving therapy with dideoxy nucleosides. *Proc Natl Acad Sci USA*, *92*, 2398-2402.

[67] Smith, B. J., McKimm-Breshkin, J. L., McDonald, M., Fernley, R. T., Varghese, J. N. & Colman, P. M. (2002). Structural studies of the resistance of influenza virus neuramindase to inhibitors. *J Med Chem*, *45*, 2207-2212.

[68] Smith, I. L., Cherrington, J. M., Jiles, R. E., Fuller, M. D., Freeman, W. R. & Spector, S. A. (1997). High-level resistance of cytomega-lovirus co ganciclovir is associated with alterations in both the UL97 and DNA polymerase genes. *J Infect Dis*, *176*, 69-77.

[69] Strasfeld, L. & Chou, S. (2010). Antiviral drug resistance, mechanisms and clinical implications. *Infect Dis Clin North Am* 24 (2), 413-437

[70] Sullivan, V., Talarico, C. L., Stanat, S. C., Davis, M., Coen, D. M. & Biron, K. K. (1992). A protein kinase homologue controls phosphorylation of ganciclovir in human cytomegalovirus infected cells. *Nature*, *358*, 162-164.

[71] Tenney, D. J., Rose, R. E., Baldick, C. J., Pokornowski, K. A., Eggers, B. J., Fang, J., Wichroski, M. J., Xu, D., Yang, J., Wilber, R. B. & Colonno, R. J. (2009). Long term monitoring shows hepatitis B virus resiistance to entecavir in nucleoside naive patients is rare through 5 years of therapy. *Hepatology*, *49*, 1503-1514.

[72] Thomas, H. C. & McDade, H. (1996). Viral dynamics in hepati- tis B virus infection. *Proc Natl Acad Sci US A*, *93*, 4398-4402.

[73] Van der Beek, M. T., Vermont, C. L., Bredius, R. G., Marijt, E. W., van der Blij-de Brouwer, C. S., Kroes, A. C., Claas, E. C. & Vossen, A. C. (2013). Persistence and antiviral resistance of varicella zoster virus in hematological patients. *Clin Infect Dis*, *56*, 335-343.

[74] Von Itzstein, M. (2007). The war against influetjza: discovery and development of sialidase inhibitors. *Nattre Rev Drug Discov*, *6*, 967-974.

[75] Wagner, R., Matrosovich, M. & Klenk, H. D. (2002). Functional balance between haemaggutinin and neuraminidase in influenza virus infections. *Rev Med Virol*, *12*, 159-166.

[76] Wang, J., Qiu, J. X., Soto, C. & DeGrado, W. F. (2011). Structural and dynamic mechanisms for the function and inhibition of the M2 proton channel from influenza A virus. *Curr Opin Struct Biol*, *21*, 68-80.

[77] Werle Lapostolle, B., Bowder, S., Locarnini, S., Wursthorn, K., Petersen, J., Lau, G., Trepo, C., Marcellin, P., Goodman, Z., Delaney, W. E. I. V., Xiong, S., Brosgart, C. L., Chen, S. S., Gibbs, C. S. & Zoulim, F. (2004). Persistence of cccDNA during the natural history of chronic hepatitis B and decline during adefovir dipivoxil therapy. *Gastroenterology*, *126*, 1750-1758.

[78] Westby, M., Lewis, M., Whitcomb, J., Youle, M., Pozniak, A. L., James, I. T., Jenkins, T. M., Perras, M. & van der Ryst, E. (2006). Emergence of CXCR4-using human-immunodeficiency virus type 1 (HIV-1) variants in a min rity of HIV-1-infected patients following treatment with! the CCR5 antagonist maraviroc is from a pretreatment CXCR4-using virus reservoir. *J Virol*, *80*, 4909-4920.

[79] Westby, M., Srnith-Burchnell, C., Mori, J., Lewis, M., Mosley, M., Stockdale, M., Dorr, P., Ciaramella, G. & Perras, M. (2007). Reduced maximal inhibition in phenotypic susceptibility assays indicates that viral strains resistant to the CCR5 antagonist maraviroc utilize inhibitor-bound receptor for entry. *J Virol*, *81*, 2359-2371.

[80] Whitcomb, J. M., Parkin, N. T., Chappey, C., Hellmann, N. S. & Petropoulos, C. J. (2003). Broad nucleoside reverse-tran-scriptase inhibitor cross-resistance in human immunodeficiency virus type 1 clinical isolates. *J Infect Dis*, *188*, 992- 1000.

[81] Yang, G., Song, Q., Charles, M., Drosopoulos, W. C., Arnold, E. & Prasad, V. R. (1996). Use of chimeric human immunodeficiency virus types 1 and 2 reverse transcriptases for structure-function analysis and for mapping suseptibilty to nonnucleoside inhibitors. *J Acquir Immune Defic Syndr Hint Retrovirol*, *11*, 326-333.

[82] Young, J., Howard, Z. & Shepler, L. (1986). Medication noncompliance in schizophrenia: codification and update**.** *Bull Am Acad Psychiatry Law*, *14***,** 105–122.

[83] Yuen, M. F. (2013). Reduction of hepatitis B surface antigen and covalently closed circular DNA by nucleos(t)ide analogues of different potency. *Clin Gastroenterol I-lepatol*, *11*, 1004-1010.

[84] Zambon, M. C. (2013). Surveillance for antiviral resistance. *Influenza Other Respir Viruses*, *7*, (Suppl 1), 37-43.

[85] Zeuzem, S., Soriano, V., Asselah, T., Bronowicki, J. P., Lohse, A. W., Mullhaupt, B., Schuchmann, M., Bourliere, M., Buti, M., Roberts, S. K., Gane, E. J., Stern, J. O., Vinisko, R., Kukolj, G., Gallivan, J. P., Bocher, W. O. & Mensa, F. J. (2013). Faldaprevir and deleobuvir for HCV genotype 1 infection. *N Engl J Med*, *369*, 630-639.

In: Antimicrobial Resistance: Opportunities and Challenges ISBN: 978-1-53617-943-9
Editors: P. Kumar Prabhakar and V. Mishra

Chapter 9

STRATEGIES FOR CIRCUMVENTING ANTIMICROBIAL RESISTANCE

***Keerti Bhardwaj*[1], *Iqbaljit Kaur*[1] *and Gopal L. Khatik*[2,3,*]**

[1]Department of Pharmaceutics, School of Pharmaceutical Sciences, Lovely Professional University, Phagwara Punjab, India

[2]Department of Pharmaceutical Chemistry, School of Pharmaceutical Sciences, Lovely Professional University, Phagwara, Punjab, India

[3]Department of Medicinal Chemistry, National Institute of Pharmaceutical Education and Research, Raebareli, Sarojini Nagar, Lucknow, India

ABSTRACT

From the 20th century onwards, a rise in antibiotic resistance has been observed among the pathogenic bacteria, causing subsequent fall in the efficacy of antibiotic therapy. That is leading to millions of deaths per year globally. Antimicrobial agents are proven to be one of the major tools used both in humans as well as animals for treatment and control of several bacterial infections. Although within the last fifty years, these medicaments have inherited a condition where they no longer are able to act upon the bacteria they were originally prepared to kill due to their excessive or incorrect use and thus unable to impart their antimicrobial action. Largely there can be two approaches to tackle this situation. One can be to keep searching for new antibiotics and another can be to innovate something useful and effective with the existing ones. Developing a new drug moiety is a long and expensive process and generally, does not comply with the financial goals of pharmaceutical companies. Herein we discussed the different approaches to circumventing antimicrobial resistance.

Keywords: antimicrobial, resistance, mutation, circumventing, multi-drug resistance

[*] Corresponding Author's E-mail: gopal_niper@rediffmail.com.

1. Introduction

Antimicrobial agents have proven to be a milestone in clinical health management since the latter half of the 20th century and were responsible for saving numerous lives from lethal bacterial infections. From the 20th century onwards, a rise in antibiotic resistance has been observed among the pathogenic bacteria, causing subsequent fall in the efficacy of antibiotic therapy. That is leading to millions of deaths per year globally [1]. The continuous elevation in the occurrence of resistance in various significant pathogens like vancomycin-resistant *Enterococcus* (VRE), methicillin-resistant *Staphylococcus aureus* (MRSA), third-generation cephalosporin-resistant *Escherichia coli* and *Klebsiella pneumonia,* imipenem-resistant *Acinetobacter baumannii* and multidrug-resistant (MDR) *Pseudomonas aeruginosa* creates a serious risk to people's health [2-4]. Extended-spectrum β-lactamase (ESBL)-producing pathogens and MRSA are endemic in many hospitals worldwide [5] and are increasingly getting discovered in the local hospitals that are dealt with the local healthcare providers who treated the patients first rather than the specialty hospitals or regional healthcare centers which has advanced diagnostics and treatment facilities [6]. For treating the infections of non-*Enterobacteriaceae* pathogens like *P. aeruginosa* and *A. baumannii,* Carbapenems are the only strong antimicrobials left [5, 7]. The resistance of carbapenems or fluoroquinolones will lead to a serious situation in the future [8, 9]. The latest reports state that the carbapenem-resistant *Salmonella enterica* and *E. coli* are also isolated from poultry and sea-food animals, which is a growing concern about antimicrobial agents [10, 11]. Tuberculosis still remains among the most prominent health issues in people and its control has been obstructed due to the rise of MDR *Mycobacterium tuberculosis* causing resistance against isoniazid and rifampicin, which are the two major drugs used in the treatment of tuberculosis [12]. It has also been reported recently that the extensively drug-resistant (XDR) *M. tuberculosis* along with being MDR, have also developed resistance towards any fluoroquinolones as well as the injectables antibiotics like capreomycin, kanamycin and amikacin [12, 13]. The chances of rendering this risk by developing any advanced antibiotic are reduced due to reduction in novel antibiotic discovery, especially for Gram-negative bacteria [14-17] and also due to the probability of emergence of resistance to the new antibiotics just like they evolved rapidly against the existing antibiotics [1, 18, 19]. Hence, for overcoming the antibiotic resistance problem, efficient approaches are needed.

2. Antimicrobial Resistance Mechanism

Antimicrobial agents are proven to be one of the major tools used both in humans as well as animals for the treatment and control of several bacterial infections. Although within the last fifty years, these medicaments have inherited a condition where they no longer are able to act upon the bacteria they were originally prepared to kill due to their excessive or incorrect use and thus unable to impart their antimicrobial action.

These antimicrobial-resistant bacteria living inside of animals not only cause a risk to the animal health but are capable of infecting other people via foodborne contaminants. Therefore, addressing the problem of antimicrobial resistance is of major concern right now for better public health.

2.1. Process

The supremacy and propagation of resistance in bacteria is a result of natural selection and has occurred as a known phenomenon following the Darwinian theory of "survival of the fittest". In a group of bacteria having similar characteristics, some of them always have certain traits that keep them alive under unfavorable conditions such as antimicrobial agent action. The susceptible bacteria will die under the action of antimicrobials while these bacteria survive due to their advantageous traits. When the same antimicrobial is used for a longer period of time, it will cause a drastic change in the bacterial communities creating more resistant microbes. This may lead to a condition where when the same antimicrobial agent will be used again, it will have no effect over the bacteria for treating the same infection.

2.2. Mechanisms

The susceptible bacteria can develop antimicrobial resistance either by undergoing genetic change like a mutation or by transferring the antimicrobial-resistant genetic material from one bacterium to another. These may happen through various biochemical actions:

1. Mutation
2. Destruction or inactivation
3. Efflux

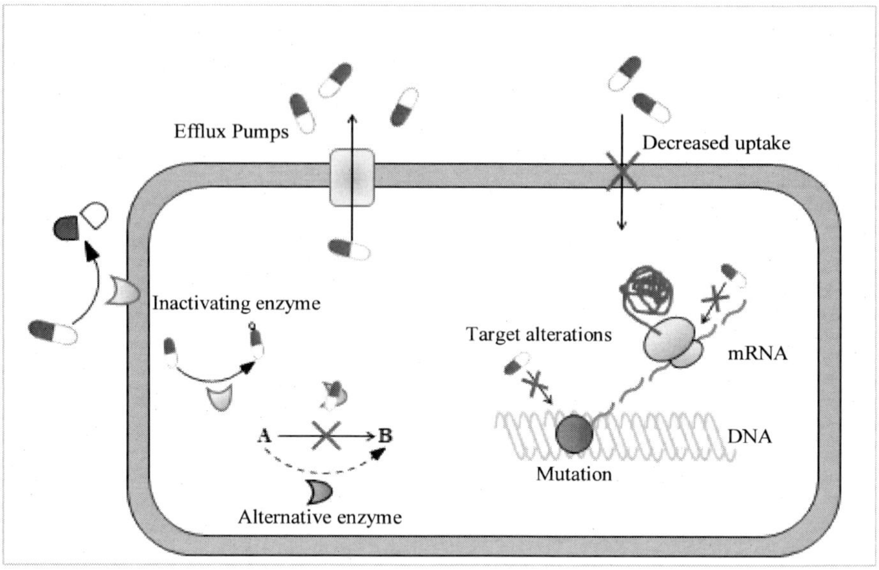

Figure 1. Diagrammatic representation of mechanisms of antimicrobial resistance.

2.2.1. Mutation

The mutation is an alteration of DNA that leads to a convert genetic material which was originally the target of an antimicrobial agent to attack. When a receptive bacteria are exposed to a therapeutically sufficient concentration of an antimicrobial agent such as fluoroquinolones, this agent tends to bind at a special enzyme like DNA gyrase. This enzyme plays an essential role in DNA replication of the given bacterium while fluoroquinolone block it, resulting in no DNA replication and ultimately, cell death. Although, when a voluntary mutation takes place in a particular portion of the genes encoding these enzymes, the antimicrobial agent will not bind properly resulting in continued DNA replication of bacteria [21, 24].

2.2.2. Destruction or Inactivation

Many bacteria contain some genes that produce certain enzymes that destruct or degrade the antimicrobial agents readily and making them ineffective.

Here, the antimicrobial is either degraded or modified by enzymatic activity before it can reach the target site of the bacterial cell [22].

2.2.3. Efflux

Some bacteria have a tendency to be resistant towards any antimicrobial through the efflux mechanism [20]. An efflux pump is typically a channel through which the antimicrobial agent has been exported out of the cell. The antimicrobial agent enters inside the bacterium from an opening known as a porin, but once entered, it is pumped back out by the efflux pump from the bacterium, leaving it intact in its intracellular integrity to impart its lethal action on the body.

Genetic material can be transferred between bacteria by several means, most often by:

a) *Conjugation:* It is mediated by a particular kind of circular DNA called a *plasmid* which replicates independently of the chromosome. Many plasmids possess the genes which are responsible for antimicrobial resistance in the bacterium. When two cells are in close proximity to each other, a hollow bridge-like structure known as pilus forms between two cells. The plasmid undergoes duplication and one of the duplicate stands of plasmid gets transferred to another bacterium through pilus. This enables susceptible bacteria to acquire resistance to a particular antimicrobial agent [25].
b) *Transformation:* In this process, the genetic material gets transferred from one bacterium to another in the form of 'naked' DNA. Whenever a bacterial cell dies and disintegrates, its DNA could be left released in the surroundings near other bacteria, floating freely, which then got scavenged by any living bacterium moving in close proximity to it and the traits carried by that genetic material gets transferred to the scavenging bacterium. One such trait could be the antimicrobial-resistance which now has been passed on to the recipient bacterial cell.
c) *Transduction:* During this process, the bacterial DNA gets transferred from one cell to another cell via a carrier that has infected those bacteria. This carrier is mostly a virus that is also known as a bacteriophage or simply a phage. When a phage infects a bacterium, it essentially takes over the bacteria's genetic processes to produce more phages. In this process, the bacterial DNA may get incorporated into the new phage DNA. Upon bacterial death and lysis or breaking apart, these new phages go on to infect other bacteria. This brings along genes from the previously infected bacterium [27].

3. Strategies of Circumventing Antimicrobial Resistance

Antimicrobial resistance (AMR) along with multidrug resistance (MDR), is increasing rapidly in several microorganisms in different healthcare centers and communities [28]. As per the suggested reports, almost around 2 million such cases of antimicrobial resistance have been reported in different types of infections in the United States (US) each year resulting in a hike of $20 billion healthcare cost directly [29]. According to the estimated studies of the European Centre for Disease Prevention and Control (ECDC) and European Medicines Agency (EMA), it has been reported that the world has experienced 25,000 deaths every year as a direct effect of the emergence of MDR infection costing about €1.5 billion [30]. The cost of treatment in Canada has been increased from $9–$14 million due to the hospitalization followed by the antimicrobial-resistance [28]. A study conducted by the Indian Network for Surveillance of Antimicrobial Resistance (INSAR) group showed an increase of methicillin-resistant *Staphylococcus aureus* (MRSA) by 41% [31]. Similarly, an elevated prevalence of Gram-negative bacterial resistance has also been observed in India [32]. It was also estimated by the World Health Organisation (WHO) that globally 3.7% new cases and 20% cases which were earlier treated have MDR-TB [33].

The journey of the development of antibiotics is exposed to many challenges due to the emergence of antimicrobial resistance. The emergence of MRSA, resistant *Pseudomonas aeruginosa* have already compromised the most effective treatments [34]. The current reports have observed a fluctuating pattern in susceptibility and prevalence of AMR [35-37] and the emergence of superbug like New Delhi metallo-β-lactamase 1 (NDM-1) positive Enterobacteriaceae are contributing negatively in treatment and managing these infections [38].

The emergence of such antimicrobial resistance needs to be addressed for maintaining public health. The strategies for controlling these resistances have been discussed below.

3.1. Conventional Approaches

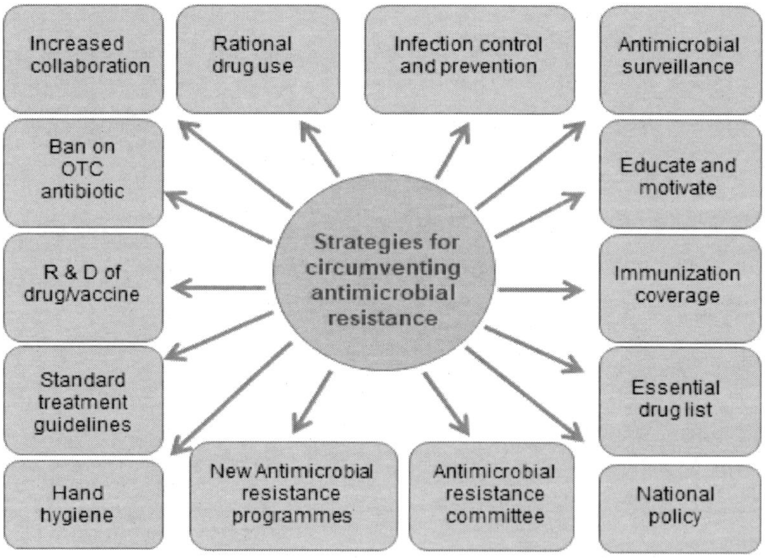

Figure 2. Strategies for combating AMR.

The threat of AMR globally needs to be addressed and collective actions need to be taken in order to control such events that put the public health at the risk. CDC has come up with 12 steps protocol for preventing antimicrobial resistance [39].

The effective ways of combating the antimicrobial-resistance have been discussed here. The collective approach from all different areas and communities at different levels is highly required at this hour of need. Several measures for circumventing AMR can be initiated right from the patient level to hospitals, community, national and international levels. Below in Figure 2, these approaches are summarized.

3.1.1. International Approach

Antimicrobial resistance has been identified as a global issue and several stakeholders have accepted its threat to mankind. In 2011, WHO drew everybody's attention towards the threat and made a theme of combating antimicrobial resistance. WHO also addressed the issue globally by stressing that a collaborative approach is needed to control the problem of antimicrobial resistance [40]. Some of the suggested approaches by WHO are listed below [41]:

a) Increased collaboration among international agencies, professional groups, governmental-nongovernmental organizations working in healthcare sectors.
b) Formation of advanced networks that keep the antimicrobial use and their respective resistance under surveillance.
c) Coming up with some international collaborative approach to circumvent the AMR.
d) Incentives to be given to R&Ds for designing new drugs or vaccines or both.
e) Working on a novel and reinforcing the already available approaches to control AMR.

3.2. National Strategies

3.2.1. National Committee with Intersectoral Coordination and Regulatory Actions

Setting up a national committee which not only monitors the emergence and subsequent effect of antimicrobial resistance but also conducts intersectoral coordination is the need of the hour. WHO suggested that these committees would then be authorized to prepare AMR policy, providing guidance on regulations, standards, training, and awareness on the proper use of antibiotics to circumvent AMR. Innovating the methods to monitor and evaluate the effect of AMR regulation and prevention would be one of the major objectives at the national level [42]. WHO suggests that all the dispensing outlets, making the availability of antimicrobial on the prescription-only basis, putting legal restrictions on all the manufacturers to report the data related to the distribution of antimicrobials and giving incentives for appropriate use of antimicrobials can be helpful in limiting AMR [43].

Implementing and establishing guidelines for national standard treatment, developing essential drug list (EDL), expanding the coverage of immunization are some additional important approaches required at the national level [41].

3.2.2 National Antimicrobial Resistance Policy, India

In the year 2011, a national policy for controlling AMR came into light. The aim of this policy was [43].

a) To comprehend the emergence, dissemination and the factors responsible for antimicrobial resistance
b) To design and execute the antimicrobial program
c) To justify the need of using antimicrobials
d) To help and promote the invention of novel, more efficient antimicrobials
e) Establishment of a surveillance system for antimicrobial resistance
f) Reinforcing the prevention and management system for infection
g) To spread awareness, train and encourage all stakeholders in reasonable antimicrobial use

As per WHO, less than 50% of all the countries are effectively applying these basic policies for proper medications [44].

3.3. Community Level Action

The communicable diseases are continued to be a major reason behind the global morbidity and mortality rate, especially, affecting those countries where proper health care services are not available [45]. According to a review on misuse of antibiotics in the community, it was stated that more than 75% of the patient were not following the antibiotic dose regimen properly and 25% of them had saved the unused antibiotics for future use, which is showing a poor behavior of antibiotic consumption [46].

In the review of the population perspective of AMR, it was concluded that, the prevention of antimicrobial resistance in the person who is suffering from that infection is the first and most important step to be taken in order to reduce the risk of prevalence of AMR in wider communities [47]. The elevating AMR has resulted in developed infections such as respiratory tract infections, typhoid fever, bacterial diarrhea that is no longer responding to their antibiotics and new antibiotics are also failing to match with them [48]. Therefore, this is a critical situation where we urgently require reforms at the community level for circumventing AMR.

Different measures directed to control and prevent AMR at community levels are needed.

3.3.1. Rational Use of Antibiotics

Reckless use of medications is a serious global issue. At a basic healthcare level, less than 30% of patients in the private sector and less than 40% of patients in the public sector are being treated as per the standard treatment guidelines in developing countries [44]. This makes it necessary to have public and professional education towards the rational use of antibiotics.

3.3.2. Over-the-Counter (OTC) Antibiotics

The precautionary measures are needed to keep the antimicrobial actions intact inclines towards the professional drug providers and health care centers by avoiding the sale or use of antibiotics without prescription. In a review of antimicrobial prescription, it was reported that the use of antibiotics without any prescription ranges from a low amount of 3% in northern Europe to 100% in African countries [49]. This indicates an immediate action to be taken on curbing the OTC consumption of antibiotics.

3.3.3. Guidelines for Local Level Antibiotics Consumption

In terms of consumption of antibiotics under common situations, a rational antibiotic therapy based on the clinical condition of the person, prevailing pathogen and its pattern of resistance in the respective locality should be designed. Suitable changes in the antibiotic moiety are needed according to the sensitivity of respective pathogens. Therefore, optimization of antibiotic guidelines is required for proper selection of antibiotics, their route of administration and suitable dose regimen [48].

3.3.4. Hygiene Standards

The alcohol-based hand sanitizers and simple hand washing have proven to be efficacious in order to prevent the infection [39, 50]. This factor is capable of limiting the infection spread, and ultimately curbs the AMR. Thus, by following high hygiene standards needs to be a top priority right now.

3.3.5. Other Approaches

Some other approaches include identification of local people having MDR infections and applying standard treatment procedures for their infection management, vaccination, prevention techniques of infection and ban over OTC availability of antibiotics [39, 43].

3.4. Health Care Setting

People are at higher risk of catching any infection when they are at any health care setting due to the presence of many common pathogens in the air. In such settings, for prevention and control of antimicrobial resistance, some of the ways are recommended by the experts as discussed below.

3.4.1. Prevention and Control of Infections in Health-Care Facilities [28, 51]

The precautionary measures are devised to limit the expansion of microbes including the resistant bacteria that may be available at various health care centers and this can reduce the risk of further spread of infection and AMR [28]. Some suitable measures are proposed below for the prevention and control of microbial infection in health-care facilities [28, 39, 43, 51-53].

a) Forming an infection prevention and control committee (IPC).
b) Practicing proper hand hygiene
c) Efficient infection diagnosis and its treatment
d) Proper use of antimicrobials
e) Close observation on the use of antibiotics and its developing resistance.
f) Enhancing the quality and supply chain of the antimicrobials.
g) Good Microbiology Practices.

3.4.2. Close Observation on the Use of Antibiotics and Its Developing Resistance

Keeping a close observation of antimicrobial resistance has proven to be very helpful worldwide. The purpose of this approach is to control the prevalence of antimicrobial resistance. This approach also acts as a tool to collect information regarding the use of antimicrobials and their resistance issue which is needed in updating the national essential diagnostic list (NEDL) and making policies for their control. This can also be helpful in improving the antimicrobial prescribing and designing appropriate therapy or suitable treatment guidelines [52, 53]. In India, National policy on AMR recommends three types of surveillance which are sentinel surveillance, point prevalence and comprehensive surveillance [43].

3.4.3 Good Microbiology Practices

Good microbial practices are proper collection, specimen handling and quick reporting with standard microbiology practices are useful to control of AMR prevalence. The good microbiology practices facilitate international standards for testing, reporting of AMR patterns to infection prevention and control committee as well as monitoring activities like disinfection and sterilization [43, 51].

3.5. At individual Level

3.5.1. Role of Physician

The physicians play a significant role in limiting AMR via the process of facilitating direct patient care, observing and providing the proper aid to infection control, complying with usage policies of antibiotics and reporting the AMR cases to IPC within the stipulated time [51]. Recognizing and preventing the conditions which may lead to bacterial growth and infection may help in controlling avoidable infections and respective AMR [39].

3.5.2. Role of Nurses/Health Care Providers

Nurses and health care providers are amongst those people who are directly related to spreading or control of infection and AMR as they are in direct contact with patients having such infections. Educating them about AMR and aseptic practices can contribute to curbing the infection spread. According to a review, initiatives taken in Thailand such as establishing a Master's programme in infection control nursing with some other short term training courses and participation of nurses in AMR prevention and control programme [54].

3.5.3. Role of Pharmacist

In the review by McCoy et al., the antibiotic stewardship program (ASPs) directed by Pharmacists has been discussed and suggested as an approach for encouraging the rational use of antibiotics. Pharmacists can explain the difference between bacterial and viral infection to the patient and making them aware that the ineffectiveness of antibacterials in viral infection. Also, some appropriate medicament could be recommended to the patient in such cases. In case of any bacterial infection diagnosed, the patient must be referred to a registered physician. Other than these the most essential role of a pharmacist is to address the concerns of antimicrobial use and spreading awareness of AMR [55]. This contribution of pharmacists may help in the reduction of avoidable use of antibiotics.

3.5.4. Patient-Level Approach

a) Follow aseptic protocol for any procedure. It has been reported that multi-drug resistance (MDR) microbes are the source of 30.2% of the blood-stream infections which traveled through catheters into the bloodstream and empirical treatment does not work on the prevention of these infections [56]. Therefore, the application of the aseptic protocol is a must to minimize the risk of any sort of infection that could be associated with any such routine or procedure.

b) Breaking the chain of infectivity [39]. By implementing simple ways like covering mouth at the time of coughing or sneezing, infection spread can be controlled.
c) Compliance with the antimicrobial regime and antibiotic. Proper compliance with the dose regimen of antibiotics can enhance the rate of infection control. Educating the patient and making him/her aware of AMR is a much-needed step [57]. Following a short-term and well-planned antibiotic dose regimen for high-risk cases can reduce the risk of AMR dramatically [58].

3.6. Other Measures

3.6.1. Pharmaceutical Promotion

WHO suggests that for marketing and promotion of medicaments, pharmaceutical firms need to follow the standards of drug promotion, advertising over the internet and direct-to-consumer advertising [41]. It is the need of the hour to search and ban any temptations given by pharmaceutical companies that in any way try to encourage improper and excessive antimicrobial use.

3.6.2. Antibiotic Use in Animals

Using antibiotic avoparin into animal foods in Europe turned out to be the reason behind Vancomycin-Resistant Enterococci (VRE) resulting in their possible colonization in the intestinal region of humans [43]. WHO exclusively called for advanced legislation to reduce the use of antimicrobials in the animal. Proper sanitation, the addition of probiotics and supplements in feed provision of vaccination for easily causing animal diseases can work towards minimizing the antimicrobial use animals [52].

3.6.3. Understanding the AMR Mechanisms

Genotyping of any bacteria and their resistant gene identification can be useful in the treatment of respective infections as well as limiting the AMR [59].

3.6.4. New Drugs and Technology Innovation

The concern for the emergence of antibiotic resistance is pointing towards the urgent need of drawing attention towards the development of new drugs and/or vaccines to circumvent AMR. Collective efforts among the national, international governments and scholastic networks are required for identifying and innovating new effective classes of antibiotics [42].

Supporting the research work of developing such new drugs by providing funds to the respective research conducting organization can help in achieving the antimicrobial advancement and prevention of AMR. In short, it is the need of the hour that substantial steps are needed to be taken by the government to instigate changes by all the responsible parties related to AMR as per the policy package for preventing AMR given by WHO [40]. This policy package suggests:

a) Work on an inclusive, funded national plan with equal responsibility for each one involved.
b) Upgrade and intensify the surveillance and laboratory facilities and capacities.
c) Assuring ceaseless wide access to necessary medicaments of good quality.
d) Spread awareness of reasonable consumption of medicines not just for humans but for animals too and proper patient care to be ensured
e) Upgrading prevention and control of infections
f) Encourage and practice research and development and novelties as new tools

3.7. Novel Approaches

Resistance to well-known antimicrobial like penicillin was reported first time in staphylococci. This stimulated by penicillinase enzyme which degrades the antibiotic. Now, over the years due to ongoing selective pressure induced by several drugs, the organisms have evolved with their resistance mechanisms leading to multidrug resistance (MDR). The most challenging MDR organisms currently identified are *Escherichia coli, Acinetobacter baumannii, Pseudomonas aeruginosa* and *Klebsiella pneumoniae* bearing extended-spectrum β-lactamases (ESBL), vancomycin-resistant MRSA, methicillin-resistant Staphylococcus aureus (MRSA), vancomycin-resistant enterococci (VRE) and extensively drug-resistant (XDR) Mycobacterium tuberculosis [60]. Another recent prevalence of resistance of carbapenem in Gram-negative Enterobacteriaceae caused by the enzyme New Delhi metallo-β-lactamase 1 (NDM-1) is proving to be a global threat and a serious issue [61]. This means that we are reaching towards the "pre-antibiotic" era because the last available antibiotics are also not protected against their resistance. Largely there can be two approaches to tackle this situation. One can be to keep searching for new antibiotics and another can be to innovate something useful and effective with the existing ones. Developing a new drug moiety is a long and expensive process and generally, does not comply with the financial goals of pharmaceutical companies. The other approach is by reinforcing the existing antibiotics with some contemporary novel techniques that are presented in Table 1.

Table 1. Novel approaches to combat the antimicrobial resistance

S. No.	Novel Strategies	Approach of action
1.	Combination drug therapy	a) Combination drugs acting on different targets in different pathways. b) Combination drugs acting on different targets in the same pathway. c) Combination drugs acting on a single target, but in different dimensions.
2.	Bacteriophage therapy	A virus which is also known as bacteriophage infects the bacterium. The protein comprised in a bacteriophage encapsulates the DNA/RNA genome and multiply inside the bacterium. This principle is used for killing the bacteriophage carrying bacterial cells.
3.	Fecal microbiota transplantation	In this process, fecal transplantation is carried out from a healthy person to the patient [62]. The purpose of this practice is to protect the gut flora as in the case of antibiotics, it could not differentiate between the bacteria and gut microbiota and destroys the needed gut commensals. Infusion of healthy fecal material could be done using different ways such as nasogastric, enema, colonoscopic and nasoduodenal routes.
4.	Antimicrobial adjuvants	Antimicrobial adjuvants counteract the mechanism applied by the pathogens to generate the antimicrobial resistance. The general approach of adjuvant action includes: Blocking the AMR elements in the bacterium Increasing the antibiotic uptake in target bacterial cell Diffusing the efflux pumps Enhancing the oxidative stress in bacteria
5.	Antimicrobial peptides	Antimicrobial peptides act in various ways which include: Stimulating the membrane permeability Neutralization of Lipopolysaccharide Hindering the cytoplasmic protein responsible for cell division or subsistence. Inhibition of the production of macromolecules Prevention of immune response and generating septic shock Anti-biofilm Activity of antimicrobial peptides against biofilm of multi-drug resistant bacteria [63, 64].

CONCLUSION

The emergence of antimicrobial resistance has proven to be a major threat to public health worldwide and circumvention of AMR is the need of the hour. Several different approaches could be employed for this purpose beginning from maintaining personal hygiene to developing new antibiotic drugs and vaccinations for the prevention of infection spread and its potential resistance. The conventional strategies are practical and feasible to be carried out. Although along with prevention, control of AMR is also an essential aspect for which new innovative drugs or reinforced existing antibiotics need to be introduced. After looking at all the aspects of inventing a new drug and rejuvenating already available antibiotics, the latter one seems to be a better and practical approach. It cannot be emphasized enough that only the production of new drug therapy will not be able to circumvent AMR but a collective approach of spreading awareness of rational use of antibiotics along with the therapies can result in circumvention of antimicrobial resistance.

REFERENCES

[1] Palmer, AC, Kishony, R. Understanding, predicting and manipulating the genotypic evolution of antibiotic resistance. *Nat. Rev. Genet.* 2013, *14*, 243–248.

[2] Meyer, E, Schwab, F, Schroeren-Boersch, B, Gastmeier, P. Dramatic increase of third-generation cephalosporin-resistant *E. coli* in German intensive care units: Secular trends in antibiotic drug use and bacterial resistance, 2001 to 2008. *Crit. Care* 2010, *14*, doi:10.1186/cc9062.

[3] Rossolini, GM, Mantengoli, E, Docquier, JD, Musmanno, RA, Coratza, G. Epidemiology of infections caused by multiresistant Gram-negatives: ESBLs, MBLs, panresistant strains. *New Microbiol.* 2007, *30*, 332–339.

[4] Spellberg, B, Guidos, R, Gilbert, D, Bradley, J, Boucher, HW., Scheld, W.M., Bartlett, J.G., Edwards, J., Jr., Infectious Diseases Society of America. The epidemic of antibiotic-resistant infections: A call to action for the medical community from the infectious diseases society of America. *Clin. Infect. Dis.* 2008, *46*, 155–164.

[5] Cantas, L, Shah, SQ., Cavaco, LM., Manaia, CM., Walsh, F, Popowska, M, Garelick, H, Burgmann, H, Sorum, H. A brief multi-disciplinary review on antimicrobial resistance in medicine and its linkage to the global environmental microbiota. *Front. Microbiol.* 2013, *4*, doi:10.3389/fmicb.2013.00096.

[6] Otter, JA., French, GL. Molecular epidemiology of community-associated meticillin-resistant *Staphylococcus aureus* in Europe. *Lancet Infect. Dis.* 2010, *10*, 227–239.

[7] Brown, S, Bantar, C, Young, HK., Amyes, SG. Limitation of *Acinetobacter baumannii* treatment by plasmid-mediated carbapenemase ARI-2. *Lancet* 1998, *351*, 186–187.

[8] Poirel, L, Nordmann, P. Carbapenem resistance in *Acinetobacter baumannii*: Mechanisms and epidemiology. *Clin. Microbiol. Infect.* 2006, *12*, 826–836.

[9] Strahilevitz, J, Jacoby, GA, Hooper, DC, Robicsek, A. Plasmid-mediated quinolone resistance: A multifaceted threat. *Clin. Microbiol. Rev.* 2009, *22*, 664–689.

[10] Fischer, J, Rodriguez, I, Schmoger, S, Friese, A, Roesler, U, Helmuth, R, Guerra, B. *Salmonella enterica* subsp. enterica producing VIM-1 carbapenemase isolated from livestock farms. *J. Antimicrob. Chemother.* 2013, *68*, 478–480.

[11] Fischer, J, Rodriguez, I, Schmoger, S, Friese, A, Roesler, U, Helmuth, R, Guerra, B. *Escherichia coli* producing VIM-1 carbapenemase isolated on a pig farm. *J. Antimicrob. Chemother.* 2012, *67*, 1793–1795.

[12] Almeida Da Silva, PE, Palomino, JC. Molecular basis and mechanisms of drug resistance in *Mycobacterium tuberculosis*: Classical and new drugs. *J. Antimicrob. Chemother.* 2011, *66*, 1417–1430.

[13] Dye, C. Doomsday postponed? Preventing and reversing epidemics of drug-resistant tuberculosis. *Nat. Rev. Microbiol.* 2009, *7*, 81–87.

[14] Freire-Moran, L, Aronsson, B, Manz, C, Gyssens, IC, So, AD, Monnet, DL, Cars, O, Group, E-EW. Critical shortage of new antibiotics in development against multidrug-resistant bacteria-time to react is now. *Drug Resist. Updat.* 2011, *14*, 118–124.

[15] Lee, JH, Jeong, SH, Cha, SS, Lee, SH. New disturbing trend in antimicrobial resistance of Gram-negative pathogens. *PLoS Pathog.* 2009, *5*, e1000221, doi:10.1371/journal.ppat.1000221.

[16] Lee, JH, Bae, IK, Lee, SH. New definitions of extended-spectrum β-lactamase conferring worldwide emerging antibiotic resistance. *Med. Res. Rev.* 2012, *32*, 216–232.

[17] Lee, JH, Jeong, SH, Cha, S-S, Lee, SH. A lack of drugs for antibiotic-resistant Gram-negative bacteria. *Nat. Rev. Drug Discov.* 2007, *6*, doi:10.1038/nrd2201-c1031.

[18] Spellberg, B, Powers, JH, Brass, EP, Miller, LG, Edwards, JE, Jr. Trends in antimicrobial drug development: Implications for the future. *Clin. Infect. Dis.* 2004, *38*, 1279–1286.

[19] Livermore, D.M. Minimising antibiotic resistance. *Lancet Infect. Dis.* 2005, *5*, 450–459.

[20] Tian ZX, Yi XX, Cho A, O'Gara F, Wang YP. CpxR Activates MexAB-OprM Efflux Pump Expression and Enhances Antibiotic Resistance in Both Laboratory and Clinical nalB-Type Isolates of *Pseudomonas aeruginosa*. *PLoS Pathog.* 2016, 12(10): e1005932. https://doi.org/10.1371/journal.ppat.1005932 PMID: 27736975

[21] Gomez JE, Kaufmann-Malaga BB, Wivagg CN, Kim PB, Silvis MR, Renedo N, et al. *Ribosomal mutations promote the evolution of antibiotic resistance in a multidrug environment.* Gilmore MS, ed. eLife. 2017, 6:e20420. https://doi.org/10.7554/eLife.20420 PMID: 28220755

[22] Sorg RA, Lin L, van Doorn GS, Sorg M, Olson J, Nizet V, et al. Collective Resistance in Microbial Communities by Intracellular Antibiotic Deactivation. *PLoS Biol.* 2016, 14(12): e2000631. https://doi.org/10. 1371/journal.pbio.2000631 PMID: 28027306.

[23] Chevereau G, DraveckaÂ M, Batur T, Guvenek A, Ayhan DH, Toprak E et al. Quantifying the Determinants of Evolutionary Dynamics Leading to Drug Resistance. *PLoS Biol.* 2015, 13(11): e1002299. https://doi.org/10.1371/journal.pbio.1002299 PMID: 26581035

[24] Moura de Sousa J, BalbontõÂn R, Durão P, Gordo I. Multidrug-resistant bacteria compensate for the epistasis between resistances. *PLoS Biol.* 2017, 15(4): e2001741. https://doi.org/10.1371/journal.pbio.2001741 PMID: 28419091

[25] Stalder T, Rogers LM, Renfrow C, Yano H, Smith Z, Top EM. Emerging patterns of plasmid-host coevolution that stabilize antibiotic resistance. *Sci Rep.* 2017,Jul 7, 7(1):4853. https://doi.org/10.1038/s41598-017-04662-0 PMID: 28687759

[26] Rolo J, Worning P, Boye Nielsen J, Sobral R, Bowden R, Bouchami O,et al. Evidence for the evolutionary steps leading to mecA-mediated β-lactam resistance in staphylococci. *PLoS Genet.* 2017, 13(4): e1006674. https://doi.org/10.1371/journal.pgen.1006674 PMID: 28394942.

[27] https://www.fda.gov/animal-veterinary/safety-health/antimicrobial-resistance, accessed on 16 Nov. 2019.

[28] The evolving threat of antimicrobial resistance. *Options for action.* World Health Organization, 2012.

[29] *Antibiotic Resistance Threats in the United States, US Department of Human and Health Services,* Centre for Disease Control and prevention, 23, 2013.

[30] ECDC/EMEA Joint Technical Report. *The bacterial challenge: time to react. European Centre for Disease Prevention and Control,* 2009. *EMEA*. doc. ref. EMEA/576176/2009.

[31] Methicillin resistant *Staphylococcus aureus* (MRSA) in India: Prevalence & susceptibility pattern. Indian Network for Surveillance of Antimicrobial Resistance (INSAR) group, India. *Indian J Med Res.* 2013, 137:363-9.

[32] Ghafur A, Mathai D, Muruganathan A, Jayalal JA, Kant R, Chaudhary D, et al. The Chennai Declaration Recommendations of A roadmap to tackle the challenge of antimicrobial resistance - A joint meeting of medical societies of India. *Indian J Cancer.* [Epub ahead of print] [Cited 2013 Oct 21]. Available from: http://www.indianjcancer.com/preprintarticle.asp?id=104065.

[33] *Antimicrobial resistance.* WHO fact sheet, Fact sheet N°194. Updated May 2013. Accessed on 17. 2013.

[34] Davies J, Davies D. Origin and evolution of antibiotic resistance. *Microbiology and Molecular Biology Reviews.* 2010,74(3):417–33

[35] Abhilash KPP, Veeraraghavan B, Abraham OC. Epidemiology and Outcome of Bacteremia Caused by Extended Spectrum Beta-Lactamase (ESBL)-producing *Escherichia coli* and *Klebsiella* Spp. in a Tertiary Care Teaching Hospital in South India. *JAPI* (supplement). 2010, 58:13-17.

[36] Patel PH, Rathod S, Chuahan B, Rathod H, Pethani J, Shah P. Changing Trend of Antibiotic Susceptibility Pattern of Common Gram Negative Bacilli Isolated From Medical Intensive Care Unit of Tertiary Care Hospital Ahmedabad, Gujarat, India. *Journal of Drug Discovery and Therapeutics.* 2013, 1(4):16-20.

[37] Deshpande VR, Karmarkar MG, Mehta PR. Prevalence of multidrug-resistant enterococci in a tertiary care hospital in Mumbai, India. *J Infect Dev Ctries.* 2013, 7(2):155-58.

[38] Nazir T, Abraham S, Islam A. Emergence of Potential Superbug Mycobacterium Tuberculosis, Lessons from New Delhi Mutant-1 Bacterial Strains. *International Journal of Health Science.* 2012, 6(1):87-94.

[39] *CDC Campaign to Prevent Antimicrobial Resistance in Healthcare Settings, 12 Steps to Prevent Antimicrobial Resistance among Long-term Care Residents.* Department of Health and Human Services Centers for Disease Control and Prevention. March 2004. Accessed on 10. 2013.

[40] Leung E, Weil DE, Raviglione M, Nakatani H. World Health Organization. World Health Day Antimicrobial Resistance Technical Working Group. The WHO policy package to combat antimicrobial resistance. *Bull World Health Organ.* 2011, 89(5):390-2.

[41] *WHO Global Strategy for Containment of Antimicrobial Resistance,* World Health Organization 2001. WHO/CDS/CSR/DRS/2001.2.

[42] *European strategic action plan on antibiotic resistance*, Regional Committee for Europe, EUR/RC61/14 EUR/RC61/Conf.Doc./7, World Health Organization Regional Office for Europe, June 2011.

[43] *National Policy for Containment of Antimicrobial Resistance,* Directorate General of Health Services, Ministry of Health & Family Welfare, India 2011.

[44] Holloway K, Dijk L. Rational Use of Medicines, *The World Medicines Situation 2011,* 3rd Edition, World Health Organization, 2011. WHO/EMP/MIE/2011.2.

[45] Raghunath D. Emerging antibiotic resistance in bacteria with special reference to India. *J Biosci.* 2008, 33(4):593–603.

[46] Kardas P, Devine S, Golembesky A, Roberts C. A systematic review and meta-analysis of misuse of antibiotic therapies in the community. *International Journal of Antimicrobial Agents.* 2005, 26(2):106-13.

[47] Lipsitch M, Samore MH. Antimicrobial use and antimicrobial resistance: A population perspective. *Emerging Infectious Diseases.* 2002, 8(4):347-54.

[48] Bhagwati A. Guidelines for antibiotic usage in common situations. *JAPI* (supplement). 2010, 58:49-50.

[49] Morgan DJ, Okeke IN, Laxminarayan R, Perencevich EN, Weisenberg S. Non-prescription antimicrobial use worldwide: a systematic review. NIH Public Access, Author Manuscript. *Lancet Infect Dis.* Author manuscript, available in PMC 2013 January 14. Accessed on 24 Oct 2013. [Published in final edited form as: Lancet Infect Dis. 2011 September, 11(9): 692–701.].

[50] WHO Guidelines on Hand Hygiene in Health Care. *World Alliance for patient safety,* World Health Organization, 2009.

[51] *Prevention of hospital-acquired infections. A practical guide.* 2nd edition. World Health Organization 2002. WHO/CDS/CSR/EPH/2002.12.

[52] Ganguly NK, Arora NK, Chandy SJ, Fairoze MN, Gill JPS, Gupta U, et al. Rationalizing antibiotic use to limit antibiotic resistance in India. Global Antibiotic Resistance Partnership (GARP) – India Working Group. *Indian J Med Res.* 2011, 134:281-94.

[53] Essack SY. Strategies for the Prevention and Containment of Antibiotic Resistance. *SA Fam Pract.* 2006, 48(1):51.

[54] Moongtui W, Picheansathian W, Senaratana W. Role of nurses in prevention of antimicrobial resistance. *Regional Health Forum.* 2011, 15(1):104-11.

[55] McCoy D, Toussaint K, Gallagher JC. The Pharmacist's Role in Preventing Antibiotic Resistance. *US Pharm.* 2011, 36(7):42-49.

[56] Parameswaran R, Sherchan JB, Varma D M, Mukhopadhyay C, Vidyasagar S. Intravascular catheter-related infections in an Indian tertiary care hospital. *J Infect Dev Ctries.* 2011, 5(6):452-58.

[57] Kardas P. Patient compliance with antibiotic treatment for respiratory tract infections. *Journal of Antimicrobial Chemotherapy.* 2002, 49:897-903.

[58] Bratzler DW, Dellinger EP, Olsen KM, Peri TM, Auwaerter PG, Bolon MK, et al. Clinical practice guidelines for antimicrobial prophylaxis in surgery. American Society of Health-System Pharmacists (ASHP) report. *Am J Health-Syst Pharm.* 2013, 70:195-283.

[59] Bergeron MG, Ouellette M. Preventing Antibiotic Resistance through Rapid Genotypic Identification of Bacteria and of Their Antibiotic Resistance Genes in the Clinical Microbiology Laboratory. *Journal of Clinical Microbiology.* 1998, 36(8): 2169–72.

[60] Alekshun MN, Levy SB (2007) Molecular Mechanisms of Antibacterial Multidrug Resistance. *Cell* 128: 1037-1050.

[61] Kumarasamy KK, Toleman MA, Walsh TR, Bagaria J, Butt F, et al. (2010) Emergence of a new antibiotic resistance mechanism in India, Pakistan, and the UK: a molecular, biological, and epidemiological study. *Lancet Infect Dis* 10: 597–602.

[62] Bakken JS, Borody T, Brandt LJ, Brill JV, Demarco DC (2011) Treating Clostridium difficile Infection With Fecal Microbiota Transplantation. *Clinical Gastroenterology and Hepatology* 9: 1044-1049.

[63] Giacometti A, Cirioni O, Ghiselli R, Mocchegiani F, Del Prete MS, et al. (2002) Potential therapeutic role of cationic peptides in three experimental models of septic shock. Antimicrob. *Agents Chemother* 46: 2132–2136.

[64] Park SC, Park Y, Hahm KS (2011) The role of antimicrobial peptides in preventing multidrug-resistant bacterial infections and biofilm formation. *Int J Mol Sci.* 12: 5971-5992.

In: Antimicrobial Resistance: Opportunities and Challenges ISBN: 978-1-53617-943-9
Editors: P. Kumar Prabhakar and V. Mishra

Chapter 10

METALLIC NANOPARTICLES IN TACKLING OF ANTIMICROBIAL RESISTANCE

Anamika Jain[1], Payal Kesharwani[2], Mohd. Hashim Mansoori[3], Vijay Mishra and C. K. Sudhakar[4,*]

[1]Sagar Institute of Pharmaceutical Sciences, Sagar, MP, India
[2]Rameesh Institute of Vocational and Technical Education, Greater Noida, UP, India
[3]School of Pharmacy, G. H. Raisoni University, Amravati, Maharastra
[4]School of Pharmaceutical Sciences, Lovely Professional University, Phagwara, Punjab, India

ABSTRACT

Nowadays, antimicrobial resistance is one of the most significant threats to human and animal health and animal production systems have been pointed out as one of the most relevant hotspots of antimicrobial resistance generation and dissemination. With the growing need to minimize or eliminate the use of environmental-risk substances, as the green chemistry principles describe, the synthesis of noble metal nanoparticles (NPs) using biological entities has received increasing attention in the last decade. They are recognized by their outstanding antimicrobial properties by themselves and have been used for centuries to treat infections. Absorbance spectroscopy is useful to characterize metal nanoparticles, because they possess bright color which is visible by naked eye. By this technique, qualitative information about the nanoparticle can be obtained. Transmission electron microscope is also widely used to characterize nanomaterials to gain information about particle size, shape, crystallinity and interparticle interaction. Different MNP, including titanium, copper, magnesium and particularly silver and gold, are known for their

* Corresponding Author's E-mail: sudhakar.20477@lpu.co.in.

antimicrobial, antiviral and antifungal capabilities. These metallic NPs are actively investigated as disinfectants, in food processing, and for use as additives in clothing and in medical devices. There are many applications of MNPs in different field.

1. Anti Microbial Resistance: An Introduction

Among the most important was the discovery of antimicrobial agents, beginning with the synthesis of arsphenamine by Paul Ehrlich as the century dawned (Moellering, 1995). With this discovery, the dreaded scourge of syphilis was brought under control, although not eradicated. However, the toxicity of the drug made it less than ideal as an antimicrobial agent. Shortly thereafter, optochin (ethyl cupreine) was tried for therapy of pneumococcal pneumonia, but it too was toxic and was not effective enough to be successful. Moreover, pneumococci with resistance to this drug were isolated from patients who failed to respond to treatment- one of the first observations of antimicrobial.

Antimicrobial resistance has been fueled by inappropriate use of antimicrobial agents- especially those directed against bacteria. Widespread industrial and agricultural use of antimicrobials has played a role, but the unwillingness of the medical profession to accept measures for the restraint or control of indiscriminate prescribing and inappropriate dosing of antibiotics needs to be addressed. Clinicians have failed to deal with a potentially solvable problem, and others are taking up the challenge. Nowadays, antimicrobial resistance is one of the most significant threats to human and animal health (Gottlieb and Nimmo, 2011; Soulsby, 2007; Stella et al., 2018) and animal production systems have been pointed out as one of the most relevant hotspots of antimicrobial resistance generation and dissemination (Catry et al., 2003; Mathew et al., 2007; Merle et al., 2012). Evidence suggests that the incorrect use of antimicrobials in animal production is the most important factor for the emergence, selection, and dissemination of antimicrobial resistance (Sayah et al., 2005; Tempini et al., 2018). Some of the practices associated to the misuse of antimicrobials are wrong diagnostics, erroneous treatment schemes, incorrect dosage, or non-compliance of required treatment periods (Maron et al., 2013).

2. Metallic Nanoparticles

With the growing need to minimize or eliminate the use of environmental-risk substances, as the green chemistry principles describe, the synthesis of noble metal nanoparticles (NPs) using biological entities has received increasing attention in the last decade (Mohanpuria et al., 2008). Due to their unique physical and chemical properties, metal NPs have attracted tremendous interest in modern chemical research, and have found applications in a wide variety of fields, such as photochemistry, electrochemistry, optics and catalysis. To gain a better understanding of their fundamental properties and to optimize their activity for various applications, NPs must be characterized precisely in terms of size, shape and composition (Shankar et al., 2004). For example, both the size and shape of NPs have been shown to affect their catalytic activity. Biosynthesis of metal NPs has gained immense interest towards researchers, and it acts as a suitable substitute for hazardous chemical methods to achieve nanostructured materials with the controlled morphology of shape and size (Vilchis-Nestor et al., 2008; Li et al., 2007; Huang et al., 2007; Shankar et al., 2003).

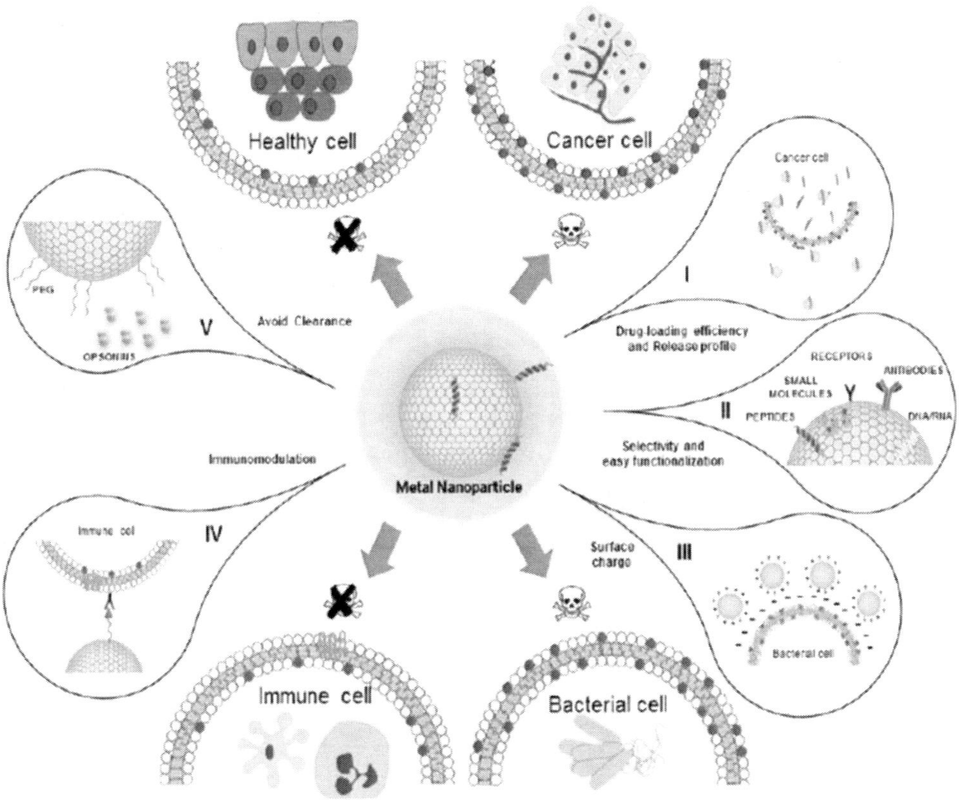

Figure 1. Strategies to deliver antimicrobial peptides (AMPs) using metal nanoparticles. Metal nanoparticles can be covered with a variety of chemical groups and (bio) molecules to enhance their delivery abilities. (I) The plasma membranes of cancer cells have anionic phospholipids exposed on the outer leaflet. The antitumor effect of MNP is determined, among other factors, by their passive targeting towards the cancer cells surface, with the help of cationic AMPs. MNP can improve the deposition and distribution of higher local doses of AMPs at the tumor site. (II) By active targeting, MNP can become selective to different types of cells or microorganisms. They can be functionalized with biomolecules or other ligands that would preferentially bind to cell surface receptors or other proteins, antibodies and/or DNA/RNA to gene delivery and silencing purposes. (III) The adsorption of MNP on bacteria membranes may lead to specific ion toxicity (membrane depolarization, perturbation of redox balance or membrane damage). Changes occur on the negatively charged cell surface, increasing its permeability. (IV) The ability of peptides, proteins or monoclonal antibodies to trigger the immune system, inducing a robust immune response, can be applied in the design of vaccines. (V) The nanoparticle surface modification with biocompatible polymers (e.g., PEG) may avoid its inactivation by the reticuloendothelial system. Furthermore, it can improve the stability, increase the solubility, decrease the cytotoxicity and enhance the bioavailability of a given drug.

Metal nanoparticles are created from nanoscale matter (1–100 nm) containing both organic and inorganic materials combined with various metals (Crooks et al., 2001; Sun et al., 2002; Iravani et al., 2011). These include transition metals, semiconductors, alkali metals, etc. The metal part of the nanoparticles plays a great role in the use of metal nanoparticles as green catalysts due to their large surface area-to-volume ratio compared to the bulk material. This enables efficient bonding, thereby allowing reactants to bind together at metal sites, and hence, reactions occur more efficiently. Also, layered transition metal dichalcogenides (TMDs) were found to serve as rare metals in combination with nanoparticles (Murphy et al., 2005). Some examples of TMDs include molybdenum disulfide (MoS_2), tungsten disulfide (WS_2), molybdenum diselenide ($MoSe_2$), molybdenum telluride ($MoTe_2$), etc. TMDs are of particular interest specifically because of their unique properties such as their excellent optical and electrical properties and large energy density (Hong et al., 2014; Orofeo et al., 2014).

Nanoparticles of noble metals have been synthesized for many applications. They are recognized by their outstanding antimicrobial properties by themselves and have been used for centuries to treat infections (Rajchakit and Surojini, 2017; Rai et al., 2009; Majdalawieh et al., 2014). Due to their small size and notably high surface area-to-volume ratio, among other properties, they are attractive conjugate platforms to improve AMP efficacy (El-sayed, 2001; Petros and Desimone, 2010). Among all metal NPs, silver (Ag) and gold (Au) NPs are extensively investigated regarding their strong antimicrobial potential (Jeong et al., 2018).

Metal-based nanoparticles (MNPs) are colloidal particles with exclusive properties. They possess specific optical behavior, electrical conductivity, and high thermal and chemical stability, which bulk forms do not possess (El-sayed, 2001; Bajaj et al., 2017). A major advantage of MNPs derives from their ability to be modified into bioconjugates (Moyano and Vincent, 2011; Cao-milan and Liz-Marzan, 2014). By using different molecules, it is possible to alter MNPs so that they can fulfill the criteria as efficient drug delivery agents (Figure 1), which implicates high stability between the drug and the NP, low toxicity and immunogenicity, affinity to the target, controlled release, and safe degradation and, ultimately, reduced health-care costs (Zhang, 2015). MNPs are generally unstable in suspension. It is important to prevent NPs from binding between each other in order to avoid aggregation phenomena (Rizzello and pompa, 2014). The problem associated with this is compromising the results due to the instability of NPs in culture media. These media contain salts that may alter the size and charge of a particle and affect ion release, leading to alterations in the NP biodistribution and irreproducible results (Rizzello and pompa, 2014).

3. Types of Mettallic Nanoparticles

3.1. Iron Oxide Nanoparticles

Laura, 2019 Iron oxide materials which are hybridized with noble metal nanoparticles (NPs) have attracted increasing interest among researchers because of their cooperative effects on combined magnetic, electronic, photonic, and catalytic activities.

Among all metal materials, noble and magnetic metals are two groups of the most promising materials. Magnetic iron oxides (mainly magnetite and maghemite) are principally attractive as components of multifunctional nanomaterials because they interact with external fields, giving them the ability to be easily recovered and reused or otherwise manipulated by magnetic means (Castillo et al., 2019). Super paramagnetic iron oxide NPs have multiple functionalities that make them very attractive for biomedical applications (Laurent et al., 2008).

Magnetic manipulation has been successfully proposed for drug targeting in specific locations of the body, to enhance cell therapy by retaining therapeutic cells on the site to be regenerated or for tissue engineering in order to reproduce the native cellular organization of tissues. In contrast, severe restrictions are related to the use of magnetic nanoparticles (MNPs) for in vivo biomedical applications (Laurent et al., 2008). Some of them are related to the organs involved in NPs metabolism (lungs, liver, spleen, and kidney), others with the chemical and physical characteristics of MNPs that can influence their pharmacokinetics (absorption, biodistribution and elimination), targeting ability, and optical properties (Dai et al., 2014). It is known that nanometric iron oxides can be synthesized by several routes, such as thermal decomposition, co-precipitation and solvothermal methods (Majidi et al., 2016).

3.2. Zeolites Nanoparticles

In this context, to preserve the unique properties of nanoparticles and subnanometric particles for various applications, it is urgently desired to explore distinct types of supports to control the size and dispersion of metal particles by protecting them with interconnected pores and/or cavities. Zeolites, with characteristic macrospores, uniform cage structures, and fine-tuned acid–base sites, appear to be most promising host materials to confine the small metal particles, especially with size of <2 nm (Riahi et al., 2002; Kim et al., 2014). On the other hand, the microspores of zeolite hosts will bring about problems in the molecule diffusion and metal site accessibility, and the applications of metal particles confined in zeolite hosts are significantly restricted if the molecule diffusion has to be considered.

Since the success of large-scale industrial production, zeolites have been widely used as catalysts and support materials (Ennaert et al., 2016; Weckhuysen and Yu, 2015). Strictly speaking, zeolites are microporous aluminosilicates built by SiO_4 and AlO_4 tetrahedral linked into corner-sharing networks, where the Si atoms bear charges of +4 and the Al atoms charges of +3 (Meng and Xiao 2014; Speybroeck et al., 2015). The Si and/or Al atoms can be substituted by other elements to construct zeolite-like materials such as phosphoaluminates. Due to the presence of aluminum in zeolite framework, electroneutrality requires the incorporation of charge compensating cations, which occupy extra-framework sites in the channels and cages. For zeolites supported metal species, three typical types can be recognized: (a) metal species loading on the outer surface of zeolite crystals, (b) metal species encapsulated within the channels or cavities of zeolites, and (c) metal species incorporated into zeolite framework (Figure 2). The samples of type (a), denoted as metal/zeolites, are usually prepared from zeolite supports and metal precursors via simple impregnation (Fodor et al., 2015).

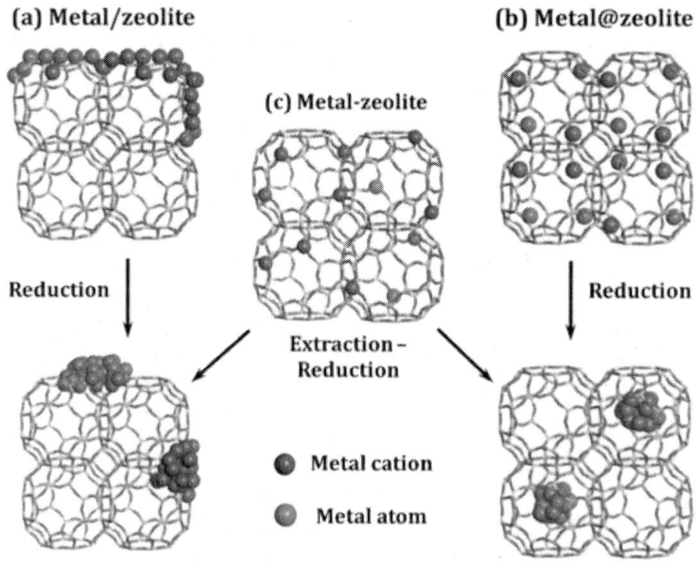

Figure 2. Three typical types of metal containing zeolites.

3.3. Platinum Group Metals (PGMs) -Based Binary Alloy Systems

The platinum group metals (PGMs) are six noble and precious metals, Ru, Rh, Pd, Os, Ir, and Pt. They are all transition metals of the d-block with groups of from 8 to 10 and periods of 5 to 7 in the periodic table of the elements; therefore, they have a comparatively high density of states (DOS) around the Fermi energy (EF) in their electronic structure (Pettifor, 1977).

3.4. Ru-Based Alloy NPs

Ru is the third most studied element in the PGMs; this is because the Ru–Pt system is an attractive combination as an active electrocatalyst with a high CO-poisoning tolerance of the methanol oxidation reaction (MOR) or similar, which is a key reaction of direct methanol fuel cells. It is well known that Pt NPs show superior catalytic activity for MOR and many kinds of Pt based binary alloys were studied (Xia et al., 2015; Qi et al., 2017). However, CO-poisoning during the reaction is a crucial problem and it was found that Ru was able to improve its CO-poisoning tolerance (Xie et al., 2016). Among many reports about this system, shape-controlled Ru–Pt nanocrystals (nanowires, nanorods, and nanocubes) were recently synthesized by a one-step solvothermal method, which revealed that the {111} facet of fcc Pt–Ru possesses superior MOR activity and CO-poisoning resistance relative to the {100} facet (Huang, et al., 2018).

3.5. Rh-Based Alloy NPs

In the Rh case, the combinations with Pd or Pt are the most studied. They show high oxidizing and reducing catalytic activity for CO, hydrocarbon, and NOx, therefore they are practically used as three-way catalysts for the purification of exhaust automobile gas (Bowker, 2008). Moreover, they are studied as electrocatalysts for the ethanol oxidation reaction (EOR) or similar (Kugai et al., 2006; Yuan et al., 2010). In addition, the other class 4 metals (Co and Ni) are also relatively well studied. As one example, RhNi was studied as a catalyst for the decomposition of hydrous hydrazine to hydrogen (Singh et al., 2009). The authors found that Rh4Ni NPs reached 100% conversion at room temperature, even though the best monometallic catalyst, Rh NPs, can convert only 43.8% at the same condition.

3.6. Gold Nanoparticles

Gold nanoparticles with a central core and protruded sharp tips from central core were normally categorised as nanoflowers and nanostars depending on the sharpness and number of tips. Both these types of nanostructures exhibit local field enhancement at the sharp protrusions and nanogaps called hot spots that could be used for a wide range of applications including surface enhanced spectroscopy (Lal et al., 2007), imaging (Kawata et al., 2009), molecular sensors (Stewart et al., 2008) and photothermal therapy (Hao et al., 2007).

Gold nanoparticles (AuNPs), of different geometric shapes, can be used for spatially and temporally controlled drug release, thereby taking advantage of their biocompatibility, size and ease of functionalization, in addition to their optic, plasmonic and photothermal properties [Eustis and El-Sayed, 2006]. Optic properties in AuNPs can be modulated by controlling their shape, size, and surface modification through synthetic approaches. A wide variety of AuNP shapes have been reported, including gold nanospheres (AuNSs), nanorods (AuNRs), nanoprisms (AuNPrs), nanoclusters (AuNCs), nanoflowers and core-shells (Alfranca et al., 2016; AL-Jawad et al., 2018; Uppal and Bose, 2017). Irradiation of AuNPs in their plasmonic band, allows them to absorb and dissipate the energy as local heat, which can be exploited in biomedical applications, including photothermal therapy and drug release (Kuo et al., 2012).

Localized surface plasmon resonance (LSPR) on AuNPs is highly dependent on their shape, size, and the surrounding medium, and in this context they are useful for biosensing and detection [Carvalho et al., 2018]. However, their unique optical properties allow not only transport and release of molecules with therapeutic properties, but also theranostic applications (Peralta et al., 2015).

3.7. Silver Nanoparticles

The nanotechnology further increases the surface-to-volume ratios and catalytic activity of AgNPs, which leads to great enhancement in their antimicrobial activity (Choi et al., 2008; Yu et al., 2013). And the antimicrobial mechanism of AgNPs is generally considered as a multi-factor, multi-way, and multi-target process (Bae et al., 2010; Kahru and Ivask 2013; Suresh et al., 2013). AgNPs can attach to the cell membranes and interact with the molecules on the membranes, which will damage the integrity and permeability of the membranes and thereby leading to the cytomorphosis and the leakage of intracellular contents (Marambio-Jones and Hoek 2010; Sondi and Salopek-Sondi 2004). And the reactive oxygen species (ROS) produced by AgNPs and silver ions released from AgNPs not only damage the cell membranes but also react with the molecules in the functional proteins and DNA, which will interfere the metabolism and DNA duplication (AshaRani et al., 2008; Choi and Hu 2008; Hackenberg et al., 2011). All these injuries to cells finally lead to the cell death.

Because the antimicrobial mechanism of AgNPs is complicated, the physical and chemical properties of AgNPs can make a great influence on their antimicrobial activity, including size, shape, coating, and surface charge (Marambio-Jones and Hoek 2010).

4. Properties of Metallic Nanoparticles

4.1. Shape and Structural Properties

During their formation, NPs minimize their total free energy by adopting facetted structures. These structures exhibit faces of different Miller indices, mainly (111) and (100) faces. It worth to note that (111) faces have lower free energy than (100) ones (i.e., (111) < (100)). Hence, the so-called "spherical" NPs possess mainly (111) faces whereas others kind of NPs (e.g., nanoplates, nanorods) present large (100) faces. The NPs shape and the kind of faces will also influence bacteria/NPs interaction and then their antibacterial activity. For instance, Hong et al., added NaCl to the reaction medium to modify the nanoparticles shape [Hong et al., 2016]. Starting from AgNO3 and polyvinylpyrrolidone (PVP k30) in ethylene glycol, they obtained "spherical" nanoparticles of 60 nm (15 nm) exhibiting mainly {111} facets. With a small amount of NaCl, truncated-nanocubes are formed. With these truncated-nanocubes, there are by-products such as bipyramids, nanospheres and nanorods. The average size of these truncated-nanocubes is 55 nm (10 nm), and the sample exhibits mainly {100} facets with {111} ones for truncated faces.

4.2. Methods of Preparation of Metallic Nanoparticles

4.2.1. Physical Methods

4.2.1.1. Ball Milling

In mechanical milling (MM), a suitable powder charge (typically, a blend of elemental) is placed in a high energy mill, along with a suitable milling medium. The objective of milling is to reduce the particle size and blending of particles in new phases. In these techniques, mixtures of elemental or pre-alloyed powders are grinded under a protective and defensive environment in equipment that can generate high-energy compressive impact forces inclusive of attrition or shaker mills (Iravani et al., 2018). The different type of ball milling can be used for synthesis of nanomaterials in which balls impact upon the powder charge (Koch and Whittenberger, 1996). The balls may roll down the surface of the chamber in a series of parallel layers or they may fall freely and impact the powder and balls beneath them. For large scale production with nano size grain, mechanical millings are more economical processes (Klaus et al., 1999). The kinetics of mechanical milling or alloying depends on the energy transferred to the powder from the balls during milling (Konishi et al., 2007). The energy transfer is governed by many parameters such as the type of mill, the powder supplied to drive the milling chamber, milling speed, size and size distribution of the balls, dry or wet milling, temperature of milling and the duration of milling (Yadav et al., 2012).

4.2.1.2. Lithography

- Vapor phase and vaporization

In the physical deposition method, material is deposited on a surface either as a thin film or as nanoparticles. Highly controlled vacuum technique such as thermal evaporation and sputtered deposition causes vaporization of material, which is further condensed on a substrate (Pandey et al., 2011). Physical vapor deposition techniques such as pulsed vapor deposition are generally used for the preparation thin film of lanthanum strontium cobalt (Park JS et al., 2017). For the pulsed laser deposition, laser ablation is employed on solid target that causes formation of plasma of ablated species and further, these ablated species are deposited on a substrate to produce a film (Gondoni et al., 2013). This method is extensively used to deposit metal nanoparticle and thin film on carbon nanotubes (Pandey et al., 2011).

Advantages

- Simple method for the formation of thin metal films.
- Disadvantage
- Expensive method.
- Generates low volume of material.

This method is used for the formation of most efficient thin-film solar cells, Cu (In,Ga) Se2 thin film using pulsed laser deposition (PLD). The femtosecond (fs)-pulsed laser deposition (Fs-PLD) derived copper indium gallium selenide (CIGS) thin films shows prominent antireflection and excellent crystalline structure (Chen et al., 2014).

4.2.1.3. Ion Sputtering

Ion sputtering method includes vaporization of a solid through sputtering with a beam of inert gas ions. Recently this method was used for the preparation of nanaoparticles from several metals using magnetronsputtering of metal targets. In this method collimated beams ofthe nanoparticles is formed and the mass nanostructured films are deposited on the silicon substrates. The entire process is performed at relatively low pressures (1 mTorr) (Swihart et al., 2003). Sputter deposition is done in evacuated vacuum chamber wheresputtering gas is admitted and working pressure (eg. 0.05 and 0.1 mbar) is maintained. A very high voltage is introduced in to the target (cathode) and free electrons are moved in spiral path using magnetic system where they collide with sputtering gas (argon) atoms and leads to ionization of gas. This continuous process produces a glow discharge (plasma) to ignite.

The positively charged gas ions attracted towards target where they continuously impinge. This event repeated occurs and approaches the surface of target with energy above the surface binding energy, an atom can be expelled. The collisions occur between metal atoms and gas molecules continuously in vacuum chamber that leads to scattering of atoms forming a diffuse cloud (Vanecht, 2012).

Advantages

- The composition of sputtered material is not altered and remains same as that of the target material (Kruis et al., 1998).
- Method of choice for refractory metals and intermetallic compounds than other methods like evaporation and laser ablation.
- Economical method as the sputtering equipment is less expensive than electron-beam lithography systems (Nie et al., 2009).
- Less impurities are generated than those created by chemical methods.
- Alloy nanoparticles can be produced with easier control on composition than other chemical reduction methods (Nie et al., 2009).
- This method is a versatile technique to synthesize ionic nanoparticles with spacious sizes and compositions that are not obtainable in solution (Johnson et al., 2016).
- Slow deposition of heavier ions or mass-selected ions gives unparalleled control of different parameters such as size, composition and charges of ions deposited onto surfaces (Johnson et al., 2016).

Disadvantages

- The nature sputtering gas (He, Ne, Ar, Kr, and Xe) can produce effect on surface morphology, composition, texture, and the optical properties of the nanocrystalline metal oxide films (Chandra et al., 2006).

4.2.1.4. Laser Ablation

In the laser ablation method, laser irradiation is used to reduce the particle size to nano level. The solid target material is placed under a thin layer and then exposed to pulsed laser irradiation. Mainly Nd:YAG (neodymium-doped yttrium aluminium garnet) laser at 106 μm output and its harmonic, Ti:Sapphire (Titanium-doped sapphire) laser and coppe vapor lasers are used (Simakin et al., 2004). The irradiation of material to laser leads to fragmentation of solid material in the form of nanoparticles, which remains in liquid that surrounds the target and produces colloidal solution. The laser pulse duration and energy determines the relative amount of ablated atoms and particles formed (Kruis et al., 1998). Several parameters such as time duration of laser pulse, wavelength, ablation time, laser fluency and effective surrounding liquid medium with or without surfactant influences ablation efficiency and characteristic of metal particle formed (Nour et al., 2010).

Advantages

- Relatively simple and effective technique for the formation of large amount of small particles (nano-size) in the form of suspension.
- Nanoparticle properties can be changed by selecting the laser parameter and nature of liquid accordingly.
- Nanoparticles formation is possible without adding surfactant in liquid media (Simakin et al., 2004).

Disadvantage

- Prolong time laser ablation leads to formation of high amount of nanoparticles in the colloidal solution which block the laser path and also laser energy is get absorbed by already formed nanoparticles instead of target surface. This overall leads to reduction in ablation rate (Ghorbani, 2014).

4.2.1.5. Spray Pyrolysis

In the spray pyrolysis method, nanoparticle precursors in vapor form are delivered into the hot reactor. For the delivery of precursor, a nebulizer is employed that directly deliver precursor in the minute small droplets form into the hot reactor. Metals such as acetate, nitrate and chloride are commonly served as metal precursor (Okuyama and Lenggoro, 2003). The apparatus used for nanoparticle preparation by spray pyrolysis method consist of three main parts 1) A fluid nebulizer for atomization of metal precursor solution 2) A thermostatically (200 °C to 1200 °C) controlled vertical tubular reactor. 3) A precipitator for collecting nanoparticles (Ghaffarian et al., 2011). The apparatus can be modified by employing atomized techniques such as two fluids nozzle or air assisted (pneumatic) pumps or sprayers, ultrasonic sound, vibrating orifice and spinning disk (Okuyama and Lenggoro, 2003).

The ultrasonic spray pyrolysis, is an improved method where ultrasound is used to produce atomized droplet from precursor solution and formed aerosol droplets are further transported from atomizer to reactor furnace by the carrier gas for nanoparticles formation that are later collected by collection system. The heavily diluted precursor solution is converted to small aerosol droplet of size 1–10 μm, which depends upon frequency of ultrasound and properties of thesolution (Majerič et al., 2015).

Advantages

- Relatively simple method.
- Low cost method (Okuyama and Lenggoro, 2003).
- The particle size can be controlled and reproducible (Ghaffarian et al., 2011).

4.2.2. *Chemical Method*

4.2.2.1. Sol-Gel Process

Livage et al. first time review this technique on sol gel chemistry of transition metal oxides (Livage et al., 1988). The sol gel method of nanoparticles synthesis involves either; a) Mixing ofpreformed colloids metal (oxide) with a sol containing the matrix forming species followed by gel formation, b) Direct mixing of metal and metal oxide or nanoparticles within a prehydrolysed silica sol. c) Complexation of metal with silone and reduction of metal before hydrolysis (Cushing BL et al., 2004). In this method, a network formation is introduced using colloidal suspension (sol) and gelatin to form a network in continuous liquid phase (gel). The ions of metal alkoxides and aloxysilanes are used as a precursor for synthesis of colloids. The tetramethoxysilane (TMS) and tetraethoxysilane are most commonly employed which forms silica gel. Metal alkoxides are organo-metallic precursor for various metals such as silica, aluminium, titanium and many other and are immiscible in water. Alcohol is used as mutual solvent. Initially in this method, homogeneous solution of one or more selected alkoxides is prepared and a catalyst is added to initiate a reaction at controlled pH. The sol-gel formation involves four mai steps; hydrolysis, condensation, particle growth and agglomeration of particle (Rajput, 2015). In direct precipitation of metal or metal oxide technique, the metal oxide particles are precipitated from silica sol usually by heat treatment at low temperature. Thin films are mainly prepared by using this technique (Cushing et al., 2004).

Advantages

- Simplest method.
- Particle size and morphology is possible to control by systematic monitoring of reaction parameters (Hasnidawani et al., 2016).

4.2.2.2. Chemical Vapor Deposition

In this method of deposition, thin film of target material is deposited on a surface through the chemical reaction of gaseous molecule containing atoms useful for film formation (Pedersen et al., 2014). The target material is released in the form of volatile molecule and works as precursor and then series of chemical reactions take place between the precursor fragment, precursor and substrate surface to produce a thin film. Generally atomic layer deposition (ALD) thin films are produced by the surface chemical reaction in this method (Miikkulainen et al., 2013).

The deposition of target material can be achieved by: 1) Thermally active chemical vapor deposition (TACVD) (Doll et al., 2010) 2) Plasma enhanced chemical vapor deposition (PECVD) and 3) Photo initiated chemical vapor deposition (PICVD) (Bedair et al., 2015).

4.2.2.3. Chemical Reduction Technique

In chemical reduction method, ionic salt is reduced in an appropriate medium in the presence of surfactant using different reducing agents (Guzmán et al., 2009). Reducing agent such as sodium borohydride is used in aqueous solution to prepare metal nanoparticles. Formed metal nanoparticles are capped by using trisodium citrate (TSC) or sodium lauryl sulphate (SLS). Sometimes stabilizing agent is used with reducing agent. The metal nanoparticles stability in the dispersion was monitored by the analysis of absorbance (Chattopadhyay and Patel, 2014). Reducing agents such as sodium borohydrate ($NaBH_4$), glucose, ethylene glycol, ethanol, citrate of sodium, and hydrazine hydrate etc. are used for silver nanoparticles synthesis (Landage et al., 2014).

Advantages

- Simplest method used for preparation of metal nanoparticles (Guzmán et al., 2009).

Disadvantages

- Several limitations associated with reducing agents such as toxicity, expensive, poor reducing ability, high costs, and impurities (Zhang et al., 2010).

4.2.2.4. Solvo-Thermal

Solvo-thermal method is used for the preparation of nanophase in presence of water or other organic chemicals like methanol, ethanol and polyol as a solvent. Reaction is produced in pressure vessel that allows solvent (water and alcohol) heating above their boiling point temperature (Yang et al., 2007). The kinetics of crystallization (crystal formation) can be increased by the one to two orders of magnitude by employing microwave assisted reactions (microwave solvo-thermal) (Komarneni et al., 2003).

Advantages

- Preparation of high quality crystallized monodispersed nanocrystals.
- This method is more preferred for the preparation of narrow size distribution of high degree of crystallization of nanocrystallites over conventional oil bath heating (Yang et al., 2007).

4.2.2.5. Co-Precipitation

However, the most common method for the production of magnetite nanoparticles is the chemical coprecipitation technique of iron salts (Yadav et al., 2012). The main advantage of the coprecipitation process is that a large amount of nanoparticles can be synthesized but with limited control on size distribution. This is mainly due to that the kinetic factors are controlling the growth of the crystal (Mathe et al., 2007). Thus the particulate magnetic contrast agents synthesized using these methods include ultrasmall particles of iron oxide (USPIO) (10–40 nm), small particles of iron oxide (SPIO) (60–150 nm). Besides, monocrystalline USPIOs are also called as monocrystalline iron oxide nanoparticles (MIONs), whereas MIONs when cross-linked with dextran they are called crosslinked iron oxide nanoparticles CLIO (10–30 nm) (Xu et al., 2015). The modification of the dextran coating by carboxylation leads to a shorter clearance half-life in blood (Simakin et al., 2004). Hence, ferumoxytol (AMAG Pharmaceuticals), a carboxyalkylated polysaccharide coated iron oxide nanoparticle, is already described as a good first-pass contrast agent but uptake by macrophages is unspecific and too fast to enhance the uptake in macrophage-rich plaques (Mody et al., 2010).

4.2.2.6. Biological Method/Biomimetic Method/Green Synthesis Method

Preparation of nanoparticles using green synthesis methods is an emerging trend of nanotechnology. These techniques are emerged to overcome problems such as reaction complications, high cost and safety issue of conventional methods. The green chemistry approaches incorporated novel techniques to the synthesis processes and numerous applications of chemical substances to reduce threat to health and environment. These approaches mainly include; 1) Clean chemistry, 2) Atom economy, 3) Environmentally benign chemistry, and 4) Benign by design chemistry (Bhosale et al., 2014).

The biological methods of nanoparticle preparation involve application of different microorganism and their enzymes, plant products like isolates and extracts. These techniques have number of advantages over other physical and chemical methods as it is a cost effective, ecofriendly method, and easily scaled up for large scale production. In addition, green synthesis does not involve use of high pressure, energy, temperature and toxic chemicals (Karnani et al., 2013).

The natural biogenic metallic nanoparticle synthesis is divided into two categories.

1) The biologically stable form using microorganisms and their enzymes. The formed metallic nanostructures are stable and inert in nature that can be safely separated from contaminated sample (Pantidos and Horsfall, 2014).
2) Biosorption: this is a unique method of nanoparticle synthesis where metal cations in aqueous media is allow to bind with organism cell wall that further leads to formation of stable nanoparticles because of cell wall or peptide interaction (Yong et al., 2002).

4.2.2.7. Bacteria

The interest for appropriate implementation of naturally occurring recourses such as microorganism for the synthesis of nanoparticles is continuously increasing worldwide. Prokaryotes have gained attention as a mean of metallic nanoparticle synthesis because of their abundance in environment and ability to adopt extreme condition. Bacteria have certain advantages like rapid multiplication, and easy to cultivate and manipulate (Pantidos et al., 2014). The growth can be controlled by controlling conditions like oxygenation, temperature and incubation time. He et al., proposed that changing the pH of growth medium during incubation produces nanoparticle of different size and shape (He et al., 2007).

4.2.2.8. Fungi

Enzyme and protein secreted by fungi functions as reducing agent that can be used for metal nanoparticle synthesis from metal salt. Metal such as Ag has ability to bind with cytoplasmic membrane due to electrostatic interaction and get reduced which further forms silver nuclei and eventually leads to accumulation of silver nuclei and nanoparticles (Ahmad et al., 2003). Commonly used fungi for nanoparticle preparation such as *Fusarium oxysporum*, *Aspergillus fumigates*, *Trichoderma reesei* and *Fusarium oxysporum* secret certain reducing agents that reduce these metal ions. Practically microorganism based nanoparticle syntheses is not feasible as they require maintenance of high aseptic condition (Pantidos et al., 2014).

Advantages

- Nanoparticle synthesis is easy to scale up and downstream processing, economical flexibility and large surface area because of mycelia.
- Productivity is more than bacteria as fungi secret more amounts of proteins (Pantidos et al., 2014).

Disadvantage

- Safety risk.
- Nanoparticles produced are of variable size (Pantidos et al., 2014).

5. Other Methods of Nanoparticle Synthesis

5.1. Electrochemical Deposition

Methods such as electrode position synthesis or template synthesis are used for the production of nanomaterials (Dikusar et al., 2009). Ion etching technique is used for the production of porous alumina membranes (Furneaux et al., 1989). In addition, for the manufacturing of porous semiconductor structures, an electrochemical deposition method such as electrochemical anode dissolving processes is preferred (Dikusar et al., 2009). The metal nanoparticles are prepared by electrochemical deposition (ED) using baths containing metal salts. The baths are either acidic or basic and use a three terminal potentiostat. In this technique an electrode, cathode is needed where the metal nanoparticles are deposited whereas another anode electrode and Ag\AgCl or calomel is a reference electrode. Slight voltage is applied for suitable time in an electrolytic bath containing metal salt. This technique is widely employed for different purposes such as cyclic voltammetry, double pulse and potential step deposition (Islam and Islam, 2013).

Advantages

- Simple, fast and inexpensive method.
- Unique method that produces nanoparticles of controlled size and morphology.

Major advantage

- The nanoparticles get directly attached to the substrate (Mohanty, 2011).

5.2. Microwave Assisted Nanoparticles Preparation

Nowadays microwave techniques are more proffered over thermal heating for the preparation nanoparticles. Microwave frequency of range 300 MHz to 300 GHz is applied that leads to orientation of polar molecule such as H_2O with the electric field. The re-orientation of dipolar molecules with an alternating electric field causes molecular friction and loss of energy in the form of heat. Recently this technique was successfully implemented for the preparation of silver nanoparticles where silver nitrate solution is irradiated with carboxymethyl chitosan, which acts as reducing agent and a stabilizer (Wang et al., 2010).

5.3. Supercritical Fluid Technology

Supercritical fluids (SCFs) are widely used in the several areas of material and chemical science. The SCFs have distinguishing feature that their physicochemical properties can be modified between gaseous and liquid states by simply altering the pressure or temperature, which provides opportunity to change reaction environment like density, viscosity, diffusivity or surface tension (Aymonier et al., 2006). The ability of SCFs to get rapid expansion is a main feature, which is selectively employed for the synthesis of fine powders such as nanoparticles (Ye et al., 2003). Many of the physical properties of SCFs vary with density. The most commonly used SCFs are carbon dioxide ($scCO_2$), nitrous oxide, water, methanol, ethanol, ethane, propane, n-hexane and ammonia (Sekhon et al., 2012).

5.4. Ultra Sound Technique

The use of ultra sound for material synthesis has been extensively increased in recent few years. Ultrasound assisted synthesis techniques include two main methods, sonochemistry and ultrasonic spray pyrolysis (Natarajan et al., 2017).

5.4.1. Characterization of MNP

Absorbance Spectroscopy: Spectroscopy is useful to characterize metal nanoparticles, because they possess bright colour which is visible by naked eye. By this technique, qualitative information about the nanoparticle can be obtained. By applying Beer's law absorbance is measured. Depending on path length (l), nanoparticle conc. (c), extinction coefficient (A) can be obtained (Kumar et al., 2018).

Infrared Spectroscopy: This method can provide information on organic layers surrounding metallic nanoparticles. It also gives valuable information to understand the surface structure of the metal nanoparticles (Kumar et al., 2018).

TEM: (Transmission electron microscopy) is also widely used to characterize nanomaterials to gain information about particle size, shape, crystallinity and interparticle interaction. TEM is a high spatial resolution structural and chemical characterization tool. It has the capability to directly image atoms in crystalline specimens at resolutions close to 0.1 nm, smaller than interatomic distance. An electron beam can also be focused to a diameter smaller than ~0.3 nm, allowing quantitative chemical analysis from a single nanocrystal (Kumar et al., 2018).

SEM: (Scanning Electron Microscopy) It is a powerful technique for imaging any material surface with a resolution down to about 1nm. The interaction of an incident electron beam with the specimen produces secondary electrons, with energies smaller than 50 ev. SEM can give information about the purity of nanoparticle sample (Kumar et al., 2018).

AFM: It is a better choice for nonconductive nanomaterials. Typically, it has vertical resolution of less than 0.1 nm and lateral resolution of around 1 nm. It gives detailed information on the atomic scale, which is important for understanding electronic structure and chemical bonding of atoms and molecules (Kumar et al., 2018).

XRD: It is useful and widely used technique for determining the crystal structures of crystalline materials. Diffraction line widths are closely related to the size and their distribution, strain in nanocrystal. The line width is broadened, as the size of the nanocrystal decreases, due to loss of long range order relative to the bulk. XRD line can be used to determine the particle size by Debye-Scherrer method (Kumar et al., 2018).

$$D = 0.9\,\lambda / b\cos\Theta$$

where, D = nanocrystal diameter

λ = light wavelength
b = full width half at max. Of the peak (radians)
Θ = Bragg angle.

FTIR: It is widely adopted techniques compared to IR spectroscopy. Functional groups attached to the metallic nanoparticle surface show different FTIR pattern than those of free groups (Kumar et al., 2018).

EXAFS: (Extended X-ray Absorption Fine Structure) this is one of the most reliable and powerful characterization technique to evaluate the structure of metallic nanoparticles; especially it is useful to determine bimetallic nanoparticles. To gain appropriate information about the structure, the sample of metallic nanoparticles should be homogeneous. This method provides the no. Of atoms surrounded the x-ray absorbing atom and their interatomic distances involved in the shells (Antoniak, 2011).

XPS: (X-ray Photoelectron Spectroscopy) it is used to provide information on metal state. Suppose the oxidation state of metal on the surface. It is often oxidized by air. So, by using this method, 0-valency of surface metal must be confirmed (Trigo, 2011).

6. Biomedical Applications

Applications of metallic NPs in the biomedical fields are numerous, and there is strong potential for continued growth in this area. Metallic NPs are widely used for their antimicrobial functionality; for example, silver NPs (AgNPs) have been incorporated into wound dressings, bone cements and implants (Chaloupka et al., 2010). Gold NPs (AuNPs) have medically relevant optical and anticancer properties. For example, Alanazi et al., described how surface plasmon absorption and surface plasmon light scattering can be used for diagnostic and therapeutic applications (Alanazi et al. 2010), and Patra et al., described the fabrication and application of AuNPs for targeted cancer therapy (Patra et al., 2010).

6.1. Antimicrobial Applications

The ongoing development of antimicrobial agents is important due to the continuous selection of antibiotic resistance traits in bacteria and other pathogens. Different metallic NPs, including titanium, copper, magnesium and particularly silver and gold, are known for their antimicrobial, antiviral and antifungal capabilities (Rai et al., 2009). These metallic NPs are actively investigated as disinfectants, in food processing, and for use as additives in clothing and in medical devices (Marambio-Jones et al., 2010).

6.2. Antibacterial Activity

AgNP exposure causes toxicity to bacteria, primarily from Ag ions released into aqueous solution following partial oxidation (Chaloupka et al., 2010). Ag ions and small NPs interact with the plasma membrane, disturbing cellular functions, including permeability and respiration, and ultimately leading to lysis. AgNP exposure can prevent DNA replication and protein synthesis by binding to DNA or by denaturizing ribosomes (Chaloupka et al., 2010).

The detailed mechanisms of AuNP antibacterial functionality against *Escherichia coli* is described by Cui et al., (Cui et al., 2012). AuNPs were shown to collapse membrane potential, strongly inhibiting ATPase activity and resulting in a decrease in cellular ATP levels. Another consequence of AuNP exposure was inhibited binding of tRNA to the ribosome subunit. Bacterial susceptibility to antimicrobial agents can depend on the cell wall structure. One of the first reports of antibacterial effects of biosynthesized NPs was published by Vigneshwaran et al., (Vigneshwaran et al., 2007). These authors used the edible mushroom Pleurotus sajor-caju for the synthesis of AgNPs and performed successful antimicrobial testing against Staphylococcus aureus (G+ve) and Klebsiella pneumoniae (G-ve) bacteria. Raheman et al., reported the extracellular synthesis of AgNPs by means of an endophytic fungus, Pestalotia sp., isolated from leaves of *Syzygium cumini* and the antimicrobial properties of AgNPs against *S. aureus* (ATCC-25923) and *Salmonella typhi* (ATCC-51812) (Raheman et al., 2011).

REFERENCES

[1] Ahmad, Absar, Priyabrata Mukherjee, Satyajyoti Senapati, Deendayal Mandal, M. Islam Khan, Rajiv Kumar, and Murali Sastry. "Extracellular biosynthesis of silver nanoparticles using the fungus *Fusarium oxysporum*." *Colloids and surfaces B: Biointerfaces* 28, no. 4 (2003): 313-318.

[2] Alanazi, Fars K., Awwad A. Radwan, and Ibrahim A. Alsarra. "Biopharmaceutical applications of nanogold." *Saudi Pharmaceutical Journal* 18, no. 4 (2010): 179-193.

[3] Alaqad, Khalid, and Tawfik A. Saleh. "Gold and silver nanoparticles: synthesis methods, characterization routes and applications towards drugs." *J. Environ. Anal. Toxicol* 6, no. 4 (2016): 525-2161.

[4] Alfranca, Gabriel, Álvaro Artiga, Grazyna Stepien, María Moros, Scott G. Mitchell, and Jesús M. de la Fuente. "Gold nanoprism–nanorod face off: comparing the heating efficiency, cellular internalization and thermoablation capacity." *Nanomedicine* 11, no. 22 (2016): 2903-2916.

[5] Alivisatos, A. Paul, Kai P. Johnsson, Xiaogang Peng, Troy E. Wilson, Colin J. Loweth, Marcel P. Bruchez Jr, and Peter G. Schultz. "Organization of 'nanocrystal molecules' using DNA." *Nature* 382, no. 6592 (1996): 609.

[6] AL-Jawad, Selma MH, Ali A. Taha, Mohammad MF Al-Halbosiy, and Lamyaa FA AL-Barram. "Synthesis and characterization of small-sized gold nanoparticles coated by bovine serum albumin (BSA) for cancer photothermal therapy." *Photodiagnosis and photodynamic therapy* 21 (2018): 201-210.

[7] Antoniak, Carolin. "Extended X-ray absorption fine structure of bimetallic nanoparticles." *Beilstein journal of nanotechnology* 2, no. 1 (2011): 237-251.

[8] AshaRani, P. V., Grace Low Kah Mun, Manoor Prakash Hande, and Suresh Valiyaveettil. "Cytotoxicity and genotoxicity of silver nanoparticles in human cells." *ACS nano* 3, no. 2 (2008): 279-290.

[9] Atiyeh, Bishara S., Michel Costagliola, Shady N. Hayek, and Saad A. Dibo. "Effect of silver on burn wound infection control and healing: review of the literature." *burns* 33, no. 2 (2007): 139-148.

[10] Aymonier, Cyril, Anne Loppinet-Serani, Helen Reverón, Yves Garrabos, and François Cansell. "Review of supercritical fluids in inorganic materials science." *The Journal of Supercritical Fluids* 38, no. 2 (2006): 242-251.

[11] Bae, Eunjoo, Hee-Jin Park, Jeongjin Lee, Younghun Kim, Jeyong Yoon, Kwangsik Park, Kyunghee Choi, and Jongheop Yi. "Bacterial cytotoxicity of the silver nanoparticle related to physicochemical metrics and agglomeration properties." *Environmental Toxicology and Chemistry* 29, no. 10 (2010): 2154-2160.

[12] Bajaj, Manish, Satish K. Pandey, Tamanna Nain, Surinder K. Brar, Prabhjot Singh, Shivakanksha Singh, Nishima Wangoo, and Rohit K. Sharma. "Stabilized cationic dipeptide capped gold/silver nanohybrids: Towards enhanced antibacterial and antifungal efficacy." *Colloids and Surfaces B: Biointerfaces* 158 (2017): 397-407.

[13] Bedair, Tarek M., Seung Jung Yu, Sung Gap Im, Bang Ju Park, Yoon Ki Joung, and Dong Keun Han. "Effects of interfacial layer wettability and thickness on the coating morphology and sirolimus release for drug-eluting stent." *Journal of colloid and interface science* 460 (2015): 189-199.

[14] Bhosale, R. R., A. S. Kulkarni, S. S. Gilda, N. H. Aloorkar, R. A. Osmani, and B. R. Harkare. "Innovative eco-friendly approaches for green synthesis of silver nanoparticles." *Int J Pharm Sci Nanotech* 7 (2014): 2328-2337.

[15] Bowker, Michael. "Automotive catalysis studied by surface science." *Chemical Society Reviews* 37, no. 10 (2008): 2204-2211.

[16] Busbee, Brantley D., Sherine O. Obare, and Catherine J. Murphy. "An improved synthesis of high-aspect-ratio gold nanorods." *Advanced Materials* 15, no. 5 (2003): 414-416.

[17] Cao-Milan, Roberto, and Luis M. Liz-Marzan. "Gold nanoparticle conjugates: recent advances toward clinical applications." *Expert opinion on drug delivery* 11, no. 5 (2014): 741-752.

[18] Carvalho, Patrícia M., Mário R. Felício, Nuno C. Santos, Sónia Gonçalves, and Marco M. Domingues. "Application of light scattering techniques to nanoparticle characterization and development." *Frontiers in chemistry* 6 (2018): 237.

[19] Castillo, Rafael R., and María Vallet-Regí. "Functional Mesoporous Silica Nanocomposites: Biomedical applications and Biosafety." *International journal of molecular sciences* 20, no. 4 (2019): 929.

[20] Catry, Boudewijn, Hans Laevens, L. A. Devriese, Geert Opsomer, and A. De Kruif. "Antimicrobial resistance in livestock." *Journal of veterinary pharmacology and therapeutics* 26, no. 2 (2003): 81-93.

[21] Chaloupka, Karla, Yogeshkumar Malam, and Alexander M. Seifalian. "Nanosilver as a new generation of nanoproduct in biomedical applications." *Trends in biotechnology* 28, no. 11 (2010): 580-588.

[22] Chandra, R., A. K. Chawla, and P. Ayyub. "Optical and structural properties of sputter-deposited nanocrystalline Cu2O films: Effect of sputtering gas." *Journal of nanoscience and nanotechnology* 6, no. 4 (2006): 1119-1123.

[23] Chattopadhyay, D. P., and B. H. Patel. "Nano metal particles: synthesis, characterization and application to textiles." *Manufacturing Nanostructures, One Central Press (OCP)* (2014): 184-215.

[24] Chen, Shih-Chen, Dan-Hua Hsieh, Hsin Jiang, Yu-Kuang Liao, Fang-I. Lai, Chyong-Hua Chen, Chih Wei Luo et al., "Growth and characterization of Cu (In, Ga) Se 2 thin films by nanosecond and femtosecond pulsed laser deposition." *Nanoscale research letters* 9, no. 1 (2014): 280.

[25] Chernousova, Svitlana, and Matthias Epple. "Silver as antibacterial agent: ion, nanoparticle, and metal." *Angewandte Chemie International Edition* 52, no. 6 (2013): 1636-1653.

[26] Chin, Ang Bee, and Iskandar Idris Yaacob. "Synthesis and characterization of magnetic iron oxide nanoparticles via w/o microemulsion and Massart's procedure." *Journal of materials processing technology* 191, no. 1-3 (2007): 235-237.

[27] Choi, Okkyoung, and Zhiqiang Hu. "Size dependent and reactive oxygen species related nanosilver toxicity to nitrifying bacteria." *Environmental science & technology* 42, no. 12 (2008): 4583-4588.

[28] Choi, Okkyoung, Kathy Kanjun Deng, Nam-Jung Kim, Louis Ross Jr, Rao Y. Surampalli, and Zhiqiang Hu. "The inhibitory effects of silver nanoparticles, silver ions, and silver chloride colloids on microbial growth." *Water research* 42, no. 12 (2008): 3066-3074.

[29] Crooks, Richard M., Mingqi Zhao, Li Sun, Victor Chechik, and Lee K. Yeung. "Dendrimer-encapsulated metal nanoparticles: synthesis, characterization, and applications to catalysis." *Accounts of chemical research* 34, no. 3 (2001): 181-190.

[30] Cui, Yan, Yuyun Zhao, Yue Tian, Wei Zhang, Xiaoying Lü, and Xingyu Jiang. "The molecular mechanism of action of bactericidal gold nanoparticles on *Escherichia coli*." *Biomaterials* 33, no. 7 (2012): 2327-2333.

[31] Cushing, Brian L., Vladimir L. Kolesnichenko, and Charles J. O'Connor. "Recent advances in the liquid-phase syntheses of inorganic nanoparticles." *Chemical reviews* 104, no. 9 (2004): 3893-3946.

[32] Dai, Xiaoyan, Hong Xu, Yongguang Li, Hongchen Gu, and Meng Wei. "Influence of the physical and chemical properties of magnetic nanoparticles on their performance in a chemiluminescence immunoassay." *Clinical biochemistry* 47, no. 3 (2014): 220-226.

[33] Dikusar, A. I., P. G. Globa, S. S. Belevskii, and S. P. Sidel'nikova. "On limiting rate of dimensional electrodeposition at meso-and nanomaterial manufacturing by template synthesis." *Surface Engineering and Applied Electrochemistry* 45, no. 3 (2009): 171-179.

[34] Doll, G. L., B. A. Mensah, H. Mohseni, and T. W. Scharf. "Chemical vapor deposition and atomic layer deposition of coatings for mechanical applications." *Journal of thermal spray technology* 19, no. 1-2 (2010): 510-516.

[35] Eichfeld, Sarah M., Lorraine Hossain, Yu-Chuan Lin, Aleksander F. Piasecki, Benjamin Kupp, A. Glen Birdwell, Robert A. Burke et al., "Highly scalable, atomically thin WSe2 grown via metal–organic chemical vapor deposition." *ACS nano* 9, no. 2 (2015): 2080-2087.

[36] El-Nour, Kholoud MM Abou, Ala'A. Eftaiha, Abdulrhman Al-Warthan, and Reda AA Ammar. "Synthesis and applications of silver nanoparticles." *Arabian journal of chemistry* 3, no. 3 (2010): 135-140.

[37] El-Sayed, Ivan H., Xiaohua Huang, and Mostafa A. El-Sayed. "Surface plasmon resonance scattering and absorption of anti-EGFR antibody conjugated gold nanoparticles in cancer diagnostics: applications in oral cancer." *Nano letters* 5, no. 5 (2005): 829-834.

[38] El-Sayed, Mostafa A. "Some interesting properties of metals confined in time and nanometer space of different shapes." *Accounts of chemical research* 34, no. 4 (2001): 257-264.

[39] Ennaert, Thijs, Joost Van Aelst, Jan Dijkmans, Rik De Clercq, Wouter Schutyser, Michiel Dusselier, Danny Verboekend, and Bert F. Sels. "Potential and challenges of zeolite chemistry in the catalytic conversion of biomass." *Chemical Society Reviews* 45, no. 3 (2016): 584-611.

[40] Eustis, Susie, and Mostafa A. El-Sayed. "Why gold nanoparticles are more precious than pretty gold: noble metal surface plasmon resonance and its enhancement of the

radiative and nonradiative properties of nanocrystals of different shapes." *Chemical society reviews* 35, no. 3 (2006): 209-217.

[41] Fodor, Daniel, Takashi Ishikawa, Frank Krumeich, and Jeroen A. van Bokhoven. "Synthesis of single crystal nanoreactor materials with multiple catalytic functions by incipient wetness impregnation and ion exchange." *Advanced Materials* 27, no. 11 (2015): 1919-1923.

[42] Furneaux, R. C., W. R. Rigby, and A. P. Davidson. "The formation of controlled-porosity membranes from anodically oxidized aluminium." *Nature* 337, no. 6203 (1989): 147.

[43] Ghaffarian, Hamid Reza, Mahboobeh Saiedi, Mohammad Ali Sayyadnejad, and Ali Morad Rashidi. "*Synthesis of ZnO nanoparticles by spray pyrolysis method.*" (2011): 1-6.

[44] Ghorbani, Hamid Reza. "A review of methods for synthesis of Al nanoparticles." *Oriental Journal of Chemistry* 30, no. 4 (2014): 1941-1949.

[45] Gondoni, Paolo, Matteo Ghidelli, Fabio Di Fonzo, Andrea Li Bassi, and Carlo S. Casari. "Fabrication of nano-engineered transparent conducting oxides by pulsed laser deposition." *JoVE (Journal of Visualized Experiments)* 72 (2013): e50297.

[46] Gottlieb, Thomas, and Graeme R. Nimmo. "Antibiotic resistance is an emerging threat to public health: an urgent call to action at the Antimicrobial Resistance Summit 2011." *Med J Aust* 194, no. 6 (2011): 281-3.

[47] Guzmán, Maribel G., Jean Dille, and Stephan Godet. "Synthesis of silver nanoparticles by chemical reduction method and their antibacterial activity." *Int J Chem Biomol Eng* 2, no. 3 (2009): 104-111.

[48] Hackenberg, Stephan, Agmal Scherzed, Michael Kessler, Silke Hummel, Antje Technau, Katrin Froelich, Christian Ginzkey, Christian Koehler, Rudolf Hagen, and Norbert Kleinsasser. "Silver nanoparticles: evaluation of DNA damage, toxicity and functional impairment in human mesenchymal stem cells." *Toxicology letters* 201, no. 1 (2011): 27-33.

[49] Hao, Feng, Colleen L. Nehl, Jason H. Hafner, and Peter Nordlander. "Plasmon resonances of a gold nanostar." *Nano letters* 7, no. 3 (2007): 729-732.

[50] Hasan, Saba. "A review on nanoparticles: their synthesis and types." *Research Journal of Recent Sciences ISSN* 2277 (2015): 2502.

[51] Hasnidawani, J. N., H. N. Azlina, H. Norita, N. N. Bonnia, S. Ratim, and E. S. Ali. "Synthesis of ZnO nanostructures using sol-gel method." *Procedia Chemistry* 19 (2016): 211-216.

[52] He, Shiying, Zhirui Guo, Yu Zhang, Song Zhang, Jing Wang, and Ning Gu. "Biosynthesis of gold nanoparticles using the bacteria Rhodopseudomonas capsulata." *Materials Letters* 61, no. 18 (2007): 3984-3987.

[53] Hong, Xuesen, Junjie Wen, Xuhua Xiong, and Yongyou Hu. "Shape effect on the antibacterial activity of silver nanoparticles synthesized via a microwave-assisted

method." *Environmental science and pollution research* 23, no. 5 (2016): 4489-4497.

[54] Hong, Xun, Juqing Liu, Bing Zheng, Xiao Huang, Xiao Zhang, Chaoliang Tan, Junze Chen, Zhanxi Fan, and Hua Zhang. "A universal method for preparation of noble metal nanoparticle-decorated transition metal dichalcogenide nanobelts." *Advanced Materials* 26, no. 36 (2014): 6250-6254.

[55] Huang, Jiale, Qingbiao Li, Daohua Sun, Yinghua Lu, Yuanbo Su, Xin Yang, Huixuan Wang et al., "Biosynthesis of silver and gold nanoparticles by novel sundried Cinnamomum camphora leaf." *Nanotechnology* 18, no. 10 (2007): 105104.

[56] Huang, Liang, Xueping Zhang, Qingqing Wang, Yujie Han, Youxing Fang, and Shaojun Dong. "Shape-control of Pt–ru nanocrystals: tuning surface structure for enhanced electrocatalytic methanol oxidation." *Journal of the American Chemical Society* 140, no. 3 (2018): 1142-1147.

[57] Iravani, Siavash, S. Thota, and D. C. Crans. "Methods for preparation of metal nanoparticles." *Metal nanoparticles: synthesis and applications in pharmaceutical sciences* (2018): 15-32.

[58] Iravani, Siavash. "Green synthesis of metal nanoparticles using plants." *Green Chemistry* 13, no. 10 (2011): 2638-2650.

[59] Islam, Md, and M. Saidul Islam. "Electro-deposition Method for Platinum Nano-particles Synthesis." *Saidul, Electro-Deposition Method for Platinum Nano-Particles Synthesis* (2013).

[60] Jeong, Woo-jin, Jiyoon Bu, Luke J. Kubiatowicz, Stephanie S. Chen, YoungSoo Kim, and Seungpyo Hong. "Peptide–nanoparticle conjugates: a next generation of diagnostic and therapeutic platforms?" *Nano convergence* 5, no. 1 (2018):

[61] Johnson, Grant E., Trevor Moser, Mark Engelhard, Nigel D. Browning, and Julia Laskin. "Fabrication of electrocatalytic Ta nanoparticles by reactive sputtering and ion soft landing." *The Journal of Chemical Physics* 145, no. 17 (2016): 174701.

[62] Kahru, Anne, and Angela Ivask. "Mapping the dawn of nanoecotoxicological research." *Accounts of Chemical Research* 46, no. 3 (2012): 823-833.

[63] Karikalan, Natarajan, Raj Karthik, Shen-Ming Chen, Chelladurai Karuppiah, and Arumugam Elangovan. "Sonochemical synthesis of sulfur doped reduced graphene oxide supported CuS nanoparticles for the non-enzymatic glucose sensor applications." *Scientific reports* 7, no. 1 (2017): 2494.

[64] Karnani, Richa L., and Abhay Chowdhary. "Biosynthesis of silver nanoparticle by eco-friendly method." *Indian J. Nanosci* 1, no. 1 (2013): 25-31.

[65] Kawata, Satoshi, Yasushi Inouye, and Prabhat Verma. "Plasmonics for near-field nano-imaging and superlensing." *Nature Photonics* 3, no. 7 (2009): 388.

[66] Kim, Jeong-Chul, Seungyeop Lee, Kanghee Cho, Kyungsu Na, Changq Lee, and Ryong Ryoo. "Mesoporous MFI zeolite nanosponge supporting cobalt nanoparticles

as a Fischer–Tropsch catalyst with high yield of branched hydrocarbons in the gasoline range." *ACS Catalysis* 4, no. 11 (2014): 3919-3927.

[67] Klaus, Tanja, Ralph Joerger, Eva Olsson, and Claes-Göran Granqvist. "Silver-based crystalline nanoparticles, microbially fabricated." *Proceedings of the National Academy of Sciences* 96, no. 24 (1999): 13611-13614.

[68] Koch, Carl C., and J. D. Whittenberger. "Mechanical milling/alloying of intermetallics." *Intermetallics* 4, no. 5 (1996): 339-355.

[69] Komarneni, Sridhar. "Nanophase materials by hydrothermal, microwave-hydrothermal and microwave-solvothermal methods." *Current science* (2003): 1730-1734.

[70] Konishi, Yasuhiro, Kaori Ohno, Norizoh Saitoh, Toshiyuki Nomura, Shinsuke Nagamine, Hajime Hishida, Yoshio Takahashi, and Tomoya Uruga. "Bioreductive deposition of platinum nanoparticles on the bacterium Shewanella algae." *Journal of biotechnology* 128, no. 3 (2007): 648-653.

[71] Kruis, F. Einar, Heinz Fissan, and Aaron Peled. "Synthesis of nanoparticles in the gas phase for electronic, optical and magnetic applications—a review." *Journal of Aerosol Science* 29, no. 5-6 (1998): 511-535.

[72] Kugai, Junichiro, Velu Subramani, Chunshan Song, Mark H. Engelhard, and Ya-Huei Chin. "Effects of nanocrystalline CeO2 supports on the properties and performance of Ni–Rh bimetallic catalyst for oxidative steam reforming of ethanol." *Journal of Catalysis* 238, no. 2 (2006): 430-440.

[73] Kumar, Anil, Bhargavi Mazinder Boruah, and Xing-Jie Liang. "Gold nanoparticles: promising nanomaterials for the diagnosis of cancer and HIV/AIDS." *Journal of Nanomaterials* 2011 (2011):

[74] Kumar, Harish, Nagasamy Venkatesh, Himangshu Bhowmik, and Anuttam Kuila. "Metallic Nanoparticle: A Review." *Biomedical Journal of Scientific & Technical Research* 4, no. 2 (2018): 3765-3775.

[75] Kuo, Wen-Shuo, Yi-Ting Chang, Keng-Chi Cho, Kuo-Chih Chiu, Chi-Hsiang Lien, Chen-Sheng Yeh, and Shean-Jen Chen. "Gold nanomaterials conjugated with indocyanine green for dual-modality photodynamic and photothermal therapy." *Biomaterials* 33, no. 11 (2012): 3270-3278.

[76] Lal, Surbhi, Stephan Link, and Naomi J. Halas. "Nano-optics from sensing to waveguiding." *Nature photonics* 1, no. 11 (2007): 641.

[77] Landage, S. M., A. I. Wasif, and P. Dhuppe. "Synthesis of nanosilver using chemical reduction methods." *International Journal of Advanced Research in Engineering and Applied Sciences* 3, no. 5 (2014): 14-22.

[78] Lansdown, Alan BG. "Silver in health care: antimicrobial effects and safety in use." In *Biofunctional textiles and the skin*, vol. 33, pp. 17-34. Karger Publishers, 2006.

[79] Laurent, Sophie, Delphine Forge, Marc Port, Alain Roch, Caroline Robic, Luce Vander Elst, and Robert N. Muller. "Magnetic iron oxide nanoparticles: synthesis,

stabilization, vectorization, physicochemical characterizations, and biological applications." *Chemical reviews* 108, no. 6 (2008): 2064-2110.

[80] Li, Shikuo, Yuhua Shen, Anjian Xie, Xuerong Yu, Lingguang Qiu, Li Zhang, and Qingfeng Zhang. "Green synthesis of silver nanoparticles using Capsicum annuum L. extract." *Green Chemistry* 9, no. 8 (2007): 852-858.

[81] Link, Stephan, and Mostafa A. El-Sayed. "Optical properties and ultrafast dynamics of metallic nanocrystals." *Annual review of physical chemistry* 54, no. 1 (2003): 331-366.

[82] Livage, Jacques, Marc Henry, and Clément Sanchez. "Sol-gel chemistry of transition metal oxides." *Progress in solid state chemistry* 18, no. 4 (1988): 259-341.

[83] Majdalawieh, Amin, Marsha C. Kanan, Oussama El-Kadri, and Sofian M. Kanan. "Recent advances in gold and silver nanoparticles: synthesis and applications." *Journal of nanoscience and nanotechnology* 14, no. 7 (2014): 4757-4780.

[84] Majerič, Peter, Bernd Friedrich, and Rebeka Rudolf. "Au-nanoparticle synthesis via ultrasonic spray pyrolysis with a separate evaporation zone." *Materiali in tehnologije* 5, no. 49 (2015): 791-796.

[85] Majidi, Sima, Fatemeh Zeinali Sehrig, Samad Mussa Farkhani, Mehdi Soleymani Goloujeh, and Abolfazl Akbarzadeh. "Current methods for synthesis of magnetic nanoparticles." *Artificial cells, nanomedicine, and biotechnology* 44, no. 2 (2016): 722-734.

[86] Marambio-Jones, Catalina, and Eric MV Hoek. "A review of the antibacterial effects of silver nanomaterials and potential implications for human health and the environment." *Journal of Nanoparticle Research* 12, no. 5 (2010): 1531-1551.

[87] Marambio-Jones, Catalina, and Eric MV Hoek. "A review of the antibacterial effects of silver nanomaterials and potential implications for human health and the environment." *Journal of Nanoparticle Research* 12, no. 5 (2010): 1531-1551.

[88] Maron, Dina Fine, Tyler JS Smith, and Keeve E. Nachman. "Restrictions on antimicrobial use in food animal production: an international regulatory and economic survey." *Globalization and health* 9, no. 1 (2013): 48.

[89] Mathew, Alan G., Robin Cissell, and S. Liamthong. "Antibiotic resistance in bacteria associated with food animals: a United States perspective of livestock production." *Foodborne pathogens and disease* 4, no. 2 (2007): 115-133.

[90] Mathew, Daliya S., and Ruey-Shin Juang. "An overview of the structure and magnetism of spinel ferrite nanoparticles and their synthesis in microemulsions." *Chemical engineering journal* 129, no. 1-3 (2007): 51-65.

[91] Meng, Xiangju, and Feng-Shou Xiao. "Green routes for synthesis of zeolites." *Chemical reviews* 114, no. 2 (2013): 1521-1543.

[92] Merle, Roswitha, Peter Hajek, Annemarie Käsbohrer, Christine Hegger-Gravenhorst, Yvonne Mollenhauer, Matthias Robanus, Fritz-R. Ungemach, and

Lothar Kreienbrock. “Monitoring of antibiotic consumption in livestock: a German feasibility study.” *Preventive veterinary medicine* 104, no. 1-2 (2012): 34-43.

[93] Miikkulainen, Ville, Markku Leskelä, Mikko Ritala, and Riikka L. Puurunen. “Crystallinity of inorganic films grown by atomic layer deposition: Overview and general trends.” *Journal of Applied Physics* 113, no. 2 (2013): 2.

[94] Mody, Vicky V., Rodney Siwale, Ajay Singh, and Hardik R. Mody. “Introduction to metallic nanoparticles.” *Journal of Pharmacy and Bioallied Sciences* 2, no. 4 (2010): 282.

[95] Moellering Jr, Robert C. “Past, present, and future of antimicrobial agents.” *The American journal of medicine* 99, no. 6 (1995): 11s-18s.

[96] Moghimi, S. Moein, A. Christy Hunter, and J. Clifford Murray. “Nanomedicine: current status and future prospects.” *The FASEB journal* 19, no. 3 (2005): 311-330.

[97] Mohanpuria, Prashant, Nisha K. Rana, and Sudesh Kumar Yadav. “Biosynthesis of nanoparticles: technological concepts and future applications.” *Journal of nanoparticle research* 10, no. 3 (2008): 507-517.

[98] Mohanty, U. S. “Electrodeposition: a versatile and inexpensive tool for the synthesis of nanoparticles, nanorods, nanowires, and nanoclusters of metals.” *Journal of applied electrochemistry* 41, no. 3 (2011): 257-270.

[99] Mohl, Melinda, Dorina Dobo, Akos Kukovecz, Zoltan Konya, Krisztian Kordas, Jinquan Wei, Robert Vajtai, and Pulickel M. Ajayan. “Formation of CuPd and CuPt bimetallic nanotubes by galvanic replacement reaction.” *The Journal of Physical Chemistry C* 115, no. 19 (2011): 9403-9409.

[100] Moyano, Daniel F., and Vincent M. Rotello. “Nano meets biology: structure and function at the nanoparticle interface.” *Langmuir* 27, no. 17 (2011): 10376-10385.

[101] Murphy, Catherine J., Tapan K. Sau, Anand M. Gole, Christopher J. Orendorff, Jinxin Gao, Linfeng Gou, Simona E. Hunyadi, and Tan Li. “Anisotropic metal nanoparticles: synthesis, assembly, and optical applications.” *Journal of Physical Chemistry*. B 109, No. 29, (2005): 3857-13870.

[102] Nie, Min, Kai Sun, and Dennis Desheng Meng. “Formation of metal nanoparticles by short-distance sputter deposition in a reactive ion etching chamber.” *Journal of Applied Physics* 106, no. 5 (2009): 054314.

[103] Okuyama, Kikuo, and I. Wuled Lenggoro. “Preparation of nanoparticles via spray route.” *Chemical engineering science* 58, no. 3-6 (2003): 537-547.

[104] Orofeo, Carlo M., Satoru Suzuki, Yoshiaki Sekine, and Hiroki Hibino. “Scalable synthesis of layer-controlled WS2 and MoS2 sheets by sulfurization of thin metal films.” *Applied Physics Letters* 105, no. 8 (2014): 083112.

[105] Pandey, Priyanka A., Gavin R. Bell, Jonathan P. Rourke, Ana M. Sanchez, Mark D. Elkin, Bryan J. Hickey, and Neil R. Wilson. “Physical vapor deposition of metal nanoparticles on chemically modified graphene: observations on metal–graphene interactions.” *Small* 7, no. 22 (2011): 3202-3210.

[106] Pantidos, Nikolaos, and Louise E. Horsfall. "Biological synthesis of metallic nanoparticles by bacteria, fungi and plants." *Journal of Nanomedicine & Nanotechnology* 5, no. 5 (2014): 1.

[107] Park, Jun-Sik, Wan-Ho Chung, Hak-Sung Kim, and Young-Beom Kim. "Rapid fabrication of chemical-solution-deposited La0. 6Sr0. 4CoO3− δ thin films via flashlight sintering." *Journal of Alloys and Compounds* 696 (2017): 102-108.

[108] Patra, Chitta Ranjan, Resham Bhattacharya, Debabrata Mukhopadhyay, and Priyabrata Mukherjee. "Fabrication of gold nanoparticles for targeted therapy in pancreatic cancer." *Advanced drug delivery reviews* 62, no. 3 (2010): 346-361.

[109] Pedersen, Henrik, and Simon D. Elliott. "Studying chemical vapor deposition processes with theoretical chemistry." *Theoretical Chemistry Accounts* 133, no. 5 (2014): 1476.

[110] Peralta, Donna V., Zahra Heidari, Srikanta Dash, and Matthew A. Tarr. "Hybrid paclitaxel and gold nanorod-loaded human serum albumin nanoparticles for simultaneous chemotherapeutic and photothermal therapy on 4T1 breast cancer cells." *ACS applied materials & interfaces* 7, no. 13 (2015): 7101-7111.

[111] Petros, Robby A., and Joseph M. DeSimone. "Strategies in the design of nanoparticles for therapeutic applications." *Nature reviews Drug discovery* 9, no. 8 (2010): 615.

[112] Pettifor, D. G. "A physicist's view of the energetics of transition metals." *Calphad* 1, no. 4 (1977): 305-324.

[113] Zhang, Q. L., Z. M. Yang, B. J. Ding, X. Z. Lan, Y. J. Guo, Preparation of copper nanoparticles by chemical reduction method using potassium borohydride, *Trans. Nonferrous Metals Soc. China* 20 (2010) s240–s244.

[114] Qi, Zhiyuan, Chaoxian Xiao, Cong Liu, Tian Wei Goh, Lin Zhou, Raghu Maligal-Ganesh, Yuchen Pei, Xinle Li, Larry A. Curtiss, and Wenyu Huang. "Sub-4 nm PtZn intermetallic nanoparticles for enhanced mass and specific activities in catalytic electrooxidation reaction." *Journal of the American Chemical Society* 139, no. 13 (2017): 4762-4768.

[115] Qin, Yimin. "Silver-containing alginate fibres and dressings." *International wound journal* 2, no. 2 (2005): 172-176.

[116] Raheman, Farkanda, Shivaji Deshmukh, Avinash Ingle, Aniket Gade, and Mahendra Rai. "Silver nanoparticles: novel antimicrobial agent synthesized from an endophytic fungus Pestalotia sp. isolated from leaves of Syzygium cumini (L)." *Nano Biomed Eng* 3, no. 3 (2011): 174-178.

[117] Rai, Mahendra, Alka Yadav, and Aniket Gade. "Silver nanoparticles as a new generation of antimicrobials." *Biotechnology advances* 27, no. 1 (2009): 76-83.

[118] Rai, Mahendra, Alka Yadav, and Aniket Gade. "Silver nanoparticles as a new generation of antimicrobials." *Biotechnology advances* 27, no. 1 (2009): 76-83.

[119] Rai, Mahendra, Alka Yadav, and Aniket Gade. "Silver nanoparticles as a new generation of antimicrobials." *Biotechnology advances* 27, no. 1 (2009): 76-83.

[120] Rajchakit, Urawadee, and Vijayalekshmi Sarojini. "Recent developments in antimicrobial-peptide-conjugated gold nanoparticles." *Bioconjugate chemistry* 28, no. 11 (2017): 2673-2686.

[121] Rajput, Namita. "Methods of preparation of nanoparticles-A review." *International Journal of Advances in Engineering & Technology* 7, no. 6 (2015): 1806.

[122] Rao, CN Ramachandra, Giridhar U. Kulkarni, P. John Thomas, and Peter P. Edwards. "Metal nanoparticles and their assemblies." *Chemical Society Reviews* 29, no. 1 (2000): 27-35.

[123] Riahi, G., D. Guillemot, M. Polisset-Thfoin, A. A. Khodadadi, and J. Fraissard. "Preparation, characterization and catalytic activity of gold-based nanoparticles on HY zeolites." *Catalysis Today* 72, no. 1-2 (2002): 115-121.

[124] Rizzello, Loris, and Pier Paolo Pompa. "Nanosilver-based antibacterial drugs and devices: mechanisms, methodological drawbacks, and guidelines." *Chemical Society Reviews* 43, no. 5 (2014): 1501-1518.

[125] Sanchez, Laura M., and Vera A. Alvarez. "Advances in Magnetic Noble Metal/Iron-Based Oxide Hybrid Nanoparticles as Biomedical Devices." *Bioengineering* 6, no. 3 (2019): 75.

[126] Sayah, Raida S., John B. Kaneene, Yvette Johnson, and RoseAnn Miller. "Patterns of antimicrobial resistance observed in *Escherichia coli* isolates obtained from domestic-and wild-animal fecal samples, human septage, and surface water." *Appl. Environ. Microbiol.* 71, no. 3 (2005): 1394-1404.

[127] Schultz, Sheldon, David R. Smith, Jack J. Mock, and David A. Schultz. "Single-target molecule detection with nonbleaching multicolor optical immunolabels." *Proceedings of the National Academy of Sciences* 97, no. 3 (2000): 996-1001.

[128] Sekhon, Bhupinder Singh. "Supercritical fluid technology: an overview of pharmaceutical applications." *International Journal of PharmTech Research* 2, no. 1 (2010): 810-826.

[129] Shankar, S. Shiv, Absar Ahmad, and Murali Sastry. "Geranium leaf assisted biosynthesis of silver nanoparticles." *Biotechnology progress* 19, no. 6 (2003): 1627-1631.

[130] Shankar, S. Shiv, Akhilesh Rai, Balaprasad Ankamwar, Amit Singh, Absar Ahmad, and Murali Sastry. "Biological synthesis of triangular gold nanoprisms." *Nature materials* 3, no. 7 (2004): 482.

[131] Sharma, Vivek, Kyoungweon Park, and Mohan Srinivasarao. "Colloidal dispersion of gold nanorods: Historical background, optical properties, seed-mediated synthesis, shape separation and self-assembly." *Materials Science and Engineering: R: Reports* 65, no. 1-3 (2009): 1-38.

[132] Simakin, A. V., V. V. Voronov, N. A. Kirichenko, and G. A. Shafeev. "Nanoparticles produced by laser ablation of solids in liquid environment." *Applied Physics A* 79, no. 4-6 (2004): 1127-1132.

[133] Singh, Sanjay Kumar, and Qiang Xu. "Complete conversion of hydrous hydrazine to hydrogen at room temperature for chemical hydrogen storage." *Journal of the American Chemical Society* 131, no. 50 (2009): 18032-18033.

[134] Sondi, Ivan, and Branka Salopek-Sondi. "Silver nanoparticles as antimicrobial agent: a case study on *E. coli* as a model for Gram-negative bacteria." *Journal of colloid and interface science* 275, no. 1 (2004): 177-182.

[135] Soulsby, Lord. "Antimicrobials and animal health: a fascinating nexus." *Journal of Antimicrobial Chemotherapy* 60, no. suppl_1 (2007): i77-i78.

[136] Stella, Pietro, Pierre-Alexandre Beloeil, Beatriz Guerra, Marta Hugas, and Ernesto Liebana. "The Role of the European Food Safety Authority (EFSA) in the Fight against Antimicrobial Resistance (AMR)." *Food Protection Trends* 38, no. 1 (2018): 72-80.

[137] Stewart, Matthew E., Christopher R. Anderton, Lucas B. Thompson, Joana Maria, Stephen K. Gray, John A. Rogers, and Ralph G. Nuzzo. "Nanostructured plasmonic sensors." *Chemical reviews* 108, no. 2 (2008): 494-521.

[138] Sun, Yugang, and Younan Xia. "Shape-controlled synthesis of gold and silver nanoparticles." *Science* 298, no. 5601 (2002): 2176-2179.

[139] Suresh, Anil K., Dale A. Pelletier, and Mitchel J. Doktycz. "Relating nanomaterial properties and microbial toxicity." *Nanoscale* 5, no. 2 (2013): 463-474.

[140] Swihart, Mark T. "Vapor-phase synthesis of nanoparticles." *Current Opinion in Colloid & Interface Science* 8, no. 1 (2003): 127-133.

[141] Tempini, P. N., Sharif S. Aly, B. M. Karle, and R. V. Pereira. "Multidrug residues and antimicrobial resistance patterns in waste milk from dairy farms in Central California." *Journal of dairy science* 101, no. 9 (2018): 8110-8122.

[142] Trigo, Pedro Gil Dias. *Oxide transistors produced by Combustion Synthesis: Influence of the PVP on the properties of the insulator*. PhD diss., 2017.

[143] Turkevich, John, Peter Cooper Stevenson, and James Hillier. "A study of the nucleation and growth processes in the synthesis of colloidal gold." *Discussions of the Faraday Society* 11 (1951): 55-75.

[144] Uppal, Abha, and Biplab Bose. "Synthesis, stability, and in vitro oral cancer cell toxicity of human serum albumin stabilised gold nanoflowers." *IET Nanobiotechnology* 12, no. 3 (2017): 292-297.

[145] Van Speybroeck, Veronique, Karen Hemelsoet, Lennart Joos, Michel Waroquier, Robert G. Bell, and C. Richard A. Catlow. "Advances in theory and their application within the field of zeolite chemistry." *Chemical Society Reviews* 44, no. 20 (2015): 7044-7111.

[146] Vanecht, Evert. *Gold Nanoparticles in Ionic Liquids Prepared by Sputter Deposition*. (2012).

[147] Veigas, Bruno, Pedro Pedrosa, Fábio F. Carlos, Liliana Mancio-Silva, Ana Rita Grosso, Elvira Fortunato, Maria M. Mota, and Pedro V. Baptista. "One nanoprobe, two pathogens: gold nanoprobes multiplexing for point-of-care." *Journal of nanobiotechnology* 13, no. 1 (2015): 48.

[148] Vigneshwaran, Nadanathangam, Arati A. Kathe, Perianambi V. Varadarajan, Rajan P. Nachane, and Rudrapatna H. Balasubramanya. "Silver– protein (core– shell) nanoparticle production using spent mushroom substrate." *Langmuir* 23, no. 13 (2007): 7113-7117.

[149] Vilchis-Nestor, Alfredo R., Victor Sánchez-Mendieta, Marco A. Camacho-López, Rosa M. Gómez-Espinosa, Miguel A. Camacho-López, and Jesús A. Arenas-Alatorre. "Solventless synthesis and optical properties of Au and Ag nanoparticles using Camellia sinensis extract." *Materials Letters* 62, no. 17-18 (2008): 3103-3105.

[150] Wang, Binghui, Xupin Zhuang, Wenjian Deng, and Bowen Cheng. "Microwave-assisted synthesis of silver nanoparticles in alkalic carboxymethyl chitosan solution." *Engineering* 2, no. 05 (2010): 387.

[151] Wang, Guangjie, Feng Jin, Nan Dai, Zhaoyang Zhong, Yi Qing, Mengxia Li, Ruo Yuan, and Dong Wang. "Signal-enhanced electrochemiluminescence immunosensor based on synergistic catalysis of nicotinamide adenine dinucleotide hydride and silver nanoparticles." *Analytical biochemistry* 422, no. 1 (2012): 7-13.

[152] Weckhuysen, Bert M., and Jihong Yu. "Recent advances in zeolite chemistry and catalysis." *Chemical Society Reviews* 44, no. 20 (2015): 7022-7024.

[153] Willner, Itamar, Ronan Baron, and Bilha Willner. "Growing metal nanoparticles by enzymes." *Advanced Materials* 18, no. 9 (2006): 1109-1120.

[154] Xia, Bao Yu, Hao Bin Wu, Nan Li, Ya Yan, Xiong Wen Lou, and Xin Wang. "One-pot synthesis of Pt–Co alloy nanowire assemblies with tunable composition and enhanced electrocatalytic properties." *Angewandte chemie international edition* 54, no. 12 (2015): 3797-3801.

[155] Xie, Jin, Qinghua Zhang, Lin Gu, Sheng Xu, Peng Wang, Jianguo Liu, Yi Ding et al., "Ruthenium–platinum core–shell nanocatalysts with substantially enhanced activity and durability towards methanol oxidation." *Nano Energy* 21 (2016): 247-257.

[156] Xu, Chunping, Sudipta De, Alina M. Balu, Manuel Ojeda, and Rafael Luque. "Mechanochemical synthesis of advanced nanomaterials for catalytic applications." *Chemical Communications* 51, no. 31 (2015): 6698-6713.

[157] Yadav, Thakur Prasad, R. Manohar Yadav, and D. Pratap Singh. "Mechanical milling: a top down approach for the synthesis of nanomaterials and nanocomposites." *Nanoscience and Nanotechnology* 2, no. 3 (2012): 22-48.

[158] Yadavalli, Tejabhiram, and Deepak Shukla. "Role of metal and metal oxide nanoparticles as diagnostic and therapeutic tools for highly prevalent viral infections." *Nanomedicine: Nanotechnology, Biology and Medicine* 13, no. 1 (2017): 219-230.

[159] Yang, Yong, Shigemasa Matsubara, Liangming Xiong, Tomokatsu Hayakawa, and Masayuki Nogami. "Solvothermal synthesis of multiple shapes of silver nanoparticles and their SERS properties." *The Journal of Physical Chemistry C* 111, no. 26 (2007): 9095-9104.

[160] Ye, Xiangrong, and C. M. Wai. "Making nanomaterials in supercritical fluids: a review." *Journal of chemical education* 80, no. 2 (2003): 198.

[161] Yong, Ping, Neil A. Rowson, J. Peter G. Farr, I. Rex Harris, and Lynne E. Macaskie. "Bioreduction and biocrystallization of palladium by Desulfovibrio desulfuricans NCIMB 8307." *Biotechnology and bioengineering* 80, no. 4 (2002): 369-379.

[162] Yu, Su-juan, Yong-guang Yin, and Jing-fu Liu. "Silver nanoparticles in the environment." *Environmental Science: Processes & Impacts* 15, no. 1 (2013): 78-92.

[163] Yuan, Qiang, Zhiyou Zhou, Jing Zhuang, and Xun Wang. "Seed displacement, epitaxial synthesis of Rh/Pt bimetallic ultrathin nanowires for highly selective oxidizing ethanol to CO2." *Chemistry of Materials* 22, no. 7 (2010): 2395-2402.

[164] Zhang, Xiaoying. "Gold nanoparticles: recent advances in the biomedical applications." *Cell biochemistry and biophysics* 72, no. 3 (2015): 771-775.

In: Antimicrobial Resistance: Opportunities and Challenges ISBN: 978-1-53617-943-9
Editors: P. Kumar Prabhakar and V. Mishra

Chapter 11

PLANT ANTIMICROBIALS: AN ALTERNATIVE FOR OVERCOMING MICROBIAL RESISTANCE

Jeena Gupta*
Department of Biochemistry, School of Bioengineering and Biosciences,
Lovely Professional University (LPU),
Phagwara, Punjab, India

ABSTRACT

The emergence of microbial resistance to conventional antibiotics is a serious concern for treating infectious diseases. The promising solution for overcoming these problems is the identification of new and novel antimicrobial molecules. Plants are always the best sources which constitute molecules with diverse structures and activities. The phytochemicals enriched in the plants have been reported to exert potential antimicrobial activities against both sensitive as well as resistant pathogens and followed different action mechanisms. In addition to being potent antimicrobial, the phytochemicals also show synergistic effects when delivered in combination with conventional antibiotics, thereby increasing their sensitivity, lowering MIC (minimum inhibitory concentration) and overcoming microbial resistance. In this chapter, we have discussed the potent natural molecules, their mechanism of action and their effect on reversing antibiotic resistance. Consideration of these facts projects phytochemicals as the most valuable molecules to provide the best therapeutic solutions for overcoming antimicrobial resistance and treating infectious diseases.

Keywords: phytochemical, antimicrobial, antifungal, microbial resistance, efflux pump inhibitor

---

* Corresponding Author's E-mail: jeena_gupta@yahoo.co.in.

INTRODUCTION

In medicine, antibiotics are amongst the most significant discoveries and emerge as powerful drugs to control infections/infectious diseases. However, the unrestricted and extensive use of antibiotics has escalated the development of resistance to antibiotics by imposing bacteria to selective pressure (Russell, 2003; Stewart, 2002; Cizman, 2003; Oluwatuyi et al., 2004). The emergence of antibiotic resistance has been the biggest challenge nowadays, which threatens human health and is responsible for causing millions of deaths worldwide annually. Global Antimicrobial Surveillance System (GLASS), which is World Health Organization (WHO)'s new prospective has revealed widespread occurrence of bacterial (suspected) infections in 22 countries in 2018 and resistance to antibiotic among 500,000 people. The evolution of resistance among infected individuals led to the ineffectiveness of existing antibacterial drugs and the spread of infections (WHO, 2014; Baym et al., 2016). Bacteria also possess the capacity to transmit the genetic determinants for resistance which exacerbated the resistance problem worldwide.

World Health Organization (WHO) also recognizes antibiotic resistance as the greatest threat for treating infectious diseases (Oluwatuyi et al., 2004; Nascimento et al., 2000; Walker and Levy, 2001; Wright, 2005, Conly, 2004). WHO has taken strategic actions to mitigate, prevent and control the spread and occurrence of resistant microbial strains and promoting the slogan "No action today, no cure tomorrow" (WHO, 2011). This plan includes several objectives: the correct use of antimicrobial drugs (right dosage and appropriate duration), enhanced environmental hygiene and infection control practices, strengthened surveillance system for monitoring antibiotic use on human health and also encourage the discovery of active new drugs (Pandra et al., 2010; Dotsch et al., 2009). Instead of the quest to discover new antimicrobials by pharmaceutical industries, only two new antibiotic classes i.e., cyclic lipopeptides and oxazolidinones, have entered the market in the past thirty years. These new classes of drugs are only effective against Gram-positive bacteria and there is no new development against Gram-negative bacteria. This low rate of new drug emergence is also evident from FDA (Food and Drug Administration) history and only 6 new antimicrobial drugs have been approved in last decade whose success is also compromised over resistance emergence (Alekshun and Levy, 2007; Chan et al., 2011; Gibbons, 2008). The discovery of active new agents that can suppress resistance mechanisms in bacteria or development of resistance modifying agents (RMA) provides an important alternative strategy to combat the antimicrobial resistance problem, the latter can also facilitate the recycling of previously established antibiotics which is often cheaper than discovering new candidate antibiotics (Sibanda et al., 2010).

Plants and their extracts are traditionally used for their antimicrobial activities and also provide an important source for discovering new chemical entities. Multiple of published reports have proved the antimicrobial activities and the therapeutic value of plant origin molecules on human health and several studies have clearly evidenced the improved therapeutic efficacy of already existing antibiotics by plant-derived products (Khameneh et al., 2019; Oluwatuyi et al., 2004; Sibanda et al., 2010). An important recommended strategy is the combination therapy of potent antibacterial phytomolecules with failing antibodies which was found to restores their activities (Brown, 2015; Rana et al., 2018; Vandevelde et al., 2016). Many reports have exhibited the potent activities of phytomolecules against bacterial resistance (Cowan, 1999; Khameneh et al., 2016; Shakeri et al., 2018; Fazly Bazzaz et al., 2018) which can either act alone or in combinatorial therapy with antibiotics for enhancing the antibacterial activity against various bacteria (Fazly Bazzaz et al., 2018; Fazly Bazzaz et al., 2016; Fazly Bazzaz et al., 2010; Betts and Wareham, 2014). But limited literature is available to highlight the mechanism of action and structure-activity relationship of natural compounds, which largely remains elusive. In this chapter, we have focused on describing the potent plant-derived antimicrobials, the relationship between their structure and their possible mechanism of action.

MECHANISMS OF ANTIBACTERIAL ACTIVITY

The various antimicrobial compounds available in the market have multiple targets in the bacterial cells through which they can inhibit the growth and proliferation of bacteria (Figure 1). The general mechanism involves is the chemical interference with either the synthesis or functioning of bacteria's vital components by interacting with these targets. These multiple targets include 1) the biosynthesis of bacterial cell wall (targeted by beta-lactam antibiotics), 2) the biosynthesis of proteins in bacteria (targeted by antibiotics like aminoglycosides, tetracyclines, macrolides, etc.), 3) the destruction of cell membrane of bacteria (targeted by polymyxin B), 4) the inhibition of bacterial metabolic pathways and 5) the replication and repair of bacterial DNA (targeted by Fluoroquinolone, etc.).

The antibacterial compounds also work by evading or bypassing the conventional antibacterial resistance mechanisms. There are various mechanisms through which the bacteria acquire resistance against antibacterial agents. Some of the bacterial species possess innate resistance against class/classes of antimicrobial agents. All the strains of these bacterial species are resistant to all the members of these class/classes of antimicrobial agents. However, the major concern is when the bacteria are initially susceptible to an antibacterial agent but later on acquires resistance (Tenover, 2006). The understanding of the mechanisms by which the bacteria develop antibiotic resistance will be helpful in solving this problem and identifying key targets for developing potent anti-microbial drugs (Walsh 2000). These mechanisms by which the bacteria develop resistance involves efflux pump activation (e.g., efflux pumps for trimethoprim and fluoroquinolones resistance in P. aeruginosa), destruction enzymes which destroys antibacterial agent (e.g., β-lactamase in Enterobacteriaceae), modifying enzymes which can modify antibacterial agent (e.g., fosfomycin or chloramphenicol in P. aeruginosa) and target alteration to a lower affinity recognition by the bacterial agent (Khameneh et al., 2016). The bacteria develop resistance either by employing one of these mechanisms or different mechanisms together. The spreading of resistance to bacterial populations mainly occurs through plasmids that can replicate independently and passed to cells of bacterial species.

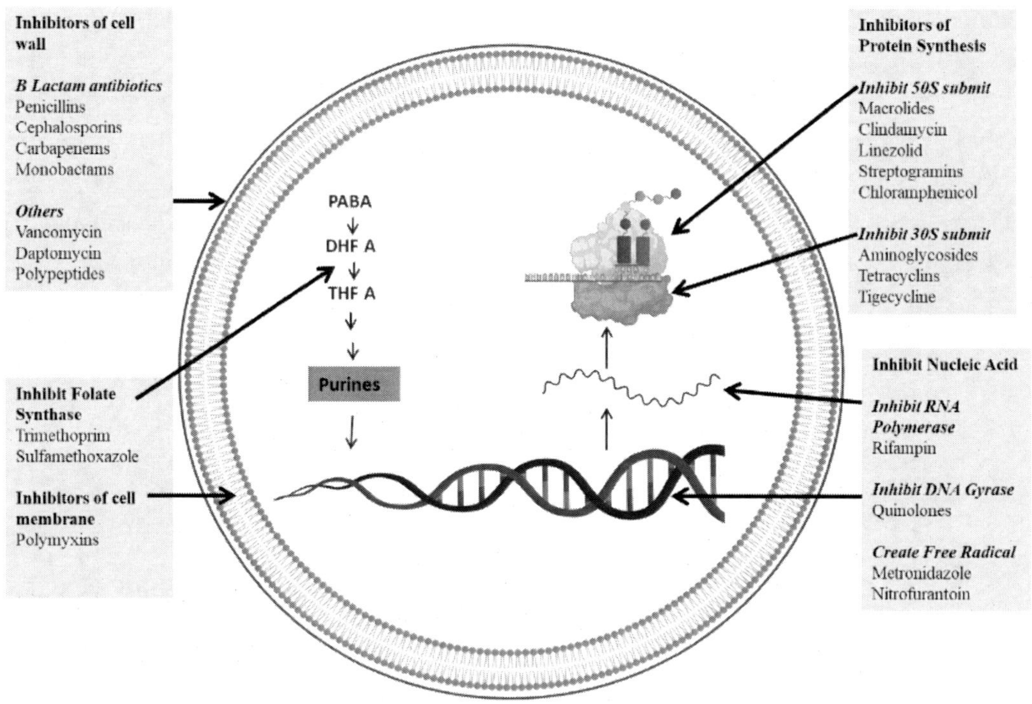

Figure 1. Antibiotic targets in the bacterial cell.
Different classes of antibiotics like β lactam antibiotics inhibit the synthesis of cell wall while Polymyxins targets bacterial cell membrane. The biosynthesis of proteins is inhibited by different antibiotic classes like aminoglycosides, macrolides, tetracyclines, etc. which target different subunits of the ribosomes. The quinolones inhibit the replication of DNA by inhibiting enzyme DNA gyrase.

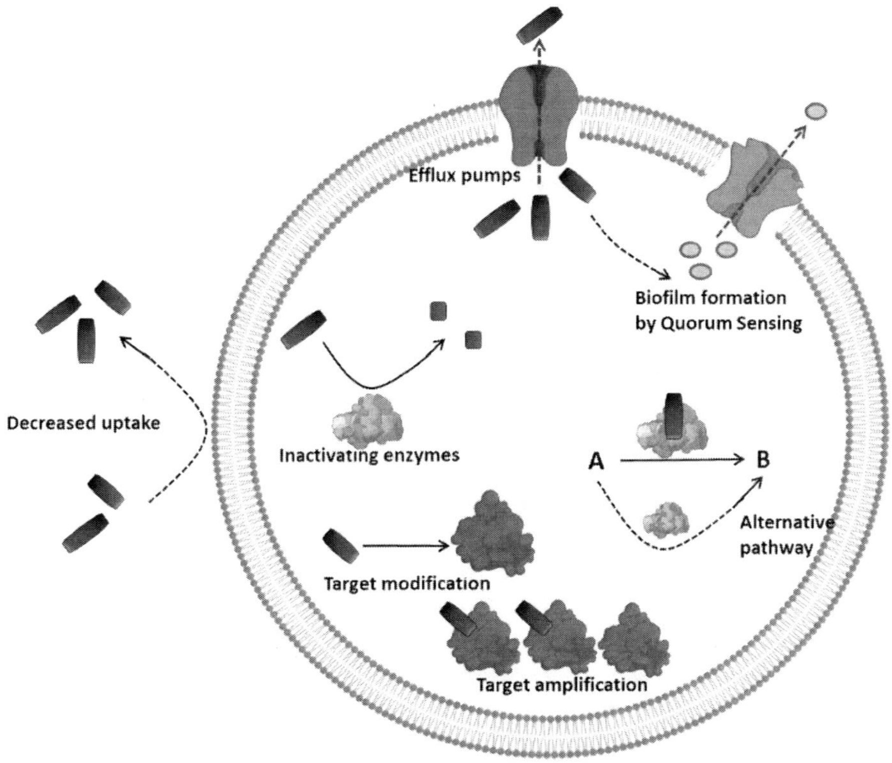

Figure 2. Multiple mechanisms of antibiotic resistance in bacteria.
The removal of antibiotics by efflux pumps in bacteria, the degradation of antibiotics by destruction enzymes, the modification and amplification of target proteins and the use of an alternative metabolic pathway.

PLANT-DERIVED ANTIMICROBIALS

Although the synthetic antimicrobial agents have got approval and have occupied the antibiotic market all over the world but researchers are also attracted to identifying the plants, animals and even microbes derived natural compounds with antimicrobial capacity (Gyawali and Ibrahim, 2014; Moloney, 2016). These compounds prove promising to overcome the bacterial resistance to antibiotics (Rossiter et al., 2017) but among all of these, the plant-derived molecules were found more potent to provide protection against bacterial infections. These plant-derived molecules are naturally occurring and belong to varied groups of chemical compounds. Their extensive existence has been implicated in providing beneficial health effects like antibacterial, antifungal, and antioxidant, etc (Barbieri et al., 2017). The combinational use of these plant-derived molecules with older antibiotics have been demonstrated to restore the potency and clinical application of these antibiotics by avoiding resistance development (Table 1). In the following sub-sections, we are discussing some of the plant's components with proven antimicrobial activities with commercialization potential.

Table 1. Potent phytochemicals with synergistic activity with known antibiotics

Phytomolecule	Class	Plant enriched	Synergistic antibiotic	Microorganism	Reference
Piperine	Alkaloid	*Piper longum Piper nigrum*	Ciprofloxacin	*S. aureus*	Khan et al., 2006
Tomatidine	Alkaloid	Solanaceous plants like potato, tomato, and eggplant	Ampicillin, Gentamicin, Ciprofloxacin, Cefepime	*Enterococcus faecalis* *S. aureus* *P. aeruginosa*	Soltani et al., 2017
Chanoclavine	Alkaloid	*Ipomoea muricata*	Tetracycline	*E. coli*	Dwivedi et al., 2019
Allicin	Organosulfur compound	*Allium sativum* (garlic)	Ciprofloxacin Cefoperazone Tobramycin	*P. aeruginosa*	Cai et al., 2008
Allyl Isothiocyanate	Organosulfur compound	*Eutrema japonicum* *Armoracia rusticana*	Erythromycin Streptomycin	*S. pyogenes* *P. aeruginosa* *E.coli*	Palaniappan et al., 2010; Saavedra et al., 2010
Baicalein	Phenol	*Scutellaria baicalensis* *Thymus vulgaris* *Scutellaria lateriflora*	β-lactam antibiotics Ciprofloxacin Tetracycline	MRSA *E.coli*	Chan et al., 2011; Fujita et al., 2005
Kaempferol rhamnoside	Phenol	*Persea lingue*	Ciprofloxacin	*S. aureus*	Holler et al., 2012
Chalcone	Phenol	*Dalea versicolor*	Erythromycin	*S. aureus*	Belofsky et al., 2004
Chebulinic acid	Phenol	*Terminalia chebula*	Quinolone	*M. tuberculosis*	Patel et al., 2015
Bergamottin epoxide	Coumarin	*Citrus paradisi*	Norfloxacin	*MRSA*	Abulrob et al., 2004
Geranyl coumarin	Coumarin		Ciprofloxacin	*S. aureus*	de Araujo et al., 2016
Gallbanic acid	Coumarin	*Ferula szowitsiana*	Ciprofloxacin	*S. aureus*	Bazzaz et al., 2010
Thymol	Terpene	*Thymus vulgaris*	Fluconazole	*Candida glabrata* *Candida krusei* *C. albicans*	Sharifzadeh et al., 2018

Alkaloids

Alkaloids are the nitrogen-containing heterocyclic compounds that possess a high degree of variability in their chemical structures. Many different studies have highlighted the anti-microbial activity of alkaloids and they have proven beneficial for treating infectious diseases (Cushnie et al., 2014). The putative mechanism for the antibacterial activity of most of the alkaloids is through the inhibition of Efflux pumps (Figure 3).

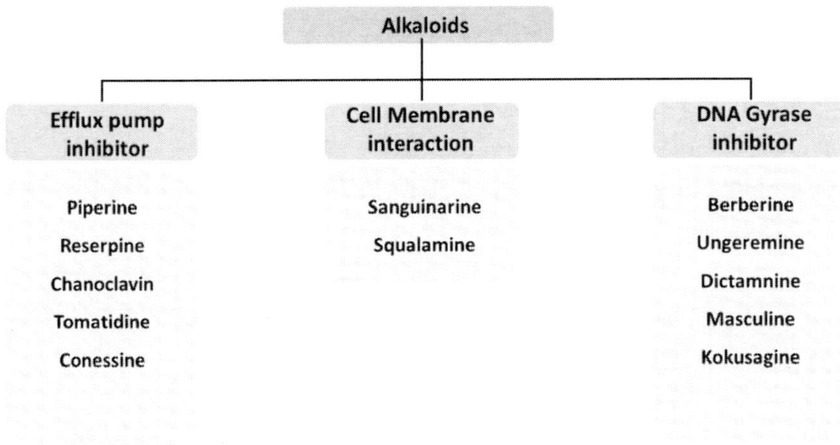

Figure 3. Alkaloids and their different mechanisms of action.

An important alkaloid is Piperine, a piperidine type molecule, which is isolated from *Piper longum* and *Piper nigrum*. Piperine when co-administered with known antibiotic ciprofloxacin, markedly reduced the Minimum Inhibitory Concentration (MIC) of ciprofloxacin for *S. aureus* and also inhibited its growth (Khan et al., 2006). The co-administration of gentamicin with piperine was also found effective against MRSA infection (Khameneh et al., 2015). The mechanistic study highlighted the inhibition of efflux pump NorA by piperine in both *S. aureus* and MRSA which is responsible for its antibiotic potentiating effects (Khameneh et al., 2015; Kumar et al., 2008).

An isoquinoline alkaloid Berberine is mainly found in the stem bark and roots of Berberis species and is the major active ingredient of *Cortex phellodendri* and *Rhizoma coptidis*, plants of importance in traditional medicine. Berberine exhibited strong antimicrobial activity against fungi, bacteria, viruses, and protozoans. The molecular mechanism involves the inhibition of cell division, intercalation with DNA, and binding with DNA gyrase, topoisomerase IV and RNA polymerase by berberine (Iwasa et al., 2001; Yi et al., 2007; Domadia et al., 2008). The inhibition of FtsZ, a cell division protein and inhibiting bacterial cell functioning by binding with cell's macromolecules, were also held responsible for its antibacterial properties (Boberek et al., 2010; Feng et al., 2016). Berberine emerges as a very strong antimicrobial agent with potency to replace conventional antibiotics and overcome antibiotic resistance. An another iso-quinoline alkaloid, ungeremine which is enriched in *Pancratium illyricum L.* bulb's methanolic extract was also found to possess potent antibacterial activities. It inhibits bacterial topoisomerase IA and causes a remarkable increase in the cleavage of DNA (Casu et al., 2011; Schrader et al., 2013). Quinoline alkaloids isolated from *Teclea afzeli's* stem bark like dictamnine, masculine and kokusagine exhibited promising antibacterial activity and functions by reducing DNA replication through inhibition of type II topoisomerases (Lin et al., 2000; Kuete et al., 2008; Heeb et al., 2011). Synthetically derived quinolone alkaloids also exhibit similar activity (Heeb et al., 2011). Some like alkyl methyl quinolones are also considered as respiratory inhibitors which reduces the consumption of O2 in bacteria (Tominaga et al., 2002).

An indole alkaloid reserpine which can be isolated from *Rauwolfia serpentine,* is highly known for its potent Efflux pump inhibitory activity (Abdelfatah et al., 2015). The co-administration of reserpine increases the antibiotic susceptibility of various bacterial species, which include *Streptococcus* spp., *Staphylococcus* spp., and *Micrococcus* spp. (Sridevi et al., 2017). It is also capable to enhance the antibiotic susceptibility of multi drug-resistant (MDR) clinical isolates belonging to *A. baumannii* species in which efflux pump AdeABC was found over-expressed (Jia et al., 2015). In another study, it was observed that *Stenotrophomonas maltophilia* gained resistance against fluoroquinolones by overexpressing efflux pumps and reserpine addition was found to decrease antibiotic resistance (Jia et al., 2015). Therefore the inhibition of efflux pumps by reserpine is the major mechanism responsible for increasing antibiotic susceptibility in both Gram-negative and Gram-positive bacteria. Some natural alkaloids like berberine and sanguinarine are also capable to alter Ftsz functionality whose inhibition is implicated in providing broad-spectrum activity to antibacterial agents (Sun et al., 2017).

Sanguinarine can be isolated from plants like *Sanguinaria canadensis, Chelidonium majus,* and *Macleaya cordata*. Sanguinarine was tested against MRSA strains and it was found to cause bacterial cell lysis by increasing the release of cell membrane-bound autolytic enzymes. The transmission-electron-microscopy of these MRSA showed septa formation throughout the cell wall highlighting its possible mechanism to increase cytoplasmic membrane deformity (Obiang-Obounou et al., 2011). Both berberine and sanguinarine also exhibit strong DNA intercalation activity and are potent inhibitors of transcription/replication (Al-Ani et al., 2015). Sanguinarine also exhibited antimycobacterial activities as demonstrated against two mycobacterial model species *Mycobacterium smegmatis* and *Mycobacterium aurum* (Newton et al., 2002).

A steroidal alkaloid tomatidine which can be isolated from solanaceous plants like potato, tomato, and eggplant, etc. displayed potential antibacterial activity against bacterium *S. aureus* either alone or in combination therapy with aminoglycosides (Jiang et al., 2016). Aminoglycosides and tomatidine showed synergistic effects against antibiotic-resistant *S. aureus* strains (Mitchell et al., 2012). Tomatidine is also considered as a potentiator for antibiotics, like ampicillin, gentamicin, ciprofloxacin, and cefepime and thereby increase the effectiveness of these antibiotics against infections caused by both Gram-negative and Gram-positive bacteria such as *Enterococcus faecalis, S. aureus and P. aeruginosa* (Soltani et al., 2017). A tricyclic ergot alkaloid, chanoclavine which can be isolated from plant *Ipomoea muricata* also exhibited synergistic effects against MDR *E. coli,* when co-administered along with tetracycline, mainly by inhibiting ATPase-dependent efflux pumps (Dwivedi et al., 2019).

A steroidal alkaloid conessine, which is enriched *Holarrhena antidysenterica* possesses therapeutic importance (Kumar et al., 2007). The extract of *Holarrhena antidysenterica* (family: Apocynaceae) has been traditionally used for treating different diseases like diarrhea, dysentery, and bacterial infections (Siriyong et al., 2017). Conessine displayed potent antibacterial activity almost as comparable to antibiotic controls and was found effective against both Gram-negative and Gram-positive bacteria. It also exerts synergistic effects in combination with conventional antibiotics (Zhou et al., 2017; Siddiqui et al., 2012). It also improves the antibiotic susceptibility of *A. baumannii* and thus can be utilized for its antibiotic resistance-modifying activity. The mechanistic insight highlights the inhibition of efflux pump AdeIJK by conessine which led to the effluxing of multiple antibiotics in bacteria *A. baumannii* (Siriyong et al., 2016; Damier-Piolle et al., 2008).

Another natural compound, Squalamine, a steroid-polyamine, is not only found in plants but was first isolated from dogfish shark. Squalamine showed broad-spectrum antimicrobial activity and mainly functions by disrupting the membrane permeability in microbes. It was found to interact with the negatively-charged phosphate groups present in the outer membrane of Gram-negative bacteria and led to its disruption. However, squalamine treatment results in leaking of cytoplasmic contents by depolarizing cytoplasmic membrane in Gram-positive bacteria leading to their rapid death (Alhanout et al., 2010).

Organosulfur Compounds

A number of sulfur-containing compounds obtained from plants thatare reported to possess potent antifungal and antibacterial activities (Kim et al., 2006; Iranshahi et al., 2008). Some of these sulfur-containing compounds like allicin, S-allyl cysteine, ajoene, isothiocyanates, dialkenyl, and dialkyl sulphides, have been reported to exert antibacterial activities against both Gram-negative and Gram-positive bacteria (Sobolewska et al., 2015). The performed investigations have highlighted that the plants which are rich in polysulphides displayed the antimicrobial activity of a wide spectrum (Asili et al., 2009; Boghrati et al., 2018).

An organosulfur compound Allicin which is also called diallyl thiosulfinate, is enriched in garlic (*Allium sativum*), belonging to family Alliaceae. Allicin's antimicrobial activities have been acknowledged for a long time, where it was observed to possess antibacterial activities against various bacteria like *Staphylococcus epidermidis, Streptococcus agalactiae,* MRSA, *P. aeruginosa*, and other oral pathogens causing periodontitis (Reiter et al., 2017). It was also found to potentiate the antibacterial activities of conventional antibiotics like ciprofloxacin, cefoperazone, and tobramycin when analyzed against *P. aeruginosa* (Cai et al., 2008).

The molecular mechanism observed for antimicrobial activity of allicin is by inhibiting the activities of sulfhydryl-dependent enzymes (Figure 4) like thioredoxin reductase, alcohol dehydrogenase, and RNA polymerase (Lanzotti et al., 2014). It was confirmed that the addition of glutathione and cysteine reduced the allicin's antimicrobial effect as these two compounds react with the disulfide bond in allicin and prevent the damage to microbial cells (Davis et al., 2005). Allicin was also found to partially inhibit the synthesis of proteins and DNA in microbial cells and was also found to target RNA (Feldberg et al., 1988).

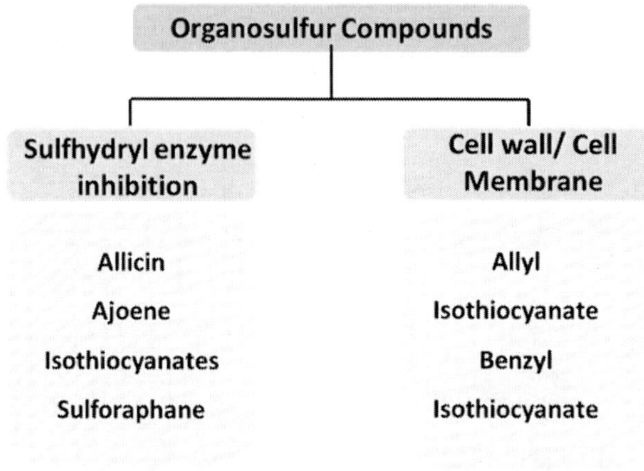

Figure 4. Organosulfur compounds and their different mechanisms of action.

Another organosulfur compound ajoene which is also found in garlic extracts have been reported to display a wide spectrum of antimicrobial activity not only against bacteria but also against fungi protozoan and viruses (more potent than allicin). The isolated compound is actually a mixture of two stereoisomers E- and Z-ajoene. The antimicrobial activity of ajoene was also found to be greatly reduced after the addition of cysteine in the media might be because of the interaction of this amino acid with ajeone's disulfide bond similar to allicin. Therefore, it can be concluded that ajeone's antibacterial mechanism is also the same as allicin by interacting with thiol-dependent enzymatic systems (Rehman et al., 2013).

The volatile Isothiocyanates (ITCs) are organosulfur compounds that are generally formed through the action of myrosinase enzyme on plant glucosinolates. After tissue disruption, myrosinase hydrolyzes glucosinolates into active compounds like thiocyanates, nitriles, and ITCs. The plants belonging to Brassicaceae family, like cabbage, cauliflower, broccoli and mustard have an abundance of these compounds but among all, ITCs showed more promising antibacterial activity and inhibits the growth of pathogenic bacteria. A study had evaluated the antimicrobial activity of ITCs extracted from horseradish root (*Armoracia rusticana*), and have analyzed them against oral pathogens where they were found to exhibit strong antimicrobial activities (Park et al., 2013). ITCs were found to inhibit the urease enzyme which results in bactericidal response against *Helicobacter pylori*, and also reduces inflammatory component responsible for causing Helicobacter infections (Fahey et al., 2013).

The underlying molecular mechanism for antimicrobial activity of ITCs is although not completely comprehended, they were proposed to react with proteins and disturb biochemical processes. The carbon atom present in the ITC group (−N=C=S) is a strong electrophile that can easily react with thiols, amines, sulfhydryl and hydroxylic groups present in amino acids of proteins (Dufour et al., 2015). They mainly react with the sulfhydryl group of cysteine residues which plays a crucial role in maintaining protein structure and function. ITCs also attack cysteine residues of bacterial P-ATPase and inhibit its ATP binding sites (Breier and Attila, 2000). Another compound that co-exists with ITCs is sulforaphane which is present in various plants like *Diplotaxis harra*. Sulforaphane is a derivative of 4-methyl sulfinyl-butyl-glucosinolate which possesses anticancer properties and also displayed potent antibacterial activities against stomach cancer-causing bacteria *H. pylori*. It was also found effective against *Listeria monocytogenes* and *S. aureus* (Benzekri et al., 2016).

Allyl ITCs (AITCs), is other organosulfur compounds with the general formula of C=C-C-N=C=S, enriched in plants belonging to Brassicaceae family like *Eutrema japonicum* and *Armoracia rusticana* and was found to possess potent antibacterial activities. AITC exert bactericidal and bacteriostatic activities against both *S. aureus* and *E. coli* (Lu et al., 2016) and was also found to effectively reduce erythromycin's MIC against *S. pyogenes* (Palaniappan et al., 2010) synergistically augment streptomycin's effect against *P. aeruginosa* and *E.coli* (Saavedra et al., 2010). There are various mechanisms reported for the observed antimicrobial potential of AITC. It was reported to damage the integrity of the cell wall when utilized in vapor state resulting in leakage of cellular content and metabolites. The creation of pores had been visualized on the bacterial cell membrane after exposure to AITC resulting in intracellular substance leakage (Lin et al., 2000). Treatment in AITC also provoked changes in internal structures as observed by electron microscopy (Lin et al., 2000; Nedorostova et al., 2009; Luciano and Holley, 2009). Studies have also suggested that AITC can oxidatively cleave disulfide bonds thereby inactivating essential intracellular-enzymes (Delaquis and Mazza, 1995).

Another isothiocyanate, benzyl ITCs have been found in *Alliaria petiolate* (Prati et al., 2004). The evaluation of this compound against 15 different isolates of MRSA showed its bactericidal activity against 11 among them, highlighting its effectiveness in suppressing MRSA (Dias et al., 2014). The chemical structure of BITC imparts it electrophilic and lipophilic properties which allows it to penetrate through the outer membrane of bacteria and disturb the integrity of bacterial cell membrane (Sofrata et al., 2011).

Phenethyl isothiocyanate (PEITC), is other ITCs type which is abundant in brassica vegetables like *Brassica rapa* and *Brassica campestris* (Jang et al., 2010). The *in vitro* evaluation of this compound against bacteria isolated from the intestinal tract of humans showed their potential antibacterial activity against Gram-positive bacteria but less against Gram-negative ones (Aires et al., 2009). It was also found to possess antifungal activity against *Alternaria brassicicola* (Drobnica et al., 1968). The factors responsible for the antifungal activity of ITCs might include ROS accumulation intracellularly, decreased oxygen consumption rate, and mitochondrial membrane depolarization (Calmes et al., 2015). Broccoli (*Brassica oleracea L.*) also contains a compound called berteroin which displayed potent bactericidal activity and low MICs values against intracellular and extracellular bacteria and was also found effective against bacterium *H. pylori* (Moon et al., 2010; Haristoy et al., 2005).

Phenolic Compounds

Phenolics constitute a wide range of natural bioactive compounds which have varied medical applications. They also play an important role in improving the antibiotic activity of conventional compounds through various mechanisms against pathogens showing resistance (Farhadi et al., 2019; Ramezani et al., 2004; Gorniak et al., 2019). The significant mechanism involves either reducing the activity of efflux pumps or functioning as an efflux pump inhibitor (Figure 5).

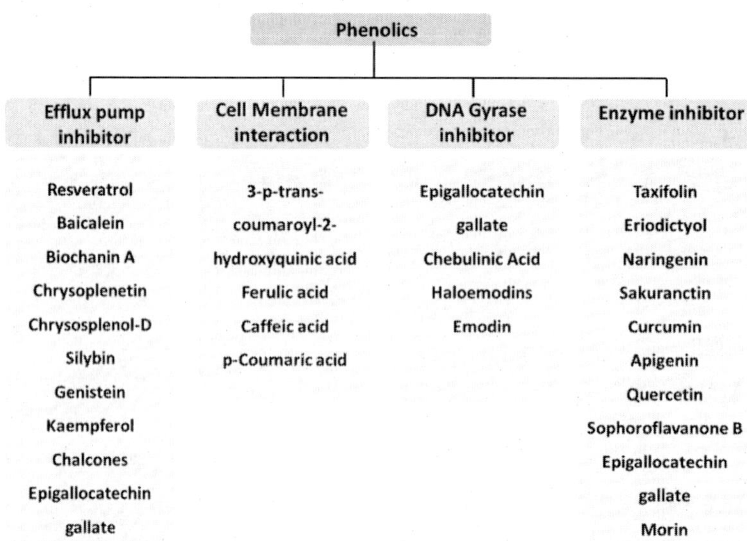

Figure 5. Phenolics and their different mechanisms of action.

A natural phenolic phytomolecule, resveratrol shows potential inhibition of efflux pump CmeABC and exhibited antimicrobial activity against bacteria like *Campylobacter jejuni* and *M. smegmatis* (Klancnik et al., 2017; Lechner et al., 2008). Resveratrol also exhibited efflux pump inhibitory activity against *Arcobacter cryaerophilus* LMG 10829 and *Arcobacter butzleri* LMG 10828 resulting in ethidium bromide accumulation in the cells (Ferreira et al., 2014).

A flavone baicalein which is isolated from roots of *Scutellaria baicalensis, Thymus vulgaris*, and *Scutellaria lateriflora* had been reported to show antibacterial activity (Lu et al., 2011). The screening of the extracts of these plants against standard strains of *Bacillus cereus, S. aureus, P. aeruginosa, C. albicans,* and *E. coli* indicate their potent antimicrobial and antioxidant properties (Bazzaz et al., 2011). The restoration of the effectiveness of antibiotics like β-lactam antibiotics, ciprofloxacin and tetracycline against MRSA by baicalein involves the inhibition of an efflux pump NorA (Chan et al., 2011). It also exerts synergistic effects with tetracycline increasing its effectiveness against E. coli which also involves the inhibition of efflux pumps (Fujita et al., 2005).

An isoflavone biochanin A showed an inhibitory effect on the MRSA efflux system and was found to reduce the expression of efflux pump NorA in this bacteria (Zou et al., 2014). It was also found to be a potent inhibitor of intracellular bacteria belonging to Chlamydia spp. (Hanski et al., 2014) and potentially inhibit efflux pump activity in Mycobacterium strains (Cannalire et al., 2017). Many other flavonoids have also demonstrated efflux pump NorA inhibitory activity. Two methoxylated flavones chrysoplenetin and chrysosplenol-D isolated from *Artemisia annua* inhibited the efflux pump NorA at sub-inhibitory berberine concentrations as a substrate (Stermitz et al., 2003). Two other phenolic compounds of classes flavonolignans and isoflavonoids could also inhibit NorA and were also reported to increase berberine and norfloxacin potency (Morel et al., 2003). A known medicinal plant *Silybum marianum* isolated flavonolignan, Silybin, and *Lupinus argenteus* isolated isoflavonoids genistein, biochanin A, and orobol were also found to potentially inhibit efflux pump NorA in *S. aureus* (Stermitz et al., 2001). The hybridization of flavonoids to antibiotics has also gained attention of many researchers as an effective technology to reduce the activity of efflux pumps. The enhanced activity and accumulation of antibiotic confirms the dual-action modes of these hybrids (Xiao et al., 2014).

Another active flavonoid, Kaempferol showed effectiveness against MRSA and *C. albicans* (fluconazole-resistant) and thus is a potential candidate against pathogenic microbes (Randhawa et al., 2016; Shao et al., 2016). It is also a potent MRSA efflux pump inhibitors similar to verapamil and mainly targets NorA efflux pump. Kaempferol's natural glycoside derivative, kaempferol rhamnoside, isolated from *Persea lingue* was found to increase ciprofloxacin antimicrobial activity by 8-folds in NorA overexpressed strain of *S. aureus* (Holler et al., 2012).

Chalcones are another type of phenolic compounds that were also found to inhibit efflux pumps and increase antibiotics activity. *Dalea versicolor* isolated 4′,6′-Dihydroxy-3′,5′-dimethyl-2′-methoxychalcone reduced erythromycin's MIC to 0.1 μg/mL from 0.4 μg/mL by inhibiting efflux pump NorA (Belofsky et al., 2004). Two synthetic forms of chalcones 4-dimethylamino-4′-dimethylaminoethoxychalcone and 4-phenoxy-4′-dimethylaminoethoxychalcone were identified during the screening of a chalcones library containing 117 synthetic and natural chalcones which showed potent inhibition of efflux pump NorA and were found equipotent as reserpine A (Holler et al., 2012). Epigallocatechin gallate (EGCG) is also a phenolic compound that exhibited antimicrobial activity against MRSA and functions by weakly inhibiting NorA efflux pump (Gibbons et al., 2004).

In addition to inhibiting efflux pumps, another antimicrobial mechanism of phenolic compounds involves the inhibition of enzyme DNA gyrase (Anderle et al., 2008). The natural phenolics like polyphenols isolated from Green tea (tannins), anthraquinones and chebulinic acid showed potent inhibition of DNA gyrase activity. Green tea EGCG can inhibit the ATP-binding site in the B-subunit of DNA gyrase (Gradisar et al., 2007). The future researchers show special interests in antimicrobial activity of EGCG as it shows multiple modes of action like efflux pump inhibition, DNA gyrase inhibition, and chromosomal penicillinase inhibition. A tannin chebulinic acid, initially extracted from *Terminalia chebula,* was found by in silico studies to inhibit DNA gyrase and can overcome quinolone resistance in *M. tuberculosis* (Patel et al., 2015). The semisynthetic derivatives of anthraquinone, called Haloemodins also showed strong inhibition of DNA gyrase in *Enterococcus faecium* (vancomycin-resistant) and MRSA. The halogenated derivatives of naturally occurring emodin can also potentially inhibited DNA gyrase in bacteria, but are weakly active against human topoisomerase I (Duan et al., 2014).

An isolated phenolic compound 3-p-trans-coumaroyl-2-hydroxyquinic acid (CHQA) from *Cedrus deodara*, was found effective in inhibiting eleven food-borne-pathogens and showed MIC in 2.5–10 mg/mL range against *S. aureus*. The action mechanism involves causing membrane hyperpolarization, damaging cytoplasmic membrane, membrane integrity loss and intracellular constituent leakage (Wu et al., 2016). The hydroxycinnamic acids like ferulic acid, caffeic acid, and p-coumaric acid interfere with the integrity of membranes and p-coumaric acid because of its lipophilic nature, shows the maximum interfering activity among them (Campos et al., 2009). Coumaric acid also shows antibacterial activity against lactic-acid-bacteria present in wine like *Lactobacillus hilgardii* and *Oenococcus oeni* (Campos et al., 2009).

A comprehensive study to identify the structure-activity relationship (SAR) of flavonoids with effects on their interaction with bacterial membranes showed a positive correlation between membrane rigidity of *E.coli* and flavonoids antibacterial activity, which suggests that flavonoids might reduce membrane rigidity. Among them, kaempferol was found to be most potent against E.coli (Wu et al., 2013).

The phenolics interaction with some of the crucial enzymes was also reported for enzymes responsible for producing the bacterial cell membrane precursors like beta-ketoacyl-acyl-carrier-protein synthase (KAS) II and III and even enzymes involved in the biosynthesis of fatty acids (mainly elongation cycle) like FabZ, FabI, and FabG. The flavonoids like taxifolin, eriodictyol, and naringenin were found potential inhibitors of KAS III enzyme of *E. faecalis* by in silico studies but however *in vivo* studies showed moderate antimicrobial activity against vancomycin-resistant bacteria and *E. faecalis*

(Jeong et al., 2009). Another flavonol 3,6-dihydroxyflavone, on the other hand, showed potent binding with KAS I and KAS III and its MIC was found to be 512 μg/mL (Farhadi et al., 2019). The covalent binding of EGCG with *E.coli* enzyme beta-ketoacyl-[acyl carrier protein] reductase (FabG) results in its aggregation and inactivation (Li et al., 2006). Naringenin's methoxy derivative, Sakuranetin isolated from *Polymnia fruticose* can competitively inhibit FabZ protein in *H. pylori* resulting in its inhibition with MIC 87.3 μM, and was found more potent than apigenin and quercetin in inhibiting this protein (Zhang et al., 2008).

Curcumin, a bioactive compound of turmeric was also found to possess bactericidal activity against E. coli and S. aureus by damaging the cell membrane. The lipophilic and amphipathic nature of curcumin allows it to cross the membrane lipid bilayer and increase its permeability (Tyagi et al., 2015). An enzyme d-Alanine:d-alanine ligase which is critical for assembling the precursors of the cell wall, was found to be inhibited by flavonoids apigenin and quercetin with quercetin being a more potent reverse inhibitor of this enzyme and competes with ATP for binding (wu et al., 2008). Another phenolic compound sophoraflavanone B which is a prenylated-flavonoid was also found to directly interact with peptidoglycan and can inhibit MRSA with MIC 15.6–31.25 μg/mL (Mun et al., 2014).

Some phenolic compounds also mechanistically inhibit enzymes like sortase, urease and dihydrofolate reductase which is responsible for their antimicrobial activity (Maresso et al., 2008; Xiao et al., 2013). EGCG was found to inhibit dihydrofolate reductase which is one of its action mechanisms against different clinical isolates (18 isolates) of *Stenotrophomonas maltophilia,* a nosocomial pathogen. For some isolates, the MIC values were found to be very low (4 μg/mL) which highlights it's potential to replace approved drug trimethoprim, a well-known dihydrofolate reductase blocker, in those patients who cannot tolerate trimethoprim side effects (Navarro-Martínez et al., 2005). *Curcuma longa* showed potential inhibitory activity against enzyme sortase A with an IC50 value of 13.8 μg/mL when evaluated against *S. aureus* while p-hydroxy mecuri-benzoic-acid, which was used as positive control, showed 40.6 μg/mL IC50. Curcumin derivatives bisdemethoxycurcumin and demethoxycurcumin can also inhibit sortase A but showed lower activity than curcumin (Park et al., 2005). A flavonol morin isolated from *Rhus verniciflua* barks also showed potent inhibition of both sortase A and B with respective IC50 of 37.39 and 8.54μM. These studies have also indicated that flavonols although unable to inhibit bacterial growth but showed fibrinogen clumping activity (Kang et al., 2006). The inhibition of enzyme urease was also observed by some phenolics which are majorly responsible for the inhibition of *H. pylori* virulence. Flavonoids 4-deoxy analogs were identified as potent inhibitors of urease by SAR studies and the IC50 of urease inhibition by 4′,7,8-trihydroxyl-2-isoflavene was 0.85 μM, which identified it as more strong urease inhibitor (20 fold activity) than its well-known inhibitor acetohydroxamic acid. Also, these synthetic urease inhibitors showed considerable side effects *in vivo* studies and natural 4′,7,8-trihydroxyl-2-isoflavene is therefore, a promising candidate (Xiao et al., 2013).

Since phenolic compounds possess diverse antibacterial action mechanisms that range from efflux pump inhibition, cell membrane interaction, cell wall biosynthesis inhibition and critical enzyme inhibition like sortase A, dihydrofolate reductase and urease, they are therefore good candidates for identifying potent antimicrobial compounds to be evaluated in clinical trials. Their synergistic activity with known antibiotics and multiple mechanisms of action makes them ideal candidates for overcoming microbial resistance.

Coumarins

Coumarins are the products of different microorganisms and plants (Smyth et al., 2009). They are known to possess various important bioactivities like anti-inflammation, analgesic, anticoagulant, estrogenic, vasodilator, anti-helminthic, sedative, hypothermic, photosensitizing, and anti-cancer (Al-Majedy et al., 2017; Wu et al., 2009; Kostova et al., 2005). Many coumarins and even their natural, as well as synthetic derivatives, were also reported to possess potent antibacterial activities (Smyth et al., 2009; Kayser and Kolodziej, 1999; Melliou et al., 2005). In a study to analyze the antibacterial activity of *Ferulago campestris* roots isolated coumarins and pyrano-coumarins; aegelinol benzoate, grandivittin and agasyllin, mostly were found to be active against Gram-negative as well as Gram-positive bacteria. Two coumarins aegelinol and agasyllin possess more potent activity against ATCC strains of *Enterobacter aerogenes*, *Salmonella enterica serovar Typhi, S. aureus,* and *Enterobacter cloacae* with MIC for aegelinol was 16 μg/mL and for agasyllin was 32 μg/mL. Their dose-dependent antibacterial activity was also found against *H. pylori* with doses in range 5 and 25 μg/mL (Basile et al., 2009).

The coumarins isolated from *Prangos hulusii* root dichloromethane extract were evaluated for their antimicrobial activity. In this study, one new coumarin 4′-senecioiloxyosthol was identified which possesses the maximum activity against *Bacillus subtilis* with MIC of 5 μg/mL. Furthermore, nine already known prenylated coumarins were also identified of which osthole was found moderately active against pathogenic bacteria *Staphylococcus aureus* (methicillin sensitive), *B. subtilis*, and *Klebsiella pneumonia,* all with MIC of 125 μg/mL (Tan et al., 2017). In another study a glucosidic aryl coumarin asphodelin A 4′-O-β-D-glucoside and its aglycon form asphodelin A were isolated from plant *Asphodelus microcarpus* and were found to possess potent antimicrobial activity against five different microorganisms *E. coli, S. aureus, C. albicans, Botrytis cinerea* and *P. aeruginosa* with MIC varies between 4 -128 μg/mL (El-Seedi et al., 2007).

The Structure Activity Relationship (SAR) of three coumarins compounds clorobiocin, novobiocin and coumermycin A1 which were derived from varied species of Streptomyces were found to display potent antimicrobial activity. Each of these compounds besides having coumarin structure also contains a noviosyl sugar-moiety which was found essential for their activity and they were found to inhibit DNA gyrase, a type II DNA topoisomerase (Maxwell, 1993). The SAR studies also reveal that the planer structure and the lipophilic characteristics, both are necessary for the antibacterial activities of coumarins (Kayser and Kolodziej, 1999). Therefore the mechanism of passive diffusion facilitates their cellular penetration mainly in Gram-positive bacteria. The free –OH groups at 6th and 7th position in the nucleus of coumarins are important for their respective antifungal and antibacterial activity (Sardari et al., 1999). The systemic analysis of SAR data also reveals that the presence of methoxy group at C-7 of coumarins, besides the presence of hydroxyl group at C-6 or either C-8 provides them invariable antibacterial activity against a wide range of bacteria. The dimethoxy aromatic arrangement in coumarins is responsible for producing promising compounds that can target microorganisms having special requirements of growth factors like beta-hemolytic *Streptococcus, Haemophilus influenza,* and *Streptococcus pneumonia* (Kayser and Kolodziej, 1999). Coumarins were also found to suppress the network of quorum-sensing in pathogenic bacteria and also prevent the production of virulence factors and formation of biofilm (Wu et al., 2009; D'Almeida et al., 2017; Zhang et al., 2018; Reen et al., 2018, 166. Gutiérrez-Barranquero et al., 2015; da SM Forezi et al., 2018).

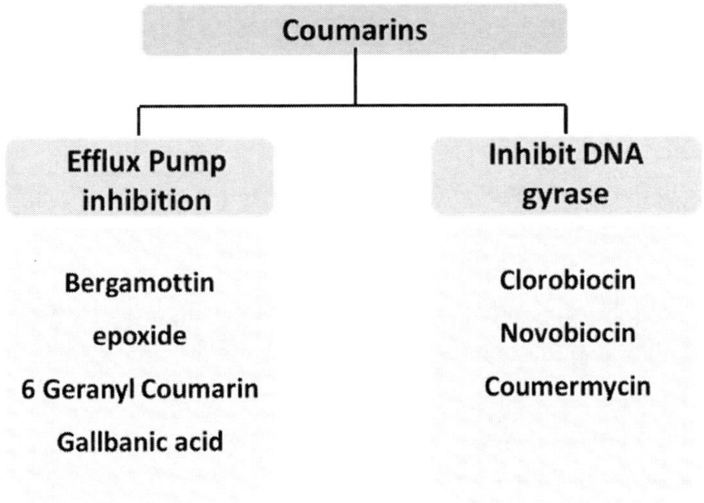

Figure 6. Coumarins and their different mechanisms of action.

The inhibition of efflux pumps was also reported by some of the coumarin derivatives in MSRA strain (Figure 6). A furanocoumarin, bergamottin epoxide isolated from grapefruit *Citrus paradisi* was found to reduce the MIC of norfloxacin by 20-folds against MRSA, and the mechanism involves efflux pump inhibition (Abulrob et al., 2004). Another study also reported that two out of seven coumarins isolated from *Mesua ferrea*, showed potential inhibition of efflux pumps in *S. aureus* clinical isolates and MRSA and also reduces the MIC of norfloxacin by 8 folds (Roy et al., 2013). The terpenoid coumarins which are formed by the fusion of plant cell isoprene units with coumarins like 6-geranyl coumarin and gallbanic acid, were also found to potentially inhibit efflux pumps with activity comparable to verapamil (well-known efflux pump inhibitor) and reduces MIC of ciprofloxacin by 8 folds (de Araujo et al., 2016; Bazzaz et al., 2010).

Terpenes

Terpenes (isoprenoids) are naturally outspread widely, present in nearly all life forms and constitute a very diverse family. They are known to perform varied functions that range from involvement in cell's primary structure like cell membranes sterols and cholesterol or cell's functions like electron transporter quinones, photosynthetic carotenoids and retinal for vision (Oldfield and Lin, 2012; Paduch et al., 2007). They are also abundantly present in fruits, flowers, and vegetables with their considerably high concentration in plant foliage and plant reproductive structures. They are also responsible for plant fragrances and are major ingredients present in herbal resins (Paduch et al., 2007). Terpenes and their derivatives are also imperative for plant defense against pathogens and herbivores (Paduch et al., 2007; Togashi et al., 2010; Lee et al., 2008; Catteau et al., 2018; Shakeri et al., 2017). The susceptibility of Gram-positive bacteria is more to terpenes than Gram-negative bacteria. The lipophilic nature of terpenes is associated with their antimicrobial activity. Monoterpenes are known to preferably interact with cell membranes, alter the topology of membranous proteins (Figure 7), thereby increasing permeability and fluidity and also disturb the electron transport chain (Paduch et al., 2007).

In a study to investigate the growth inhibitory effect of different carbon chain length containing terpene alcohols against *S. aureus,* identified that 12 carbon farnesol and 10 carbon nerolidol were most effective with MBC of 20 μg/mL and 40 μg/mL, respectively. These terpene alcohols were found to interact with the cell membrane of bacteria and increase K^+ ion leakage from cells. The hydrocarbon chain length to which the hydroxyl group is attached was therefore found to play an important role and terpene alcohols with 10-12 carbon chains are most potent against *S. aureus* (Togashi et al., 2010). A resin acid and terpene dehydroabietic acid and its derivatives were also found antibacterial against *S. aureus* (Söderberg et al., 1996; Feio et al., 1999; Savluchinske-Feio et al., 2006; Tapia et al., 1997).

Carvone is another important terpenes with anti-infective potential (Paduch et al., 2007). The optical isomers of carvone show diverse activities as (4R)-(−)-carvone was found active against *E. coli, Campylobacter jejuni,* and *E. faecium* while (4S)-(+)-carvone was found effective against *L. monocytogenes* (De Carvalho et al., 2006). Both the optical isomers of carvone also possess the potent antifungal activity and it inhibits the yeast form to filamentous form transformation of *C. albicans* and thus decreases fungus pathogenicity (Paduch et al., 2007).

Another antifungal terpene is thymol which showed potent antifungal activity against *Candida glabrata, Candida krusei,* and *C. albicans*, clinical isolates with MIC of 51.25, 70 and 49.37 µg/ml, respectively. It also shows synergistic activity with fluconazole against all tested *Candida* species (Sharifzadeh et al., 2018). The terpenes with potent antifungal activity, like thymol, carvacrol, eugenol, and menthol are also suggested as good alternatives for food industries to replace synthetic fungicides. They were reported to be effective against different food-decaying fungi like *Fusarium oxysporum, Aspergillus niger, Penicillium chrysogenum, Aspergillus fumigatus, Cladosporium spp., Aspergillus flavus, Penicillium citrinum, Aspergillus ochraceus, Botrytis cinerea, Alternaria alternata,* and *Rhizopus oryzae* (Abbaszadeh et al., 2014; Marinelli et al., 2018).

Thymol and carvacrol constitute the major components of plant *Thymus capitatus,* which when evaluated against *S. aureus, E. coli, P. aeruginosa,* and *Enterobacter aerogenes* were found potent with MIC for thymol between 0.005–0.008 mg/mL and for carvacrol between 0.007–0.008 mg/mL (Althunibat et al., 2016). Both these compounds were also found to inhibit the formation of biofilm by *Salmonella* spp. (*Salmonella enteritidis, Salmonella typhimurium,* and *Salmonella saintpaul*), on the polypropylene surface (Amaral et al., 2015). The major mechanism for thymol was suggested to disrupt the integrity of the membrane when evaluated against *S. typhimurium* (Chauhan et al., 2014). Another study had also hypothesized the concentration-dependent efflux pump inhibition by carvacrol and thymol, as increased ethidium bromide accumulation was observed in foodborne pathogens possibly through inhibition of cell efflux of ethidium bromide (Miladi et al., 2016).

Two penta-cyclic triterpenes, α-amyrin, and ursolic acid possess antibacterial activity against a broad spectrum of bacteria. The interaction of these two compounds was modeled by Langmuir-monolayer-technique where they display disorganizing effects on the membrane of E.coli (Broniatowski et al., 2015).

Cinnamaldehyde and eugenol are other important terpenoids that are enriched in several plants essential oils and also demonstrated activity against a wide range of pathogenic microorganisms. These two compounds when comprehensively studied against 30 different *H. pylori* strains that are responsible for causing duodenal and gastric ulcers, were found to prevent its growth without the development of resistance against these compounds (Ali et al., 2005). The notable bioactivity of eugenol was also observed against biofilms of clinical strains MSSA and MRSA. Eugenol was found to inhibit the construction of biofilm, disrupt pre-established biofilms, disrupt cell-cell communication, and kill bacterial cells in biofilms. At sub-inhibitory concentration, it was also found to reduce the gene expression of enterotoxin production and biofilm-related genes (Yadav et al., 2015). The similar effects of eugenol were also observed against *P. aeruginosa* where it inhibits the synthesis of virulence factor and formation of biofilm (Rathinam et al., 2017).

Cinnamaldehyde was also found to mechanistically damage the bacterial cell wall, decrease membrane potential and inhibit bacterial growth as observed by scanning electron microscopic study against *S. aureus* and *E. coli* (Zhang et al., 2015).

The number of derivatives of terpenoids also exhibited potent anti-mycobacterial activity like agelasine F, costunolide, 1,10-epoxycostunolide, reynosin, puupehenone, deschloroelatol, allolaurinterol, sandaracopimaric acid, (+)-Totarol, elisapterosin B, parthenolide, santamarine, alantolactone, elatol, debromolaurinterol, and aureole when evaluated against *M. tuberculosis*. The lipophilic nature of terpene structure was found to facilitate the penetration into the cell wall of mycobacteria (Copp, 2003).

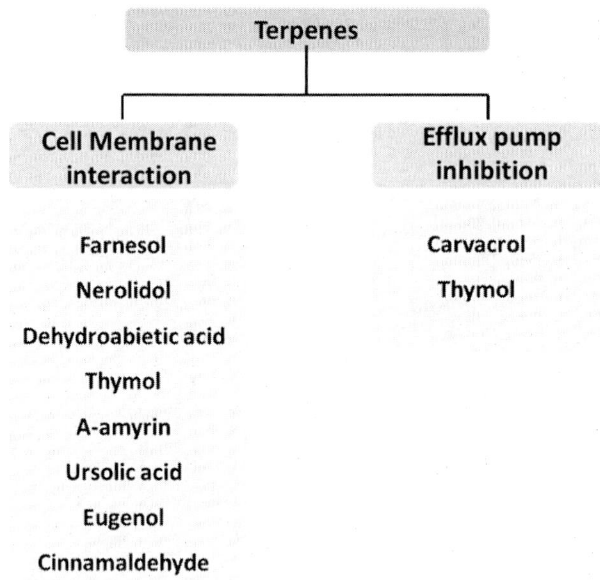

Figure 7. Terpenes and their different mechanisms of action.

Conclusion

A number of studies has suggested the effectiveness of medicinal plants for treating infectious diseases. They hold great promise for identifying new antimicrobial agents which can replace the conventional antibiotics and also overcome their microbial resistance. Being cheap, readily available and having almost no side effects, phytochemicals showed remarkable antimicrobial activity against pathogenic microorganisms and possesses lots of potentials to treat infectious diseases. Many of these compounds show synergistic activity when administered in combination with known antibiotics and also reduces their MIC values, thereby overcoming antibiotic resistance. The co-administration of natural compounds with conventional antibiotics follows different mechanisms and increases the range of their target sites providing superior efficacy and suppressing resistance development. Therefore understanding their molecular mechanism is necessary for developing novel therapeutic approaches.

The general action mechanisms of these compounds include damaging cell membranes of bacteria, efflux pump inhibition and inhibition of protein and DNA biosynthesis. For instance, during co-administration, the efflux pump inhibition by natural compound led to the accumulation of the conventional antibiotics in the bacterial cell resulting in their increased antibiotic efficacy against resistant strains. This highlights the need to conduct more mechanistic studies on potent phytochemicals for ensuring their safety and efficacy. The major challenge is the less tissue penetration and low bioavailability of these natural products. Once inside the human body they are readily acted upon by different enzymes which drastically affect their plasma concentration. Therefore, continued researches are also required for better understanding of their pharmacodynamic and pharmacokinetic properties.

REFERENCES

Abbaszadeh, S., A. Sharifzadeh, H. Shokri, A. R. Khosravi, and A. Abbaszadeh. "Antifungal efficacy of thymol, carvacrol, eugenol and menthol as alternative agents to control the growth of food-relevant fungi." *Journal de mycologie medicale* 24, no. 2 (2014): e51-e56.

Abdelfatah, Sara AA, and Thomas Efferth. "Cytotoxicity of the indole alkaloid reserpine from *Rauwolfia serpentina* against drug-resistant tumor cells." *Phytomedicine* 22, no. 2 (2015): 308-318.

Abulrob, Abedel-Nasser, Marc TE Suller, Mark Gumbleton, Claire Simons, and A. Denver Russell. "Identification and biological evaluation of grapefruit oil components as potential novel efflux pump modulators in methicillin-resistant *Staphylococcus aureus* bacterial strains." *Phytochemistry* 65, no. 22 (2004): 3021-3027.

Aires, A., V. R. Mota, M. J. Saavedra, E. A. S. Rosa, and R. N. Bennett. "The antimicrobial effects of glucosinolates and their respective enzymatic hydrolysis products on bacteria isolated from the human intestinal tract." *Journal of Applied Microbiology* 106, no. 6 (2009): 2086-2095.

Alekshun, Michael N., and Stuart B. Levy. "Molecular mechanisms of antibacterial multidrug resistance." *Cell* 128, no. 6 (2007): 1037-1050.

Alhanout, Kamel, Soazig Malesinki, Nicolas Vidal, Vincent Peyrot, Jean Marc Rolain, and Jean Michel Brunel. "New insights into the antibacterial mechanism of action of squalamine." *Journal of antimicrobial chemotherapy* 65, no. 8 (2010): 1688-1693.

Ali, Shaik Mahaboob, Aleem A. Khan, Irshad Ahmed, M. Musaddiq, Khaja S. Ahmed, H. Polasa, L. Venkateswar Rao, Chittoor M. Habibullah, Leonardo A. Sechi, and Niyaz Ahmed. "Antimicrobial activities of Eugenol and Cinnamaldehyde against the human gastric pathogen *Helicobacter pylori*." *Annals of clinical microbiology and antimicrobials* 4, no. 1 (2005): 20.

Al-Majedy, Yasameen K., Abdul Amir H. Kadhum, Ahmed A. Al-Amiery, and Abu Bakar Mohamad. "Coumarins: the antimicrobial agents." *Systematic Reviews in Pharmacy* 8, no. 1 (2017): 62.

Althunibat, Osama Y., Haitham Qaralleh, Sati Yassin Ahmed Al-Dalin, Muayad Abboud, Khaled Khleifat, Ibrahim S. Majali, H. K. Aldal'in, Walid A. Rayyan, and Ahmad Jaafraa. "Effect of thymol and carvacrol, the major components of *Thymus capitatus* on the growth of *Pseudomonas aeruginosa*." *J Pure Appl Microbiol* 10 (2016): 367-74.

Amaral, Viviane CS, Patrícia R. Santos, Alex Fiori da Silva, Adriele Rodrigues dos Santos, Miguel Machinski Jr, and Jane M. Graton Mikcha. "Effect of carvacrol and thymol on *Salmonella* spp. biofilms on polypropylene." *International journal of food science & technology* 50, no. 12 (2015): 2639-2643.

Anderle, Christine, Martin Stieger, Matthew Burrell, Stefan Reinelt, Anthony Maxwell, Malcolm Page, and Lutz Heide. "Biological activities of novel gyrase inhibitors of the aminocoumarin class." *Antimicrobial agents and chemotherapy* 52, no. 6 (2008): 1982-1990.

Asili, Javad, Amirhossein Sahebkar, Bibi Sedigheh Fazly Bazzaz, Sirus Sharifi, and Mehrdad Iranshahi. "Identification of essential oil components of Ferula badrakema fruits by GC-MS and 13C-NMR methods and evaluation of its antimicrobial activity." *Journal of Essential Oil Bearing Plants* 12, no. 1 (2009): 7-15.

Barbieri, Ramona, Erika Coppo, Anna Marchese, Maria Daglia, Eduardo Sobarzo-Sanchez, Seyed Fazel Nabavi, and Seyed Mohammad Nabavi. "Phytochemicals for human disease: An update on plant-derived compounds antibacterial activity." *Microbiological research* 196 (2017): 44-68.

Basile, Adriana, Sergio Sorbo, Vivienne Spadaro, Maurizio Bruno, Antonella Maggio, Nicoletta Faraone, and Sergio Rosselli. "Antimicrobial and antioxidant activities of coumarins from the roots of *Ferulago campestris* (Apiaceae)." *Molecules* 14, no. 3 (2009): 939-952.

Baym, Michael, Laura K. Stone, and Roy Kishony. "Multidrug evolutionary strategies to reverse antibiotic resistance." *Science* 351, no. 6268 (2016): aad3292.

Bazzaz, Bibi Sedigheh Fazly, Bahman Khameneh, Mohammad Reza Zahedian Ostad, and Hossein Hosseinzadeh. "*In vitro* evaluation of antibacterial activity of verbascoside, lemon verbena extract and caffeine in combination with gentamicin against drug-resistant Staphylococcus aureus and *Escherichia coli* clinical isolates." *Avicenna journal of phytomedicine* 8, no. 3 (2018): 246.

Bazzaz, Bibi Sedigheh Fazly, Mohamad Hassanzadeh Khayat, Seyed Ahmad Emami, Javad Asili, Amirhossein Sahebkar, and Elahe Javadi Neishabory. "Antioxidant and antimicrobial activity of methanol, dichloromethane, and ethyl acetate extracts of *Scutellaria litwinowii*." *Science Asia* 37 (2011): 327-334.

Bazzaz, Bibi Sedigheh Fazly, Sahar Sarabandi, Bahman Khameneh, and Hossein Hosseinzadeh. "Effect of Catechins, Green tea Extract and Methylxanthines in Combination with Gentamicin Against *Staphylococcus aureus* and *Pseudomonas aeruginosa*:-Combination therapy against resistant bacteria." *Journal of pharmacopuncture* 19, no. 4 (2016): 312.

Bazzaz, Bibi Sedigheh Fazly, Zahra Memariani, Zahra Khashiarmanesh, Mehrdad Iranshahi, and Mahbobeh Naderinasab. "Effect of galbanic acid, a sesquiterpene coumarin from *Ferula szowitsiana*, as an inhibitor of efflux mechanism in resistant clinical isolates of *Staphylococcus aureus*." *Brazilian Journal of Microbiology* 41, no. 3 (2010): 574-580.

Bazzaz, BS Fazly, M. Iranshahi, M. Naderinasab, S. Hajian, Z. Sabeti, and E. Masumi. "Evaluation of the effects of galbanic acid from *Ferula szowitsiana* and conferol from *F. badrakema*, as modulators of multi-drug resistance in clinical isolates of *Escherichia coli* and Staphylococcus aureus." *Research in pharmaceutical sciences* 5, no. 1 (2010): 21.

Belofsky, Gil, Diamond Percivill, Kim Lewis, George P. Tegos, and Julie Ekart. "Phenolic Metabolites of Dalea v ersicolor that Enhance Antibiotic Activity against Model Pathogenic Bacteria." *Journal of natural products* 67, no. 3 (2004): 481-484.

Benzekri, Roudaina, Lamjed Bouslama, Adele Papetti, Mejdi Snoussi, Imen Benslimene, Majdi Hamami, and Ferid Limam. "Isolation and identification of an antibacterial compound from *Diplotaxis harra* (Forssk.) Boiss." *Industrial Crops and Products* 80 (2016): 228-234.

Betts, Jonathan W., and David W. Wareham. "*In vitro* activity of curcumin in combination with epigallocatechin gallate (EGCG) versus multidrug-resistant *Acinetobacter baumannii*." *BMC microbiology* 14, no. 1 (2014): 172.

Boberek, Jaroslaw M., Jem Stach, and Liam Good. "Genetic evidence for inhibition of bacterial division protein FtsZ by berberine." *PloS one* 5, no. 10 (2010): e13745.

Boghrati, Zahra, and Mehrdad Iranshahi. "Ferula species: a rich source of antimicrobial compounds." *Journal of Herbal Medicine* 16 (2019): 100244.

Breier, Albert, and Attila Ziegelhoffer. "Lysine is the Lord," thought some scientists in regard to the group interacting with fluorescein isothiocyanate in ATP-binding sites of P-type ATPases. But, is it not cysteine? (Minireview)." *General physiology and biophysics* 19, no. 3 (2000): 253-264.

Broniatowski, Marcin, Patrycja Mastalerz, and Michał Flasiński. "Studies of the interactions of ursane-type bioactive terpenes with the model of *Escherichia coli* inner membrane—Langmuir monolayer approach." *Biochimica et Biophysica Acta (BBA)-Biomembranes* 1848, no. 2 (2015): 469-476.

Brown, David. "Antibiotic resistance breakers: can repurposed drugs fill the antibiotic discovery void?." *Nature reviews Drug discovery* 14, no. 12 (2015): 821.

Cai, Yun, Rui Wang, Mao-Mao An, Bei-Bei Liang, and Yi Fang. "In vitro bactericidal activity of allicin combined with cefoperazone, tobramycin and ciprofloxacin." *International journal of antimicrobial agents* 2, no. 31 (2008): 179-180.

Calmes, Benoit, Guillaume N'Guyen, Jérome Dumur, Carlos A. Brisach, Claire Campion, Béatrice Iacomi, Sandrine Pigné et al. "Glucosinolate-derived isothiocyanates impact mitochondrial function in fungal cells and elicit an oxidative stress response necessary for growth recovery." *Frontiers in plant science* 6 (2015): 414.

Campos, F. M., J. A. Couto, A. R. Figueiredo, I. V. Tóth, António OSS Rangel, and T. A. Hogg. "Cell membrane damage induced by phenolic acids on wine lactic acid bacteria." *International journal of food microbiology* 135, no. 2 (2009): 144-151.

Cannalire, Rolando, Diana Machado, Tommaso Felicetti, Sofia Santos Costa, Serena Massari, Giuseppe Manfroni, Maria Letizia Barreca et al. "Natural isoflavone biochanin A as a template for the design of new and potent 3-phenylquinolone efflux inhibitors against Mycobacterium avium." *European journal of medicinal chemistry* 140 (2017): 321-330.

Casu, Laura, Filippo Cottiglia, Marco Leonti, Alessandro De Logu, Emanuela Agus, Yuk-Ching Tse-Dinh, Valentina Lombardo, and Claudia Sissi. "Ungeremine effectively targets mammalian as well as bacterial type I and type II topoisomerases." *Bioorganic & medicinal chemistry letters* 21, no. 23 (2011): 7041-7044.

Catteau, Lucy, Li Zhu, Françoise Van Bambeke, and Joëlle Quetin-Leclercq. "Natural and hemi-synthetic pentacyclic triterpenes as antimicrobials and resistance modifying agents against *Staphylococcus aureus*: a review." *Phytochemistry reviews* 17, no. 5 (2018): 1129-1163.

Chan, Ben C. L., Margaret I. p., Clara B. S. Lau, S. L. Lui, Claude Jolivalt, Carine Ganem-Elbaz, Marc Litaudon et al. "Synergistic effects of baicalein with ciprofloxacin against NorA over-expressed methicillin-resistant *Staphylococcus aureus* (MRSA) and inhibition of MRSA pyruvate kinase." *Journal of ethnopharmacology* 137, no. 1 (2011): 767-773.

Chauhan, Anil Kumar, and Sun Chul Kang. "Thymol disrupts the membrane integrity of Salmonella ser. typhimurium *in vitro* and recovers infected macrophages from oxidative stress in an ex vivo model." *Research in microbiology* 165, no. 7 (2014): 559-565.

Čižman, Milan. "The use and resistance to antibiotics in the community." *International journal of antimicrobial agents* 21, no. 4 (2003): 297-307.

Conly, John M. "Antimicrobial resistance–Judicious use is the key." *Canadian Journal of Infectious Diseases and Medical Microbiology* 15, no. 5 (2004): 249-251.

Copp, Brent R. "Antimycobacterial natural products." *Natural product reports* 20, no. 6 (2003): 535-557.

Cowan, Marjorie Murphy. "Plant products as antimicrobial agents." *Clinical microbiology reviews* 12, no. 4 (1999): 564-582.

Cushnie, TP Tim, Benjamart Cushnie, and Andrew J. Lamb. "Alkaloids: an overview of their antibacterial, antibiotic-enhancing and antivirulence activities." *International Journal of Antimicrobial Agents* 44, no. 5 (2014): 377-386.

da S. M. Forezi, Luana, Thamires Q. Froes, Mariana F. C. Cardoso, Maciel de Oliveira, Carine Assuncao, Gianluca G. Nicastro, Regina L. Baldini et al. "Synthesis and Biological Evaluation of Coumarins Derivatives as Potential Inhibitors of the Production of *Pseudomonas aeruginosa* Virulence Factor Pyocyanin." *Current Topics in Medicinal Chemistry* 18, no. 2 (2018): 149-156.

D'Almeida, R. E., R. D. I. Molina, C. M. Viola, M. C. Luciardi, C. Nieto Peñalver, A. Bardon, and M. E. Arena. "Comparison of seven structurally related coumarins on the inhibition of quorum sensing of *Pseudomonas aeruginosa* and *Chromobacterium violaceum*." *Bioorganic chemistry* 73 (2017): 37-42.

Damier-Piolle, Laurence, Sophie Magnet, Sylvie Brémont, Thierry Lambert, and Patrice Courvalin. "AdeIJK, a resistance-nodulation-cell division pump effluxing multiple antibiotics in *Acinetobacter baumannii*." *Antimicrobial agents and chemotherapy* 52, no. 2 (2008): 557-562.

Davis, Stephen R. "An overview of the antifungal properties of allicin and its breakdown products–the possibility of a safe and effective antifungal prophylactic." *Mycoses* 48, no. 2 (2005): 95-100.

de Araújo, Rodrigo Santos Aquino, José Maria Barbosa-Filho, Marcus Tullius Scotti, Luciana Scotti, Ryldene Marques Duarte da Cruz, Vivyanne dos Santos Falcão-Silva, José Pinto de Siqueira-Júnior, and Francisco Jaime Bezerra Mendonça-Junior. "Modulation of drug resistance in staphylococcus aureus with coumarin derivatives." *Scientifica* 2016 (2016).

De Carvalho, Carla C. C. R., and M. Manuela R. Da Fonseca. "Carvone: Why and how should one bother to produce this terpene." *Food Chemistry* 95, no. 3 (2006): 413-422.

Delaquis, P. J., and G. Mazza. "Antimicrobial properties of isothiocyanates in food preservation." *Food Technol* 49, no. 11 (1995): 73-84.

Dias, Carla, Alfredo Aires, and Maria Saavedra. "Antimicrobial activity of isothiocyanates from cruciferous plants against methicillin-resistant *Staphylococcus aureus* (MRSA)." *International journal of molecular sciences* 15, no. 11 (2014): 19552-19561.

Domadia, Prerna N., Anirban Bhunia, J. Sivaraman, Sanjay Swarup, and Debjani Dasgupta. "Berberine targets assembly of *Escherichia coli* cell division protein FtsZ." *Biochemistry* 47, no. 10 (2008): 3225-3234.

Dötsch, Andreas, Tanja Becker, Claudia Pommerenke, Zofia Magnowska, Lothar Jänsch, and Susanne Häussler. "Genomewide identification of genetic determinants of antimicrobial drug resistance in *Pseudomonas aeruginosa*." *Antimicrobial agents and chemotherapy* 53, no. 6 (2009): 2522-2531.

Drobnica, Ľ., M. Zemanova, P. Nemec, K. Antoš, P. Kristian, A. Martvoň, and E. Zavodska. "Antifungal Activity of Isothiocyanates and Related Compounds: III. Derivatives of Diphenyl, Stilbene, Azobenzene, and Several Polycondensed Aromatic Hydrocarbons." *Appl. Environ. Microbiol.* 16, no. 4 (1968): 582-587.

Duan, Feixia, Xiaohong Li, Suping Cai, Guang Xin, Yanyan Wang, Dan Du, Shiliang He et al. "Haloemodin as novel antibacterial agent inhibiting DNA gyrase and bacterial topoisomerase I." *Journal of medicinal chemistry* 57, no. 9 (2014): 3707-3714.

Dufour, Virginie, Martin Stahl, and Christine Baysse. "The antibacterial properties of isothiocyanates." *Microbiology* 161, no. 2 (2015): 229-243.

Dwivedi, Gaurav Raj, Anupam Maurya, Dharmendra Kumar Yadav, Vigyasa Singh, Feroz Khan, Mahendra Kumar Gupta, Mastan Singh, Mahendra P. Darokar, and Santosh Kumar Srivastava. "Synergy of clavine alkaloid 'chanoclavine'with tetracycline against multi-drug-resistant E. coli." *Journal of Biomolecular Structure and Dynamics* 37, no. 5 (2019): 1307-1325.

El-Seedi, Hesham R. "Antimicrobial arylcoumarins from *Asphodelus microcarpus*." *Journal of natural products* 70, no. 1 (2007): 118-120.

Fahey, Jed W., Katherine K. Stephenson, Kristina L. Wade, and Paul Talalay. "Urease from *Helicobacter pylori* is inactivated by sulforaphane and other isothiocyanates." *Biochemical and biophysical research communications* 435, no. 1 (2013): 1-7.

Farhadi, Faegheh, Bahman Khameneh, Mehrdad Iranshahi, and Milad Iranshahy. "Antibacterial activity of flavonoids and their structure–activity relationship: An update review." *Phytotherapy research* 33, no. 1 (2019): 13-40.

Feio, S. Savluchinske, B. Gigante, J. Carlos Roseiro, and M. J. Marcelo-Curto. "Antimicrobial activity of diterpene resin acid derivatives." *Journal of Microbiological Methods* 35, no. 3 (1999): 201-206.

Feldberg, Ross S., Stephen C. Chang, Alan N. Kotik, Marcus Nadler, Zeev Neuwirth, David C. Sundstrom, and Nathan H. Thompson. "*In vitro* mechanism of inhibition of bacterial cell growth by allicin." *Antimicrobial Agents and Chemotherapy* 32, no. 12 (1988): 1763-1768.

Feng, Ruizhang, Jing Qu, Wanhai Zhou, Qin Wei, Zhongqiong Yin, Yonghua Du, Kuan Yan et al. "Antibacterial activity and mechanism of berberine on avian Pasteurella multocida." *Int J Clin Exp Med* 9, no. 11 (2016): 22886-22892.

Ferreira, Susana, Filomena Silva, João A. Queiroz, Mónica Oleastro, and Fernanda C. Domingues. "Resveratrol against *Arcobacter butzleri* and *Arcobacter cryaerophilus*: activity and effect on cellular functions." *International journal of food microbiology* 180 (2014): 62-68.

Fujita, Mai, Sumiko Shiota, Teruo Kuroda, Tsutomu Hatano, Takashi Yoshida, Tohru Mizushima, and Tomofusa Tsuchiya. "Remarkable Synergies between Baicalein and Tetracycline, and Baicalein and β-Lactams against Methicillin-Resistant *Staphylococcus aureus*." *Microbiology and immunology* 49, no. 4 (2005): 391-396.

Gibbons, Simon, Elisabeth Moser, and Glenn W. Kaatz. "Catechin gallates inhibit multidrug resistance (MDR) in *Staphylococcus aureus*." *Planta medica* 70, no. 12 (2004): 1240-1242.

Gibbons, Simon. "Phytochemicals for bacterial resistance-strengths, weaknesses and opportunities." *Planta medica* 74, no. 06 (2008): 594-602.

Górniak, Ireneusz, Rafał Bartoszewski, and Jarosław Króliczewski. "Comprehensive review of antimicrobial activities of plant flavonoids." *Phytochemistry Reviews* 18, no. 1 (2019): 241-272.

Gradišar, Helena, Primož Pristovšek, Andreja Plaper, and Roman Jerala. "Green tea catechins inhibit bacterial DNA gyrase by interaction with its ATP binding site." *Journal of medicinal chemistry* 50, no. 2 (2007): 264-271.

Gutiérrez-Barranquero, José A., F. Jerry Reen, Ronan R. McCarthy, and Fergal O'Gara. "Deciphering the role of coumarin as a novel quorum sensing inhibitor suppressing virulence phenotypes in bacterial pathogens." *Applied microbiology and biotechnology* 99, no. 7 (2015): 3303-3316.

Gyawali, Rabin, and Salam A. Ibrahim. "Natural products as antimicrobial agents." *Food control* 46 (2014): 412-429.

Hanski, Leena, Natalja Genina, Hanna Uvell, Kristina Malinovskaja, Åsa Gylfe, Timo Laaksonen, Ruzica Kolakovic et al. "Inhibitory activity of the isoflavone biochanin A on intracellular bacteria of genus Chlamydia and initial development of a buccal formulation." *PLoS One* 9, no. 12 (2014): e115115.

Haristoy, Xavier, Jed W. Fahey, Isabelle Scholtus, and Alain Lozniewski. "Evaluation of the antimicrobial effects of several isothiocyanates on Helicobacter pylori." *Planta medica* 71, no. 04 (2005): 326-330.

Heeb, Stephan, Matthew P. Fletcher, Siri Ram Chhabra, Stephen P. Diggle, Paul Williams, and Miguel Cámara. "Quinolones: from antibiotics to autoinducers." *FEMS microbiology reviews* 35, no. 2 (2011): 247-274.

Holler, Jes Gitz, Hans-Christian Slotved, Per Mølgaard, Carl Erik Olsen, and Søren Brøgger Christensen. "Chalcone inhibitors of the NorA efflux pump in Staphylococcus aureus whole cells and enriched everted membrane vesicles." *Bioorganic & medicinal chemistry* 20, no. 14 (2012): 4514-4521.

Holler, Jes Gitz, S. Brøgger Christensen, Hans-Christian Slotved, Hasse B. Rasmussen, Alfonso Gúzman, Carl-Erik Olsen, Bent Petersen, and Per Mølgaard. "Novel inhibitory activity of the Staphylococcus aureus NorA efflux pump by a kaempferol rhamnoside isolated from Persea lingue Nees." *Journal of Antimicrobial Chemotherapy* 67, no. 5 (2012): 1138-1144.

Iranshahi, Mehrdad, Mohammad Hassanzadeh-Khayat, Bibi Sedigheh Fazly Bazzaz, Zahra Sabeti, and Fatemeh Enayati. "High content of polysulphides in the volatile oil of Ferula latisecta Rech. F. et Aell. fruits and antimicrobial activity of the oil." *Journal of Essential Oil Research* 20, no. 2 (2008): 183-185.

Issam, AL-Ani, Stefan Zimmermann, Jürgen Reichling, and Michael Wink. "Pharmacological synergism of bee venom and melittin with antibiotics and plant secondary metabolites against multi-drug resistant microbial pathogens." *Phytomedicine* 22, no. 2 (2015): 245-255.

Iwasa, Kinuko, Masataka Moriyasu, Takao Yamori, Takashi Turuo, Dong-Ung Lee, and Wolfgang Wiegrebe. "In vitro cytotoxicity of the protoberberine-type alkaloids." *Journal of natural products* 64, no. 7 (2001): 896-898.

Jang, Miran, Eunyoung Hong, and Gun-Hee Kim. "Evaluation of antibacterial activity of 3-butenyl, 4-pentenyl, 2-phenylethyl, and benzyl isothiocyanate in Brassica vegetables." *Journal of food science* 75, no. 7 (2010): M412-M416.

Jeong, Ki-Woong, Jee-Young Lee, Dong-Il Kang, Ju-Un Lee, Song Yub Shin, and Yangmee Kim. "Screening of flavonoids as candidate antibiotics against Enterococcus faecalis." *Journal of natural products* 72, no. 4 (2009): 719-724.

Jia, Wei, Caiyun Li, Haiyun Zhang, Gang Li, Xiaoming Liu, and Jun Wei. "Prevalence of genes of OXA-23 carbapenemase and AdeABC efflux pump associated with multidrug resistance of Acinetobacter baumannii isolates in the ICU of a comprehensive hospital of Northwestern China." *International journal of environmental research and public health* 12, no. 8 (2015): 10079-10092.

Jia, Wei, Jiayuan Wang, Haotong Xu, and Gang Li. "Resistance of Stenotrophomonas maltophilia to fluoroquinolones: prevalence in a university hospital and possible mechanisms." *International journal of environmental research and public health* 12, no. 5 (2015): 5177-5195.

Jiang, Qi-Wei, Mei-Wan Chen, Ke-Jun Cheng, Pei-Zhong Yu, Xing Wei, and Zhi Shi. "Therapeutic potential of steroidal alkaloids in cancer and other diseases." *Medicinal research reviews* 36, no. 1 (2016): 119-143.

Kang, Sam Sik, Jae-Gyu Kim, Tae-Hoon Lee, and Ki-Bong Oh. "Flavonols inhibit sortases and sortase-mediated Staphylococcus aureus clumping to fibrinogen." *Biological and Pharmaceutical Bulletin* 29, no. 8 (2006): 1751-1755.

Kayser, Oliver, and Herbert Kolodziej. "Antibacterial activity of simple coumarins: structural requirements for biological activity." *Zeitschrift für Naturforschung C* 54, no. 3-4 (1999): 169-174.

Khameneh, Bahman, Milad Iranshahy, Morteza Ghandadi, Davod Ghoochi Atashbeyk, Bibi Sedigheh Fazly Bazzaz, and Mehrdad Iranshahi. "Investigation of the antibacterial activity and efflux pump inhibitory effect of co-loaded piperine and gentamicin nanoliposomes in methicillin-resistant *Staphylococcus aureus*." *Drug development and industrial pharmacy* 41, no. 6 (2015): 989-994.

Khameneh, Bahman, Milad Iranshahy, Vahid Soheili, and Bibi Sedigheh Fazly Bazzaz. "Review on plant antimicrobials: a mechanistic viewpoint." *Antimicrobial Resistance & Infection Control* 8, no. 1 (2019): 118.

Khameneh, Bahman, Roudayna Diab, Kiarash Ghazvini, and Bibi Sedigheh Fazly Bazzaz. "Breakthroughs in bacterial resistance mechanisms and the potential ways to combat them." *Microbial Pathogenesis* 95 (2016): 32-42.

Khan, Inshad Ali, Zahid Mehmood Mirza, Ashwani Kumar, Vijeshwar Verma, and Ghulam Nabi Qazi. "Piperine, a phytochemical potentiator of ciprofloxacin against Staphylococcus aureus." *Antimicrobial agents and chemotherapy* 50, no. 2 (2006): 810-812.

Kim, Seokwon, Roman Kubec, and Rabi A. Musah. "Antibacterial and antifungal activity of sulfur-containing compounds from *Petiveria alliacea* L." *Journal of ethnopharmacology* 104, no. 1-2 (2006): 188-192.

Klančnik, A., M. Šikić Pogačar, K. Trošt, M. Tušek Žnidarič, B. Mozetič Vodopivec, and S. Smole Možina. "Anti-Campylobacter activity of resveratrol and an extract from waste Pinot noir grape skins and seeds, and resistance of Camp. jejuni planktonic and biofilm cells, mediated via the Cme ABC efflux pump." *Journal of applied microbiology* 122, no. 1 (2017): 65-77.

Kostova, Irena. "Synthetic and natural coumarins as cytotoxic agents." *Current Medicinal Chemistry-Anti-Cancer Agents* 5, no. 1 (2005): 29-46.

Kuete, V., J. D. Wansi, A. T. Mbaveng, M. M. Kana Sop, A. Tcho Tadjong, V. Penlap Beng, F. X. Etoa, J. Wandji, J. J. Marion Meyer, and N. Lall. "Antimicrobial activity of the methanolic extract and compounds from *Teclea afzelii* (Rutaceae)." *South African Journal of Botany* 74, no. 4 (2008): 572-576.

Kumar, Ashwani, Inshad Ali Khan, Surrinder Koul, Jawahir Lal Koul, Subhash Chandra Taneja, Intzar Ali, Furqan Ali et al. "Novel structural analogues of piperine as inhibitors of the NorA efflux pump of *Staphylococcus aureus*." *Journal of Antimicrobial Chemotherapy* 61, no. 6 (2008): 1270-1276.

Kumar, Neeraj, Bikram Singh, Pamita Bhandari, Ajai Prakash Gupta, and Vijay Kumar Kaul. "Steroidal Alkaloids from Holarrhena antidysenterica (L.) W ALL." *Chemical and pharmaceutical bulletin* 55, no. 6 (2007): 912-914.

Lanzotti, Virginia, Felice Scala, and Giuliano Bonanomi. "Compounds from Allium species with cytotoxic and antimicrobial activity." *Phytochemistry reviews* 13, no. 4 (2014): 769-791.

Lechner, Doris, Simon Gibbons, and Franz Bucar. "Plant phenolic compounds as ethidium bromide efflux inhibitors in Mycobacterium smegmatis." *Journal of Antimicrobial chemotherapy* 62, no. 2 (2008): 345-348.

Lee, Ngan H., and John W. Ho. "Celastrol and terpenes as anti-infective agents." *Anti-Infective Agents in Medicinal Chemistry (Formerly Current Medicinal Chemistry-Anti-Infective Agents)* 7, no. 2 (2008): 97-100.

Li, Bing-Hui, Rui Zhang, Ya-Tao Du, Ying-Hui Sun, and Wei-Xi Tian. "Inactivation mechanism of the β-ketoacyl-[acyl carrier protein] reductase of bacterial type-II fatty acid synthase by epigallocatechin gallate." *Biochemistry and cell biology* 84, no. 5 (2006): 755-762.

Lin, Chia-Min, James F. Preston III, and Cheng-I. Wei. "Antibacterial mechanism of allyl isothiocyanate." *Journal of food protection* 63, no. 6 (2000): 727-734.

Li-Na, Zhou, Ge Xiao-Lei, D. O. N. G. Ting-Ting, G. A. O. Hui-Yuan, and Sun Bo-Hang. "Antibacterial steroidal alkaloids from *Holarrhena antidysenteriaca*." *Chinese journal of natural medicines* 15, no. 7 (2017): 540-545.

Lu, Yingjian, Rolf Joerger, and Changqing Wu. "Study of the chemical composition and antimicrobial activities of ethanolic extracts from roots of *Scutellaria baicalensis* Georgi." *Journal of agricultural and food chemistry* 59, no. 20 (2011): 10934-10942.

Lu, Zhongjing, Christopher R. Dockery, Michael Crosby, Katherine Chavarria, Brett Patterson, and Matthew Giedd. "Antibacterial Activities of Wasabi against *Escherichia coli* O157: H7 and *Staphylococcus aureus*." *Frontiers in microbiology* 7 (2016): 1403.

Luciano, Fernando B., and Richard A. Holley. "Enzymatic inhibition by allyl isothiocyanate and factors affecting its antimicrobial action against *Escherichia coli* O157: H7." *International journal of food microbiology* 131, no. 2-3 (2009): 240-245.

Malléa, Monique, Abdallah Mahamoud, Jacqueline Chevalier, Sandrine Alibert-Franco, Pierre Brouant, Jacques Barbe, and Jean-Marie Pagès. "Alkylaminoquinolines inhibit the bacterial antibiotic efflux pump in multidrug-resistant clinical isolates." *Biochemical Journal* 376, no. 3 (2003): 801-805.

Maresso, Anthony W., and Olaf Schneewind. "Sortase as a target of anti-infective therapy." *Pharmacological reviews* 60, no. 1 (2008): 128-141.

Marinelli, Lisa, Antonio Di Stefano, and Ivana Cacciatore. "Carvacrol and its derivatives as antibacterial agents." *Phytochemistry reviews* 17, no. 4 (2018): 903-921.

Maxwell, Anthony. "The interaction between coumarin drugs and DNA gyrase." *Molecular microbiology* 9, no. 4 (1993): 681-686.

Melliou, Eleni, Prokopios Magiatis, Sofia Mitaku, Alexios-Leandros Skaltsounis, Efrosini Chinou, and Ioanna Chinou. "Natural and synthetic 2, 2-dimethylpyranocoumarins with antibacterial activity." *Journal of natural products* 68, no. 1 (2005): 78-82.

Miladi, Hanene, Tarek Zmantar, Yassine Chaabouni, Kais Fedhila, Amina Bakhrouf, Kacem Mahdouani, and Kamel Chaieb. "Antibacterial and efflux pump inhibitors of thymol and carvacrol against food-borne pathogens." *Microbial pathogenesis* 99 (2016): 95-100.

Mitchell, Gabriel, Myriame Lafrance, Simon Boulanger, David Lalonde Séguin, Isabelle Guay, Mariza Gattuso, Éric Marsault, Kamal Bouarab, and François Malouin. "Tomatidine acts in synergy with aminoglycoside antibiotics against multiresistant *Staphylococcus aureus* and prevents virulence gene expression." *Journal of antimicrobial chemotherapy* 67, no. 3 (2011): 559-568.

Moloney, Mark G. "Natural products as a source for novel antibiotics." *Trends in pharmacological sciences* 37, no. 8 (2016): 689-701.

Moon, Joon-Kwan, Jun-Ran Kim, Young-Joon Ahn, and Takayuki Shibamoto. "Analysis and anti-*Helicobacter* activity of sulforaphane and related compounds present in broccoli (*Brassica oleracea* L.) sprouts." *Journal of agricultural and food chemistry* 58, no. 11 (2010): 6672-6677.

Morel, Cécile, Frank R. Stermitz, George Tegos, and Kim Lewis. "Isoflavones as potentiators of antibacterial activity." *Journal of agricultural and food chemistry* 51, no. 19 (2003): 5677-5679.

Mun, Su-Hyun, Dae-Ki Joung, Sung-Bae Kim, Sung-Joo Park, Yun-Soo Seo, Ryong Gong, Jang-Gi Choi et al. "The mechanism of antimicrobial activity of sophoraflavanone B against methicillin-resistant *Staphylococcus aureus*." *Foodborne pathogens and disease* 11, no. 3 (2014): 234-239.

Nascimento, Gislene G. F., Juliana Locatelli, Paulo C. Freitas, and Giuliana L. Silva. "Antibacterial activity of plant extracts and phytochemicals on antibiotic-resistant bacteria." *Brazilian journal of microbiology* 31, no. 4 (2000): 247-256.

Navarro-Martínez, María Dolores, Enma Navarro-Perán, Juan Cabezas-Herrera, Joaquín Ruiz-Gómez, Francisco García-Cánovas, and José Neptuno Rodríguez-López. "Antifolate activity of epigallocatechin gallate against *Stenotrophomonas maltophilia*." *Antimicrobial agents and chemotherapy* 49, no. 7 (2005): 2914-2920.

Nedorostova, Lenka, Pavel Kloucek, Ladislav Kokoska, Miluse Stolcova, and Josef Pulkrabek. "Antimicrobial properties of selected essential oils in vapour phase against foodborne bacteria." *Food control* 20, no. 2 (2009): 157-160.

Newton, Sandra M., Clara Lau, Sudagar S. Gurcha, Gurdyal S. Besra, and Colin W. Wright. "The evaluation of forty-three plant species for *in vitro* antimycobacterial activities; isolation of active constituents from Psoralea corylifolia and Sanguinaria canadensis." *Journal of ethnopharmacology* 79, no. 1 (2002): 57-67.

Obiang-Obounou, Brice W., Ok-Hwa Kang, Jang-Gi Choi, Joon-Ho Keum, Sung-Bae Kim, Su-Hyun Mun, Dong-Won Shin et al. "The mechanism of action of sanguinarine against methicillin-resistant *Staphylococcus aureus*." *The Journal of toxicological sciences* 36, no. 3 (2011): 277-283.

Oldfield, Eric, and Fu-Yang Lin. "Terpene biosynthesis: modularity rules." *Angewandte Chemie International Edition* 51, no. 5 (2012): 1124-1137.

Oluwatuyi, Moyosoluwa, Glenn W. Kaatz, and Simon Gibbons. "Antibacterial and resistance modifying activity of Rosmarinus officinalis." *Phytochemistry* 65, no. 24 (2004): 3249-3254.

Paduch, Roman, Martyna Kandefer-Szerszeń, Mariusz Trytek, and Jan Fiedurek. "Terpenes: substances useful in human healthcare." *Archivum immunologiae et therapiae experimentalis* 55, no. 5 (2007): 315.

Palaniappan, Kavitha, and Richard A. Holley. "Use of natural antimicrobials to increase antibiotic susceptibility of drug resistant bacteria." *International journal of food microbiology* 140, no. 2-3 (2010): 164-168.

Pandra, Sonia Bindu, David Banji, Otilia J. F. Banji, Ranjith Kumar Aavula, and Swetha Merugu. "Antibiotic resistance: Overview and mechanisms." *Inter J Pharm Sci Res* 1, no. 12 (2010): 17-27.

Park, Byeoung-Soo, Jae-Gyu Kim, Mi-Ran Kim, Sung-Eun Lee, Gary R. Takeoka, Ki-Bong Oh, and Jeong-Han Kim. "*Curcuma longa* L. constituents inhibit sortase A and Staphylococcus aureus cell adhesion to fibronectin." *Journal of agricultural and food chemistry* 53, no. 23 (2005): 9005-9009.

Park, Ho-Won, Kyu-Duck Choi, and Il-Shik Shin. "Antimicrobial activity of isothiocyanates (ITCs) extracted from horseradish (*Armoracia rusticana*) root against oral microorganisms." *Biocontrol science* 18, no. 3 (2013): 163-168.

Patel, Kunal, Chetna Tyagi, Sukriti Goyal, Salma Jamal, Divya Wahi, Ritu Jain, Navneeta Bharadvaja, and Abhinav Grover. "Identification of chebulinic acid as potent natural inhibitor of M. tuberculosis DNA gyrase and molecular insights into its binding mode of action." *Computational biology and chemistry* 59 (2015): 37-47.

Prati, Daniel, and Oliver Bossdorf. "Allelopathic inhibition of germination by Alliaria petiolata (Brassicaceae)." *American Journal of Botany* 91, no. 2 (2004): 285-288.

Ramezani, M. O. H. A. M. M. A. D., B. S. Fazli-Bazzaz, F. Saghafi-Khadem, and A. Dabaghian. "Antimicrobial activity of four Artemisia species of Iran." *Fitoterapia* 75, no. 2 (2004): 201-203.

Rana, Ritika, Ruchika Sharma, and Anoop Kumar. "Repurposing of Existing Statin Drugs for Treatment of Microbial Infections: How Much Promising?." *Infectious Disorders-Drug Targets (Formerly Current Drug Targets-Infectious Disorders)* 19, no. 3 (2019): 224-237.

Randhawa, Harmandeep K., Kanwarpreet K. Hundal, Pallavi N. Ahirrao, Sanjay M. Jachak, and Hemraj S. Nandanwar. "Efflux pump inhibitory activity of flavonoids isolated from Alpinia calcarata against methicillin-resistant *Staphylococcus aureus*." *Biologia* 71, no. 5 (2016): 484-493.

Rathinam, Prasanth, H. S. Vijay Kumar, and Pragasam Viswanathan. "Eugenol exhibits anti-virulence properties by competitively binding to quorum sensing receptors." *Biofouling* 33, no. 8 (2017): 624-639.

Reen, F. Jerry, José A. Gutiérrez-Barranquero, and María L. Parages. "Coumarin: a novel player in microbial quorum sensing and biofilm formation inhibition." *Applied microbiology and biotechnology* 102, no. 5 (2018): 2063-2073.

Rehman, F., and S. A. M. Y. A. Mairaj. "Antimicrobial studies of allicin and ajoene." *Int J Pharm Bio Sci* 4, no. 2 (2013): 1095-1105.

Reiter, Jana, Natalja Levina, Mark van der Linden, Martin Gruhlke, Christian Martin, and Alan Slusarenko. "Diallylthiosulfinate (allicin), a volatile antimicrobial from garlic (*Allium sativum*), kills human lung pathogenic bacteria, including MDR strains, as a vapor." *Molecules* 22, no. 10 (2017): 1711.

Rossiter, Sean E., Madison H. Fletcher, and William M. Wuest. "Natural products as platforms to overcome antibiotic resistance." *Chemical reviews* 117, no. 19 (2017): 12415-12474.

Roy, Somendu K., Neela Kumari, Sonika Pahwa, Udai C. Agrahari, Kamlesh K. Bhutani, Sanjay M. Jachak, and Hemraj Nandanwar. "NorA efflux pump inhibitory activity of coumarins from Mesua ferrea." *Fitoterapia* 90 (2013): 140-150.

Russell, A. D. "Biocide use and antibiotic resistance: the relevance of laboratory findings to clinical and environmental situations." *The Lancet infectious diseases* 3, no. 12 (2003): 794-803.

Saavedra, Maria J., Anabela Borges, Carla Dias, Alfredo Aires, Richard N. Bennett, Eduardo S. Rosa, and Manuel Simões. "Antimicrobial activity of phenolics and glucosinolate hydrolysis products and their synergy with streptomycin against pathogenic bacteria." *Medicinal Chemistry* 6, no. 3 (2010): 174-183.

Sardari, Soroush, Yoki Mori, Kiyoshi Horita, Ronald G. Micetich, Sansei Nishibe, and Mohsen Daneshtalab. "Synthesis and antifungal activity of coumarins and angular furanocoumarins." *Bioorganic & medicinal chemistry* 7, no. 9 (1999): 1933-1940.

Savluchinske-Feio, Sonia, Maria João Marcelo Curto, Bárbara Gigante, and J. Carlos Roseiro. "Antimicrobial activity of resin acid derivatives." *Applied microbiology and Biotechnology* 72, no. 3 (2006): 430-436.

Schrader, Kevin K., Fabiana Avolio, Anna Andolfi, Alessio Cimmino, and Antonio Evidente. "Ungeremine and its hemisynthesized analogues as bactericides against *Flavobacterium columnare*." *Journal of agricultural and food chemistry* 61, no. 6 (2013): 1179-1183.

Shakeri, Abolfazl, Javad Akhtari, Vahid Soheili, Seyedeh Faezeh Taghizadeh, Amirhossein Sahebkar, Rezvan Shaddel, and Javad Asili. "Identification and biological activity of the volatile compounds of *Glycyrrhiza triphylla* Fisch. & CA Mey." *Microbial pathogenesis* 109 (2017): 39-44.

Shakeri, Abolfazl, Mohammad J. Sharifi, Bibi S. Fazly Bazzaz, Ahmad Emami, Vahid Soheili, Amirhossein Sahebkar, and Javad Asili. "Bioautography detection of antimicrobial compounds from the essential oil of *Salvia pachystachys*." *Current Bioactive Compounds* 14, no. 1 (2018): 80-85.

Shao, Jing, MengXiang Zhang, TianMing Wang, Yue Li, and ChangZhong Wang. "The roles of CDR1, CDR2, and MDR1 in kaempferol-induced suppression with fluconazole-resistant *Candida albicans*." *Pharmaceutical biology* 54, no. 6 (2016): 984-992.

Sharifzadeh, A., A. R. Khosravi, H. Shokri, and H. Shirzadi. "Potential effect of 2-isopropyl-5-methylphenol (thymol) alone and in combination with fluconazole against clinical isolates of *Candida albicans*, *C. glabrata* and *C. krusei*." *Journal de mycologie medicale* 28, no. 2 (2018): 294-299.

Sibanda, Timothy, Leonard V. Mabinya, Ntsikelelo Mazomba, David A. Akinpelu, Kim Bernard, Ademola O. Olaniran, and Anthony I. Okoh. "Antibiotic producing potentials of three freshwater actinomycetes isolated from the Eastern Cape Province of South Africa." *International journal of molecular sciences* 11, no. 7 (2010): 2612-2623.

Siddiqui, Bina S., Syed Tahir Ali, Ghazala H. Rizwani, Sabira Begum, Saima Tauseef, and Aqeel Ahmad. "Antimicrobial activity of the methanolic bark extract of *Holarrhena pubescens* (Buch. Ham), its fractions and the pure compound conessine." *Natural product research* 26, no. 11 (2012): 987-992.

Siriyong, Thanyaluck, Potjanee Srimanote, Sasitorn Chusri, Boon-ek Yingyongnarongkul, Channarong Suaisom, Varomyalin Tipmanee, and Supayang Piyawan Voravuthikunchai. "Conessine as a novel inhibitor of multidrug efflux pump systems in *Pseudomonas aeruginosa*." *BMC complementary and alternative medicine* 17, no. 1 (2017): 405.

Siriyong, Thanyaluck, Sasitorn Chusri, Potjanee Srimanote, Varomyalin Tipmanee, and Supayang Piyawan Voravuthikunchai. "*Holarrhena antidysenterica* extract and its steroidal alkaloid, conessine, as resistance-modifying agents against extensively drug-resistant *Acinetobacter baumannii*." *Microbial Drug Resistance* 22, no. 4 (2016): 273-282.

Smyth, T., V. N. Ramachandran, and W. F. Smyth. "A study of the antimicrobial activity of selected naturally occurring and synthetic coumarins." *International journal of antimicrobial agents* 33, no. 5 (2009): 421-426.

Sobolewska, Danuta, Irma Podolak, and Justyna Makowska-Wąs. "*Allium ursinum*: botanical, phytochemical and pharmacological overview." *Phytochemistry reviews* 14, no. 1 (2015): 81-97.

Söderberg, Thor A., Anders Johansson, and Rolf Gref. "Toxic effects of some conifer resin acids and tea tree oil on human epithelial and fibroblast cells." *Toxicology* 107, no. 2 (1996): 99-109.

Sofrata, Abier, Ellen M. Santangelo, Muhammad Azeem, Anna-Karin Borg-Karlson, Anders Gustafsson, and Katrin Pütsep. "Benzyl isothiocyanate, a major component from the roots of Salvadora persica is highly active against Gram-negative bacteria." *PLoS One* 6, no. 8 (2011): e23045.

Soltani, Rasool, Hossein Fazeli, Rahim Bahri Najafi, and Alireza Jelokhanian. "Evaluation of the synergistic effect of tomatidine with several antibiotics against standard and clinical isolates of *Staphylococcus aureus*, *Enterococcus faecalis*, *Pseudomonas aeruginosa* and *Escherichia coli*." *Iranian journal of pharmaceutical research: IJPR* 16, no. 1 (2017): 290.

Sridevi, D., C. Shankar, P. Prakash, Joon Hong Park, and K. Thamaraiselvi. "Inhibitory effects of reserpine against efflux pump activity of antibiotic resistance bacteria." *Chemical Biology Letters* 4, no. 2 (2017): 69-72.

Stermitz, Frank R., Kevin K. Cashman, Kathleen M. Halligan, Cecile Morel, George P. Tegos, and Kim Lewis. "Polyacylated neohesperidosides from *Geranium caespitosum*: bacterial multidrug resistance pump inhibitors." *Bioorganic & medicinal chemistry letters* 13, no. 11 (2003): 1915-1918.

Stermitz, Frank R., Teresa D. Beeson, Paul J. Mueller, Jen-Fang Hsiang, and Kim Lewis. "Staphylococcus aureus MDR efflux pump inhibitors from a Berberis and a *Mahonia* (sensu strictu) species." *Biochemical systematics and ecology* 29, no. 8 (2001): 793-798.

Stewart, Philip S. "Mechanisms of antibiotic resistance in bacterial biofilms." *International Journal of Medical Microbiology* 292, no. 2 (2002): 107-113.

Sun, Ning, Ruo-Lan Du, Yuan-Yuan Zheng, Bao-Hua Huang, Qi Guo, Rui-Fang Zhang, Kwok-Yin Wong, and Yu-Jing Lu. "Antibacterial activity of N-methylbenzofuro [3, 2-b] quinoline and N-methylbenzoindolo [3, 2-b]-quinoline derivatives and study of their mode of action." *European journal of medicinal chemistry* 135 (2017): 1-11.

Tan, Nur, Merve Bilgin, Emir Tan, and Mahmut Miski. "Antibacterial activities of pyrenylated coumarins from the roots of Prangos hulusii." *Molecules* 22, no. 7 (2017): 1098.

Tapia, Aníbal A., Martha D. Vallejo, Silvia C. Gouiric, Gabriela E. Feresin, Pedro C. Rossomando, and Daniel A. Bustos. "Hydroxylation of dehydroabietic acid by Fusarium species." *Phytochemistry* 46, no. 1 (1997): 131-133.

Tenover, Fred C. "Mechanisms of antimicrobial resistance in bacteria." *The American journal of medicine* 119, no. 6 (2006): S3-S10.

Togashi, Naoko, Hajime Hamashima, Akiko Shiraishi, Yoshihiro Inoue, and Akihito Takano. "Antibacterial activities against Staphylococcus aureus of terpene alcohols with aliphatic carbon chains." *Journal of Essential Oil Research* 22, no. 3 (2010): 263-269.

Tominaga, Kazunari, Kazuhide Higuchi, Norio Hamasaki, Masaki Hamaguchi, Takashi Takashima, Tetsuya Tanigawa, Toshio Watanabe et al. "In vivo action of novel alkyl methyl quinolone alkaloids against Helicobacter pylori." *Journal of Antimicrobial Chemotherapy* 50, no. 4 (2002): 547-552.

Tyagi, Poonam, Madhuri Singh, Himani Kumari, Anita Kumari, and Kasturi Mukhopadhyay. "Bactericidal activity of curcumin I is associated with damaging of bacterial membrane." *PloS one* 10, no. 3 (2015): e0121313.

Vandevelde, Nathalie M., Paul M. Tulkens, and Francoise Van Bambeke. "Modulating antibiotic activity towards respiratory bacterial pathogens by co-medications: a multi-target approach." *Drug discovery today* 21, no. 7 (2016): 1114-1129.

Walker, Elaine S., and Foster Levy. "Genetic trends in a population evolving antibiotic resistance." *Evolution* 55, no. 6 (2001): 1110-1122.

Walsh, Christopher. "Molecular mechanisms that confer antibacterial drug resistance." *Nature* 406, no. 6797 (2000): 775.

WHO Regional Committee for Europe, *European strategic action plan on antibiotic resistance*, Copenhagen, 2011.

WHO. *Antimicrobial resistance: Global report on surveillance*, vol. 2014. Geneva: WHO; 2014.

Wright, Gerard D. "Bacterial resistance to antibiotics: enzymatic degradation and modification." *Advanced drug delivery reviews* 57, no. 10 (2005): 1451-1470.

Wu, Dalei, Yunhua Kong, Cong Han, Jing Chen, Lihong Hu, Hualiang Jiang, and Xu Shen. "D-Alanine: D-alanine ligase as a new target for the flavonoids quercetin and apigenin." *International journal of antimicrobial agents* 32, no. 5 (2008): 421-426.

Wu, L., X. Wang, W. Xu, F. Farzaneh, and R. Xu. "The structure and pharmacological functions of coumarins and their derivatives." *Current medicinal chemistry* 16, no. 32 (2009): 4236-4260.

Wu, Ting, Mengying He, Xixi Zang, Ying Zhou, Tianfu Qiu, Siyi Pan, and Xiaoyun Xu. "A structure–activity relationship study of flavonoids as inhibitors of E. coli by membrane interaction effect." *Biochimica et Biophysica Acta (BBA)-Biomembranes* 1828, no. 11 (2013): 2751-2756.

Wu, Yanping, Jinrong Bai, Kai Zhong, Yina Huang, Huayi Qi, Yan Jiang, and Hong Gao. "Antibacterial activity and membrane-disruptive mechanism of 3-p-trans-coumaroyl-2-hydroxyquinic acid, a novel phenolic compound from pine needles of Cedrus deodara, against *Staphylococcus aureus*." *Molecules* 21, no. 8 (2016): 1084.

Xiao, Zhu-Ping, Xu-Dong Wang, Peng-Fei Wang, Yin Zhou, Jing-Wen Zhang, Lei Zhang, Jiao Zhou et al. "Design, synthesis, and evaluation of novel fluoroquinolone–flavonoid hybrids as potent antibiotics against drug-resistant microorganisms." *European journal of medicinal chemistry* 80 (2014): 92-100.

Xiao, Zhu-Ping, Zhi-Yun Peng, Jing-Jun Dong, Juan He, Hui Ouyang, Yu-Ting Feng, Chun-Lei Lu et al. "Synthesis, structure–activity relationship analysis and kinetics study of reductive derivatives of flavonoids as *Helicobacter pylori* urease inhibitors." *European journal of medicinal chemistry* 63 (2013): 685-695.

Yadav, Mukesh Kumar, Sung-Won Chae, Gi Jung Im, Jae-Woo Chung, and Jae-Jun Song. "Eugenol: a phyto-compound effective against methicillin-resistant and methicillin-sensitive Staphylococcus aureus clinical strain biofilms." *PLoS One* 10, no. 3 (2015): e0119564.

Yi, Zhi-Biao, Yan Yu, Yi-Zeng Liang, and Bao Zeng. "Evaluation of the antimicrobial mode of berberine by LC/ESI-MS combined with principal component analysis." *Journal of pharmaceutical and biomedical analysis* 44, no. 1 (2007): 301-304.

Zhang, Liang, Yunhua Kong, Dalei Wu, Haitao Zhang, Jian Wu, Jing Chen, Jianping Ding, Lihong Hu, Hualiang Jiang, and Xu Shen. "Three flavonoids targeting the β-hydroxyacyl-acyl carrier protein dehydratase from *Helicobacter pylori*: Crystal structure characterization with enzymatic inhibition assay." *Protein Science* 17, no. 11 (2008): 1971-1978.

Zhang, Yun-bin, Xiaoyu Liu, Pingping Jiang, Wei-di Li, and Yi-fei Wang. "Mechanism and antibacterial activity of cinnamaldehyde against *Escherichia coli* and *Staphylococcus aureus*." *Modern Food Science and Technology* 31, no. 5 (2015): 31135.

Zhang, Yunhui, Andrea Sass, Heleen Van Acker, Jasper Wille, Bruno Verhasselt, Filip Van Nieuwerburgh, Volkhard Kaever, Aurélie Crabbé, and Tom Coenye. "Coumarin reduces virulence and biofilm formation in *Pseudomonas aeruginosa* by affecting quorum sensing, type Ⅲ secretion and c-di-GMP levels." *Frontiers in microbiology* 9 (2018): 1952.

Zou, D., K. Xie, H. Wang, Y. Chen, and M. Xie. "Inhibitory effects of biochanin A on the efflux pump of methicillin-resistant *Staphylococcus aureus* (MRSA)." *Wei sheng wu xue bao= Acta microbiologica Sinica* 54, no. 10 (2014): 1204-1211.

In: Antimicrobial Resistance: Opportunities and Challenges ISBN: 978-1-53617-943-9
Editors: P. Kumar Prabhakar and V. Mishra

Chapter 12

ALGAE: NOVEL CELL FACTORIES FOR ANTIMICROBIAL COMPOUNDS TO OVERCOME ANTIBIOTIC RESISTANCE

Pratibha Vyas*
Department of Microbiology, College of Basic Sciences and Humanities,
Punjab Agricultural University, Ludhiana, Punjab, India

ABSTRACT

Antibiotic resistance, the ability of microorganism to multiply in the presence of an antibiotic is a natural phenomenon, whereby, the intrinsic genes of the organism encode resistance towards antibiotics. Resistance to antibiotics in bacteria has been reported for almost every antibiotic once it is used. Many studies have reported the emergence of antibiotic-resistant microorganisms due to irrational use of antibiotics leading to a serious threat to mankind worldwide. The unexplored substantial diversity of algae residing in different habitats is a large resource for the discovery of novel bioactives indispensable for treating antibiotic-resistant microbial infections. Different bioactives including deschloroelatol, fucophlorethol, fucoxanthin, loliolide etc. have been isolated from different algal species. In the present chapter, diversity of algae cell factories for bioactive compounds to overcome antibiotic resistance has been discussed. The bioactive compounds of algal origin with high availability have the potential of developing into successful drugs against various infectious diseases.

Keywords: algae, antibiotic resistance, antimicrobial compounds, bioactives, cell factories

* Corresponding Author's E-mail: vyasp2000@pau.edu; vyasp2000@yahoo.co.in.

INTRODUCTION

Resistance is the unresponsiveness of a microorganism towards an antimicrobial agent in standard doses. All antimicrobial agents have the potential to select drug-resistant subpopulations of microorganisms. Antibiotic resistance has become a word wide concern with the continuous and irrational use of broad-spectrum antibiotics, imparting a negative and economic impact on health due to the long illness duration, long treatment methods including treatment with expensive drugs and high mortality. The Centers for Disease Control and Prevention (CDC) has published a report entitled "Antibiotic resistance threats in the United States" enlisting 18 antibiotic-resistant threats in the U.S. in 2013. The report mentioned that out of about two million people in U.S. acquiring antibiotic resistance, atleast 23,000 die, which is quite very alarming. Resistance is accelerated through inappropriate and irrational use of antimicrobials which include non-availability of the standard treatment guidelines to physicians or non-adherence of the guidelines, inappropriate prescription, availability of antibiotics/drugs without prescription, accessibility of poor quality of antibiotics, poor compliance and insufficient monitoring or self- prescription or administration. Many hospital-acquired infections are reported to be caused by multidrug-resistant (MDR) pathogenic microorganisms leading to slow and difficult treatment (Fridkin et al., 2014). Patients act as a reservoir of resistant organisms which are passed to community and health-care workers. Microorganisms use various mechanisms to develop resistance against drugs including enzymatic inhibition, alteration of target sites/target modifications, modification of cell wall protein/porin mutations, efflux pumps, and target changes. Resistance is a public health priority requiring a collaborative global approach.

Natural products have been a rich source of compounds with structural and chemical diversity for drug discovery. Exploration of new strategies to combat infectious diseases has become essential. Algae are the untapped resources for the discovery of bioactive compounds with the ability to produce a large array of secondary metabolites with antimicrobial activity, which might provide useful leads in developing new and novel pharmaceutical agents. The potential of the different bioactive compounds isolated from various algal genera has been reported highlighting their capability of producing metabolites of high complexity and unlimited diversity (del Rio et al., 2001; Seddek et al., 2019).

Algae are a diverse group of unicellular and multicellular eukaryotic autotrophic organisms. They are ubiquitous and have diverse habitats including oceans, fresh water, brackish water, hot springs, soil, rocks, deserts and buildings; however, mainly inhabit in aquatic environments. They also live as epiphytes or parasites on animals and plants. Majority of algae are autotrophic, thereby, play an important role in the productivity of oceans and represent the basis of food chain in aquatic habitats. In addition, some algae including *Chlorella protothecoides*, *Cochlodinium polykrikoides* and *Dinophysis acuminata* show mixotrophic metabolism (gain energy from photosynthesis as well as from the uptake of organic carbon). Additionally, some heterotrophic forms of algae such as *Crypthecodinium cohnii*, *Helicosporidium* sp., *Prototheca* sp., and *Schizochytrium* sp. are also known (Diao et al., 2018; Suzuki et al., 2018).

Algae belonging to various genera are considered as important cell factories producing bioactive compounds with the potential to act against pathogenic microorganisms including multi-drug resistant bacteria. Various studies have reported the antagonistic ability of algae of diverse habitats against different pathogenic microorganisms including Gram positive bacteria, Gram negative bacteria, yeasts and molds (Kolanjinathan and Stella, 2009; Mugilan and Sivakami, 2016; Zouaoui and Zhalem, 2017). Display of antimicrobial activities is considered as an indicator of the ability of algae to synthesize bioactive compounds as secondary metabolites. Algae of different genera and habitats have been reported to produce a diverse array of bioactive molecules including amino acids, alkenes terpenoids, cyclic polysulphides, halogenated ketones, phenolic compounds, phlorotannins and steroids with antibacterial, antifungal, antiviral, antitumor, antioxidant and anti-inflammatory activities. Bioactive compounds like isoprenoids, polyketides, polyunsaturated fatty acids, alkaloids, terpenes, carbohydrates, glycolipids, lipoproteins, bromophenols, lectins and tannins of algal origin have shown activity against human pathogens. The present chapter reviews the mechanisms of antibiotic resistance in microorganisms and the role of algal cell factories producing bioactive compounds in inhibiting the drug resistant pathogenic microorganisms including bacteria, yeasts and molds.

ANTIBIOTIC RESISTANCE

Antimicrobial resistance, an emerging public health concern is the resistance of microorganisms towards an antimicrobial agent to which they were sensitive earlier. It has serious implications leading to an economic loss to the patients and community in view of the longer duration of illness and treatment. Expensive and toxic drugs are generally used to treat infections with antibiotic-resistant microorganisms leading to an extra burden on the health system. Antibiotic resistance also increases the risk of the transmittance of resistant microorganisms across borders through humans and food-chain. Well known examples including outbreaks of methicillin-resistant *Staphylococcus aureus* (MRSA) in the Canada hospitals being originated in North India, multidrug-resistant (MDR) *Salmonella schwarzengrund* in Denmark and USA through Thai food, MDR typhoid fever in the United States of America originating in six countries including India, multidrug resistant *Mycobacterium* and malaria at Thai-Cambodia border and resistance in H1N1 and HIV are of global concerns.

A large number of microorganisms have acquired resistance towards antibiotics which has been generated and transmitted in many ways. Some of the ways include natural resistance attained by vertical transmission of chromosomally located genes, horizontal gene transfer of transposons and intrinsic resistance due genes generating a resistance phenotype. Resistance to antibiotics in bacteria is not new as this mechanism appeared during the beginning of the antibiotic era and a number of dangerous resistant strains have emerged since then which is of global concern. This is generally because of their overuse due to a lack of awareness and knowledge despite recent regulations and control on the prescription and purchase of antibiotics worldwide. In addition, self-medication is also posing a great threat affecting the quality of the therapy and correct diagnosis of the illness by a medical specialist. Also, the use of antibiotics in animal feedstock is a serious issue contributing to the spread of resistance (Orhan et al., 2006).

ANTIBIOTIC RESISTANCE MECHANISMS

Antibiotic resistance in bacteria can occur by different mechanisms; however, five mechanisms most frequently observed are enzymatic inhibition, modifications of penicillin-binding proteins (PBPs), porin mutations/modification of cell wall protein, alteration of targets, and efflux pumps (Figure 1).

Enzymatic Inhibition

Enzymatic inhibition is the most common mechanism used by bacteria for antibiotic resistance. It is based on the strategies used by bacteria to inactivate the antibacterial compounds including hydrolysis (mainly in β-lactams), transference of functional groups present in antibiotics like aminoglycoside, chloramphenicol, lincosamide and rifamycin, and some chemical modifications of antibiotics like rifamycin, tetracycline and streptogramin (Bratu et al., 2005). β-lactamases, including penicillinases (conferring resistance against penicillins, ampicillin, and cephalosporins) the main enzymes able to modify antibacterial compounds, pose a major problem in treating infections of Gram-negative bacteria (Bonnin et al., 2013). Extended-spectrum β-lactamases provide resistance against penicillins and cephalosporins by hydrolyzing these antibiotics. In general, β-lactamases do not pose any serious problem in Gram-positive bacteria, but penicillinases have been reported to have enzyme inhibition mechanism conferring resistance (Rai et al., 2012).

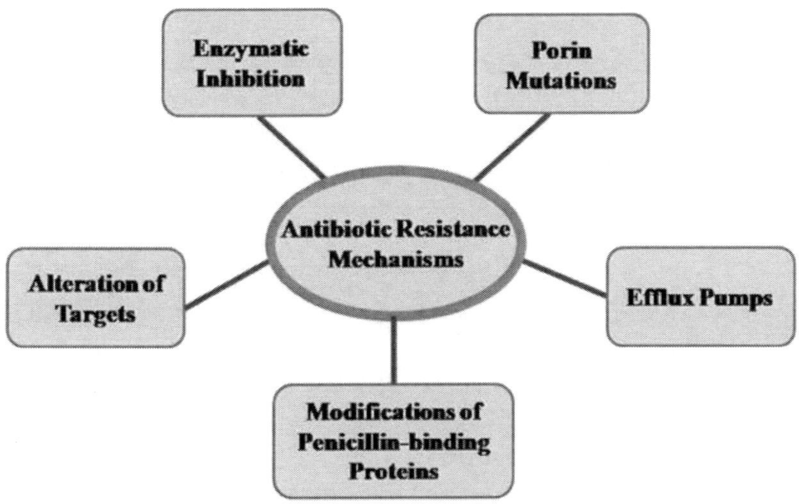

Figure 1. Mechanisms used by antibiotic resistant bacteria.

Modifications of Penicillin-binding Proteins (PBPs)

PBPs, important proteins involved in the synthesis of peptidoglycan are a major constituent of bacterial cell walls. The function of these proteins is to carry out transglycosylation and transpeptidation leading to the elongation of the glycan strand and cross-linkage between glycan chains, respectively. The target of β-lactam agents is the transpeptidase active site, whereby, D-Ala-D-Ala dipeptide in peptidoglycan is mimicked leading to the formation of a stable acyl-enzyme complex, thereby, inactivating the enzyme (Lukacik et al., 2012). Modified PBPs alter the active site, leading to the loss of affinity of β-lactams for target protein and thus, promoting resistance. Among the different modified PBPs, the most alarming is PBP2a/PBP20, which confers resistance to cephalosporins and penicillins. One such example is antibiotic-resistant staphylococci resulting due to the attainment of a gene encoding modified enzyme, PBP2a with lower affinity than the wild-type enzyme for antibiotics.

Alteration of Targets

Target modifications are one of the important mechanisms of acquired resistance towards antibiotics resulting from mutations or movement of resistance genes between different species and genera or between strains of the same species.

Once the active concentrations of antibiotics are accumulated in the bacterial cell, they bind to a specific target usually an essential protein required for the bacteria. Mutation processes leading to the synthesis of modified target proteins with low affinity to the antibiotic have resulted in generation of large number of antibiotic-resistant bacterial strains. Generally, the plasmids contain resistance genes, but these genes can also be located on transposons or integrons leading to their movement from one plasmid to another within the chromosome of the same cell or between different cells. Even the substitution of single amino acid of the protein leads to changes in binding property without influencing the other biological functions (Hopwood, 2007).

Porin Modifications

Changes in bacterial cytoplasmic membrane could also result in antibiotic resistance. Gram-negative bacteria have outer membrane consisting of hydrophobic lipopolysaccharide making the passage of hydrophilic compounds difficult across the membrane. Outer membrane porins (Omps) are the proteins aiding in the transport of hydrophilic solutes across lipid membrane (Livermore and Woodford, 2006; Vila et al., 2007). A specific porin is produced by a bacterial species and in general, antibiotic-resistant bacterial strains are not able to synthesize one or more porins. One common factor in establishing resistance is the loss or impairment of one or more Omps, for example the loss of OprD in *Pseudomonas aeruginosa* provides resistance to meropenem and imipenem (Goossens, 2003). Similarly, loss of OmpF can lead to the generation of multidrug-resistant microorganisms (Mah and O'Toole, 2001). Likewise, it was observed that the minimum inhibitory concentrations for penicillins and cephalosporins in *Escherichia coli* are inversely proportional to expression levels of OmpC and OmpF porins. High level of resistance was associated with loss of OmpF in comparison to OmpC due to the participation of the former in the formation of larger pores (Gutmann et al., 1984; Kojima and Nikaido, 2013).

Efflux Pumps

Production of a proton dependent system - an efflux pump, is an efficient mechanism of antibiotic-resistance due to an active removal of the antibiotics from inside the cell (Wright, 2011). Five families of efflux proteins including major facilitators, small multidrug resistance, ATP-binding cassette, resistance nodulation cell division, and multidrug and toxic compound extrusion are known (Kobayashi et al., 2001; Stavri et al., 2007).

Efflux systems in Gram positive bacteria are simple and generally include a single cytoplasmic membrane transporter of the families, belonging to major facilitators, small multidrug resistance and ATP-binding cassette, while the system is more complex in Gram-negative bacteria due to outer lipopolysaccharide membrane (Stavri et al., 2007). Efflux pumps either have high specificity for a particular antibiotic or can transport a number of drugs, leading to the development of multidrug-resistant microbial phenotypes.

ALGAE AS ANTIMICROBIAL AGENTS

The emergence of antibiotic resistance in microorganisms has led to research focussed on natural compounds with antimicrobial properties. The interest in algae showing antimicrobial activity has increased after the study demonstrating the ability of *Chlorella* against Gram positive and Gram negative bacteria by Pratt and co-workers (1944). Thereafter, many studies have been conducted to ascertain the potential of algae of different habitats to inhibit the pathogenic microorganisms (Table 1).

Algae including *Caulerpa Mexicana*, *Caulerpa racemosa*, *Halimeda opumtia*, *Halimeda tuna*, *Valonia aegrophila* and *Ulva pertusa* have been screened for antimicrobial activity against *Bacillus subtilis*, *Candida albicans*, *Escherichia coli* and *Staphylococcus aureus* using disc diffusion method (Mtolera and Semesi, 1996). A range of algae including *Acanthophora nojadiformis*, *Ceramium rubrum*, *Colpomenia sniosa*, *Cystoseira mediterranea*, *Dictyota linearis*, *Dictyopteris membranacea*, *Ectocarpus siliculosus*, *Enteromorpha linza*, *Gracilaria gracilis*, *Padina pavonica*, and *Ulva rigida* have been evaluated for antimicrobial activity against *Candida* sp., *Enterobacter faecalis*, *Escherichia coli*, *Pseudomonas aeruginosa*, *Staphylococcus aureus* and *S. epidermidis* (Tuney et al., 2006). Likewise, red algae *Ceramium rubrum*, *Sargassum vulgare*, *S. fusiforme* and *Padina pavonia* collected from Egypt, were tested against multidrug resistant Gram positive and Gram negative bacteria (El Shafay et al., 2016). Recently, *Halimeda* sp. from the coastal area of Peninsular Malaysia has been screened against a range of microorganisms including *Aspergilus niger*, *Bacillus subtilis*, *B. cereus*, *Candida albicans*, *Escherichia coli*, *Pseudomonas aeruginosa* and *Staphylococcus aureus* using disc diffusion method (War et al., 2018). Methanolic extracts of *Helimeda* sp. were found to exhibit activity against *Bacillus subtilis*, *B. cereus* and *Staphylococcus aureus*.

Table 1. Antimicrobial activity of algae

Algal species	Against microorganism	Reference
Amansia smultifida, *Caulerpa cupressoides*, *C. prolifera*, *Gracilaria domingensis*, *Gracilaria* sp., *Ulva fasciata*	*Bacillus subtilis*, *Citobacter freundii*, *Enterobacter aerogenes*, *Escherichia coli*, *Klebsiella pneumoniae*, *Morganella morganii*, *Pseudomonas aeruginosa*, *Salmonella typhi*, *S. typhimurium*, *S. enteritidis*, *S. choleraesuis*, *Serratia marcescens*, *Staphylococcus epidermidis*, *S. aureus*, *Vibrio cholera*	Lima-Filho et al., 2002
Cystoseira tamariscifolia	*Saccharomyces cerevisiae*, *Candida albicans*, *Debaryomyces* sp., *Kluyveromyces* spp. *Aspergillus flavus*, *Penicillium* sp.	Zineb et al., 2004
Spirulina platensis	*Bacillus subtilis*, *Candida albicans*, *Enterobacter aerogenes*, *E. cloacae*, *Escherichia coli*, *Pseudomonas aeruginosa*, *P. vulgaris Salmonella typhimurium*, *Staphylococcus aureus*, *S. epidermidis*, *Streptococcus faecalis*	Ozdemir et al., 2004
Gracilaria tikvahiae, *Sargassum fluitans* *Ulva fasciata*, *Ulva lactuca*	*Candida albicans*, *Escherichia coli*, *Salmonella enteriditis*, *Pseudomonas aeruginosa*, *Staphylococcus aureus*, *S. epidermidis*, *Streptococcus faecalis*	Oranday et al., 2004
Gracilaria corticata, *Ulva fasciata*, *Enteromorpha compressa*	*Pseudomonas aeruginosa*, *P. fluorescens*, *Aeromonas hydrophila*, *Edwardsiella tarda*, *Vibrio alginolyticus*	Choudhury et al., 2005
Dictyopteris membranaceae, *Cystoseira barbata*	*Bacillus subtilis*, *Candida albicans*, *Enterobacter cloacae*, *Escherichia coli*, *Pseudomonas aeruginosa*, *Salmonella typhimurium*, *Staphylococcus aureus*, *S. epidermidis*, *Streptococcus faecalis*,	Ozdemir et al., 2006
Enteromorpha linza	*Candida albicans*, *Staphylococcus aureus*, *S. epidermidis*, *Streptococcus fecalis*, *Bacillus subtilis*, *Salmonella typhimurium*, *Pseudomonas aeruginosa*, *Enterobacter cloacae*, *Escherichia coli*	Sukatar et al., 2006
Acanthophora nojadiformis, *Ceramium rubrum*, *Colpomenia sniosa*, *Cystoseira mediterranea*, *Dictyota linearis*, *Dictyopteris membranacea*, *Ectocarpus siliculosus*, *Enteromorpha linza*, *Gracilaria gracilis*, *Padina pavonica*, *Ulva rigida*	*Candida* sp., *Enterobacter faecalis*, *Escherichia coli*, *Pseudomonas aeruginosa*, *Staphylococcus aureus*, *S. epidermidis*	Tuney et al., 2006
Cladostephus spongiosus f. *verticillatus*, *Corallina officinalis*, *Cystoseira barbata*, *Dictyota dichotoma*, *Halopteris filicina*, *Ulva rigida*	*Escherichia coli*, *Enterobacter aerogenes*, *Enterococcus faecalis*, *Micrococcus luteus*, *Staphylococcus aureus*	Taskin et al., 2007

Algal species	Against microorganism	Reference
Caulerpa racemosa, *C. scalpelliforrmis*, *Ceramium indica*, *Corallina officinalis*, *Cystoseira* sp., *Enteromorpha intestinalis*, *Gelidiella acerosa*, *G. gracilis*, *Gelidium* sp., *Halimeda tuna*, *Heterosiphonia* sp., *Gracilaria* corticata, *Galaxaura oblongata*, *Grateloupia* sp., *Iyengaria stellata*, *Laurentia pedicullata*, *Padina gymnospora*, *Sargassum* sp., *Stoechospermum marginatum*, *Sargassum* sp., *Spathoglossum variabile*, *Udotea indica*, *Gastroclonium iyengarii*, *Ulva lactuca*	*Bacillus cereus*, *Klebsiella pneumoniae*, *Pseudomonas. testosteroni*, *Citrobacter freundii*	Nair et al., 2007
Sargassum dentifolium, *Laurencia papillosa*, *Jania corniculata*	*Bacillus subtilis*, *Staphylococcus albus*, *Streptococcus faecalis*, *Escherichia coli*, *Candida albicans*, *Aspergillus flavus*	Shanab, 2007
Jania rubens	*Bacillus cereus*, *B. subtilis*, *Candida albicans*, *Escherichia coli*, *Enterobacter cloacae*, *Pseudomonas aeruginosa*, *Streptococcus faecalis*, *Salmonella typhimurium*, *Staphylococcus aureus*, *S. epidermidis*	Karabay-Yavasoglu et al., 2007
Dictyota acutiloba	*Aspergillus niger*, *Bacillus subtilis*, *Candida albicans*, *Enterobacter* sp., *Klebsiella pneumoniae*, *Pseudomonas aeruginosa*, *Salmonella typhi*, *Staphylococcus aureus*	Solomon and Santhi, 2008
Sargassum sp.	*Bacillus subtilis*, *Escherichoia coli*, *Staphylococcus aureus*	Patra et al., 2008
Polysiphonia virgata	*Mycobacterium tuberculosis*, *M. smegmatis*	Saravanakumar et al., 2008
Halimeda gracilis, *Hypnea musiformis*, *Gracilaria edulis*, *Sargassum myricystum*, *Turbinaria conoides*, *Ulva lactua*	*Enterococcus faecalis*, *Escherichia coli*, *Klebsiella pneumoniae*, *Pseudomonas aeruginosa*, *Staphylococcus aureus*	Kolanjinathan and Stella, 2009
Gracilaria changii	*Acinetobacter calcoaceticus*, *Aspergillus niger*, *A. flavus*, *Bacillus cereus*, *B. licheniformis*, *B. subtilis*, *Candida albicans*, *Citrobacter preundii*, *Cryptococcus neoformans*, *Enterobacter enterocolitica*, *E. aerogenes*, *Erwinia* sp., *Escherichia coli*, *Fusarium* sp., *F. oxysporium*, *Klebsiella pneumoniae*, *Micrococcus* sp., *Mucor* sp., *Penicillium* sp., *Pseudomonas aeruginosa*, *P. mirabilis*, *Rhodotorula rubra*, *Rhizophus* sp., *Saccharomyces cerevisiae*, *Salmonella typhi*, *S. paratyphyi*, *Serratia marcescens*, *Staphylococcus aureus*, *S. epidermidis*, *Trichoderma viride*, *Trichophyton mentagrophytes*	Sasidharan et al., 2009

Table 1. (Continued)

Algal species	Against microorganism	Reference
Gracilaria edulis, *Padina gymnospora*, *Sargassum wightii*, *Ulva lactuca*	*Klebsiella pneumonia*, *Pseudomonas aeruginosa*, *Shigella dysentriae*, *Salmonella bodii*, *Staphylococcus aureus*, *Vibrio cholerae*, *S. paratyphi*	Vallinayagam et al., 2009
Trichodesmium erythraeum	*Bacillus cinerea*, *B. subtilis*, *Enteobacter faecalis*, *Erwinia amylovora*, *Escherichia coli*, *Klebsiella pneumoniae*, *Pseudomonas aeruginosa*, *Proteus* vulgaris, *Salmonella typhimurium*, *Staphylococcus aureus*, *Trichophyton rubrum*, *T. mentagrophytes*, *T. simii*, *Scopulariopsis* sp. 101/01, *Aspergillus flavus*, *Aspergillus niger*, *Candida* albicans	Thillairajasekar et al., 2009
Eucheuma denticulatum	*Staphylococcus aureus*, *S. pyogenes*	Al-Haj et al., 2009
Kappaphycus sp., *Padina* sp.	*Bacillus megaterium*, *Citroacitro* sp., *Enterobacter* sp., *Escherichia coli*, *Pseudomonas aeruginosa*, *P. fluorescens*, *Proteus mirabilis*, *P. vulgaris*, *Klebsiella pneumoniae*, *Salmonella fexneri*, *S. typhimurium*, *Staphylococcus aureus*, *Vibrio cholerae*	Rajasulochana et al., 2009
Colpomenia sinuosa, *Dictyota dichotoma*, *Dictyota dichotoma* var. *implexa*, *Petalonia fascia*, *Scytosiphon lomentaria*	*Bacillus subtilis*, *Candida albicans*, *Escherichia coli*, *Enterobacter aerogenes*, *Staphylococcus aureus*, *Proteus vulgaris*, *Salmonella typhimurium*	Demirel et al., 2009
Ulva fasciata	*Vibrio parahaemolyticus*, *V. alginolyticus*, *V. vulnificus*	Chakraborty et al., 2010
Enteromorpha intestinalis, *Chaetomorpha aerea*, *Ulva fasciata*	*Escherichia coli*, *Klebsiella pneumonia*, *Pseudomonas aeruginosa*, *Staphylococcus aureus*	Seenivasan et al., 2010
Chlorella pyrenoidosa, *Nostoc muscorum*, *Spirulina platensis*	*Escherichia coli*, *Pseudomonas aeruginosa*, *Staphylococcus aureus*	Arun et al., 2012
Dunaliella salina and *Pseudokirchneriella subcapitata*	*Escherichia coli*, *Klebsiella* spp., *Pseudomonas aeruginosa*, *Staphylococcus aureus*	Pane et al., 2015
Lyngbya	*Chlorella vulgaris*, *Scenedesmus quadricauda*, *Selenastrum capricornatum*, *Anabaena variabilis*, *Bacillus subtilis*, *B. pumilus*	Devi and Mehta, 2015
Chlorella sp. and *Ulva lactuca*	*Bacillus subtilis*, *Candida albicans Escherichia coli*, *Klebsiella pneumonia*, *Pseudomonas aeruginosa*, *Staphylococcus aureus*, *Salmonella typhi*	Chowdhury et al., 2015
Cyclotella comta, *Nostoc linckia Oscillatoria limosa*,*Spirogyra quinina*, *Spirulina major*, *Zygnema stellinum*	*Aeromonas hydrophila*, *Aspergillus niger*, *Bacillus subtilis*, *Candida albicans*, *Clostridium perfringens*, *Escherichia coli*, *Proteus vulgaris*, *Salmonella typhi*, *Streptococcus faecalis*	Mugilan and Sivakami, 2016

Algal species	Against microorganism	Reference
Dictyota dichotoma, *Corallina elongate*, *Ulva lactuca*	*Escherichia coli, Staphylococcus aureus, Salmonella* sp., *Candida albicans, Penicillium* sp.	Zouaoui and Zhalem, 2017
Sargassum swartzii, Jania rubens, Stoechospermum marginatum	*Bacillus subtilis, Enterococcus faecalis, Escherichia coli, Klebsiella pneumoniae, Pseudomonas aeruginosa, Proteus vulgaris, Staphylococcus aureus, Salmonella typhi, Streptococcus pyogenes, Aspergillus niger, Candida albicans, Cryptococcus neoformans, Penicillium* sp., *Trichophyton* sp.	Sasikala and Geetha, 2017
Chlorella vulgaris	*Escherichia coli*, *Bacillus subtilis* and *Pseudomonas aeruginosa*, *Staphylococcus aureus*, *Candida albicans*	Mashhadinejad et al., 2017
Ulva reticulata	*Staphylococcus aureus*	Nainu et al., 2018
Halimeda sp.	*Bacillus subtilis*, *B. cereus*, *Escherichia coli*, *Pseudomonas aeruginosa*, *Staphylococcus aureus*, *Candida albicans*, *Aspergilus niger*	War et al., 2018
Anabaena oryzae, *Oscillatoria* sp., *Stigonema ocellatum*	*Candida albicans*, *Serratia marcescens*	Seddek et al., 2019

Table 2. Algal cell factories for antimicrobial bioactive compounds

Algal species	Bioactive compound	Against microorganism	Reference
Eucheuma serra, *Galaxaura marginata*	Lectins	*Vibrio pelagius*, *V. vulnificus*	Liao et al., 2003
Solieria filiformis	lectins	*Enterobacter aerogenes*, *Klebsiella pneumoniae*, *Pseudomonas aeruginosa*, *Proteus* sp., *Salmonella typhi*, *Serratia marcescens*	Holanda et al., 2005
Colpomenia sinuosa, *Dictyota dichotoma*, *Petalonia fascia* and *Scytosiphon lomentaria*	Phlorotannins	*Bacillus subtilis*, *Enterobacter aerogenes*, hemorrhagic *Escherichia coli*, methicillin-oxacillin-resistant *Staphylococcus aureus*, *Proteus vulgaris Salmonella typhimurium*	Demirel et al., 2009
Phaeodactylum tricornutum	Fatty acids	*Staphylococcus aureus*	Desbois et al., 2009

Table 2. (Continued)

Algal species	Bioactive compound	Against microorganism	Reference
Stoecheospermum marginatum, *Padina tetrastromatica*, *P. gymnospora*, *Chnoospora implexa*, *Asparagopsis taxiformis*, *Amphiroa fragilissima*, *Jania reubens*, *Caulerpa racemosa*, *Caulerpa peltata*, *Caulerpa taxifolia*, *Codium fragile*, *Chlorodesmis fastigiata*	Phenols, tannins, saponinis, cardiac glycosides, terpenoides, alkaloids, anthraquinones and flavonoides	*Bacillus subtilis*, *Candida albicans*, *Escherichia coli*, *Klebsiella pneumoneae*, *Pseudomonas* sp., *Proteus vulgaris*, *Staphylococcus aureus*, *Streptococcus pyrogenes*, *Serrratia marganii*	Kotnala et al., 2009
Synechocystis sp., *Himanthalia elongate*	Fatty acids and volatile compounds	*Staphylococcus aureus*, *Escherichia coli*, *Aspergillus niger*, *Candida albicans*	Plaza et al., 2010
Chaetoceros mulleri	Triglycerides and docosa-pentaenoic acid (DPA))	*Escherichia coli*, *Bacillus subtilis Staphylococcus aureus*	Sánchez-Saavedra et al., 2010
Chaetomorpha aerea	Sulphated galactan	*Staphylococcus aureus*	Pierre et al., 2011
Sargassum swartzii	Sulphated polysaccharides	*Escherichia coli*	Vijayabaskar et al., 2012
Himanthalia elongata	fucoxanthin	*Listeria monocytogenes*	Rajauria and Abu-Ghannam, 2013
Ascophyllum nodosum, *Laminaria hyperborea*	Laminarin (polysacch)	*Escherichia coli*, *Staphylococcus aureus*	*Kadam* et al., 2015
Marine algae	Fucoidan	*Escherichia coli, Staphylococcus aureus*	Yu et al., 2015
Ulva lactuca	Loliolide, phytol, neophytadiene, hexadecanoic acid methyl ester, 8 heptadecene and Bis 2 (ethyl hexyl) phathalate, palmitic acid, oleic acid	*Alternaria solani*, *Fusarium solani*, *Rhizoctonia solani*, *Sclerotinia sclerotiorum*, *Phytophthora infestans* and *Botrytis cinerea*	Selim et al., 2015
Phaeodactylum tricornutum	Eicosapentaenoic acid	*Staphylococcus aureus*	Rodolfi et al., 2017
Chlorella marina, *Nannochloropsis occulata*, *Chaetoceros affinis*	Fatty acids	*Enterobacter* spp., *Escherichia coli*, *Klebsiella pneumoniae*, *Streptococcus*, *Pseudomonas aeruginosa*, *Proteus* spp., *Salmonella typhi*, *Shigella sonnaie*, *Staphylococcus aureus*, *Vibrio cholerae*	Saritha et al., 2017

Algal species	Bioactive compound	Against microorganism	Reference
Sargassum oligocystum and *Sargassum crassifolium*	flavonoids, tannins, phenolics, sterols and terpenoids	*Aeromonas hydrophila, Bacillus subtilis, Escherichia coli, Micrococcus luteus, Psuedomonas aeruginosa, P. fluorescens, Staphylococcus aureus, Streptococcus mutans, Aspergillus niger, A. parasiticus, Candida tropicalis Penicillium expansum, Sacchromyces cerevisiae*	Baleta et al., 2017
Ulva reticulata		*Staphylococcus aureus*	Nainu et al., 2018
Ulva rigida	fatty acids, mainly oleic, linoleic, palmitic and stearic	*Staphylococcus aureus, Enterococcus faecalis*	Ismail et al., 2018
Dunaliella sp.	Stearic acid, lutein, α-carotene, zeaxanthin, ferulic acid	*Bacillus subtilis, Escherichia coli, Enterobacter cloacae, Staphylococcus aureus*	Kilic et al., 2018
Anabaena oryzae, *Oscillatoria* sp. and *Stigonema ocellatum*	Diacetone alcohol, acetic acid butyl ester mesityl oxide and heptadecane	*Serratia marcescens* and *Candida albicans*	Seddek et al., 2019

ALGAL CELL FACTORIES FOR ANTIMICROBIAL BIOACTIVE COMPOUNDS

Algae have received great interest as microbial cell factories in view of their ability to produce a large array of distinctive compounds with wide applications. Algal extracts have been reported to contain bioactive compounds including polyphenols, vitamin C, mycosporine-like amino acids and tocopherols with antioxidants properties and pigments such as carotenoids, chlorophylls, and phycobilins with anti-bacterial, anti-fungal, anti-viral, anti-inflammatory anti-oxidative and anti-tumor properties (Metzger et al., 2002; Kuda et al., 2007). The extracts and active compounds of algae have shown a broad-range of antimicrobial activity (Table 2). Exhibition of antimicrobial activity is an indicator of the algal ability to synthesize bioactive compounds as secondary metabolites.

Many studies have reported the isolation of bioactive metabolites from algae with antibacterial, antifungal and antivirals properties (Kotnala et al., 2009; Kilic et al., 2018; Seddek et al., 2019). Algae have been found to accumulate secondary metabolites exhibiting different levels of antimicrobial activity in different algal cultures (Kotnala et al., 2009; Saritha et al., 2017; Seddek et al., 2019). Algal synthesis of bioactive compounds like alkaloids, bromophenols, carbohydrates, glycolipids, isoprenoids, lectins, lipoproteins, polyketides, polyunsaturated fatty acids, tannins and terpenes exhibit antibacterial activity against human pathogens (Table 2). Many studies have reported the broad-spectrum antimicrobial activity of the crude extracts obtained from different species of algae, which is considered necessary for producing new and novel antibiotics.

Bioactive compounds including deschloroelatol, fucophlorethol, fucoxanthin, loliolide, and have been obtained from *Ascophyllum nodosum*, *Myagropsis myagroides Laurencia rigida* and *Undaria pinnatifida* (Shannon and Ghannam, 2016; Karpiński and Adamczak, 2019). In addition, parsiguine and tjipanazoles isolated from the alga *Tolypothrix tjipanasensis* have shown antimicrobial activity (Sakthivel and Kathiresan, 2012). Many studies have also reported the isolation of antimicrobial substances active against multidrug-resistant (MDR) bacteria. One such example includes the isolation of eicosapentaenoic acid from *Phaeodactylum tricornutum* active against MDR *Staphylococcus aureus*.

Fatty acids occur either as glycolipids and phospholipids in cells forming an integral part of cell membranes or storage products in the form of triglycerides for carbon and energy (Mendiola et al., 2007). Fatty acids have been reported to be produced in large amounts by cyanobacteria as well as eukaryotic microalgae. Oleic, linoleic, palmitic and stearic acids produced by algae have been reported to have broad-spectrum antimicrobial activity against *Enterobacter* spp., *Escherichia coli*, *Klebsiella pneumonia*, *Vibrio cholerae*, *Streptococcus*, *Proteus* spp., *Pseudomonas aeruginosa, Salmonella typhi, Staphylococcus aureus, Shigella sonnaie* etc. (Table 2).

Lectins, carbohydrate-binding proteins have high specificity towards glycolipids, glycoproteins and oligosaccharides. Lectins produced by algae are monomeric low molecular weight proteins with high content of acidic amino acids and do not require any metal ions for their activity. They are considered as cell markers for diagnostic purposes and also play an important role in detecting the modifications of glycan synthesis. Algae including *Eucheuma serra*, *Galaxaura marginata*, *Solieria filiformis*, and *Chlorella minutissima* have been reported to produce lectins as secondary metabolites with antimicrobial activity (Table 2).

Widely distributed in nature, terpenoids, derived from the precursor isopentenyl pyrophosphate, are reported to be produced by marine algae in enormous amounts (Table 2). They are divided into seven groups as hemiterpenoids, sesquiterpenoids, monoterpenoids, diterpenoids, sesterterpenoids, triterpenoids and polyterpenoids whereby monoterpenoids, sesterterpenoids and sesquiterpenoids are the major terpenoids with antimicrobial and antiviral activities (Abad et al., 2011). Sesquiterpenoid hydroquinones including acetylmajapolene A and B, and tiomanene isolated from *Laurencia* sp. have been reported for antimicrobial activity (Vairappan et al., 2008). Similarly, a bioactive compound sargafuran with strong activity against *Propionibacterium acne* has been isolated from the brown sea alga *Sargassum macrocarpum* (Kamei et al., 2009). Likewise, *Peyssonnelia* sp. has been reported to produce sesquiterpene hydroquinones and peyssonoic acid A and B inhibiting the growth of bacterial and fungal pathogens (Lane et al., 2010).

Phenolic compounds, constituting the largest group of bioactive secondary metabolites range from simple structures to highly complex polymeric molecules, have an aromatic ring with one or more hydroxyl groups. They have the ability to exhibit varied antimicrobial potential against different bacteria and fungi either by altering cell permeability and loss of internal macromolecules or by interfering with functions of cell membrane (Abu-Ghannam et al., 2013). Monodictyoquinone, an anthraquinon (1,8-dihydroxy2-methoxy-6-methyl anthraquinone) with significant antimicrobial activity against bacteria and fungi has been isolated from *Monodictys* sp. (El-Beih et al., 2007). Recently *Sargassum oligocystum* and *Sargassum crassifolium* have been shown to produce phenolics in addition to flavonoids, tannins, sterols and terpenoids with broad-spectrum antimicrobial activity against Gram positive bacteria, Gram negative bacteria, molds and yeasts (Baleta et al., 2017).

In addition, polyphenols synthesized by algae have shown activity against microorganisms (Table 2). Polyphenols can be divided into two groups as phloroglucinol and phlorotannins where former have an aromatic phenyl ring containing three hydroxyl groups and the latter are polymers of phloroglucinol containing an additional hydroxyl group or halogen (Fernando et al., 2016). Phlorotannins produced by *Colpomenia sinuosa*, *Dictyota dichotoma*, *Petalonia fascia* and *Scytosiphon lomentaria* have shown antioxidant and antibacterial activity (Demirel et al., 2009).

Apart from above mentioned compounds, new and novel polysaccharides are also being explored from prokaryotic and eukaryotic algae in view of their remarkable antimicrobial activities. Algae belonging to *Pterocladia capillacea* and *Dictyopteris membranacea* have shown the production of polysaccharides (Zeid et al., 2014). Nostoflan, an acidic polysaccharide, isolated from *Nostoc flagelliforme* has been reported to show strong antiviral activity against HSV-1 virus (Kanekiyo et al., 2007). Synergistic effect of a complex sulphated polysaccharide, fucoidan present in the cell wall of edible brown algae with antibiotics against pathogenic bacteria has been studied (Lee et al., 2013).

Alkaloids, chemical compounds containing basic nitrogen atoms and deriving from amino acids have potent therapeutic efficiency. Many alkaloids and some other nitrogenous heterocyclic compounds have been obtained from algae with antagonistic effects against large number of microorganisms (Table 2).

CONCLUSION

Irrational use of antibiotics has led to the emergence of antibiotic-resistant microorganisms posing a serious threat to mankind worldwide. Increase in the antibiotic resistance in microorganisms posing a threat of bioterrorism attack with multi-drug resistant organisms has led to the research on the development of antimicrobial drugs. It becomes essential to explore new strategies for combating infectious diseases. Natural products obtained for microorganisms are a rich source of diverse compounds for drug discovery. Algae are the untapped resources for the discovery of novel bioactives that can be used to replace the antimicrobial agents against which microbes have developed resistance. The interest in algal -based antimicrobial compounds has increased over the past few years in view of their large diversity and ability to produce different secondary metabolites. They and are present everywhere ranging from fresh water to saline aquatic habitats, soils, rocks, deserts, hot water springs, and also live in association with plants and animals. Algae belonging to various genera and different habitats have shown their potential for the production of various bioactive compounds and inhibition of pathogenic microorganisms. Secondary metabolites including alkaloids, terpenoids, polyphenols, fatty acids, lectins, terpenoids, polysaccharides etc. of algal origin have shown promising antimicrobial activity against multi drug resistant strains. Taking into consideration the vast potential of algae for the production of novel bioactive compounds, it is required to isolate and screen new species residing in different habitats to introduce new technologies to overcome the challenge of antibiotic resistance.

REFERENCES

Abad, M. J., Bedoya, L. M., & Bermejo, P. (2011). Marine compounds and their antimicrobial activities. *Science against microbial pathogens: Communicating Current Research and Technological Advances*, 51, 1293-1306.

Abu-Ghannam, N., & Rajauria, G. (2013). Antimicrobial activity of compounds isolated from algae. In *Functional ingredients from algae for foods and nutraceuticals* (pp. 287-306). Woodhead Publishing.

Al-Haj, N. A., Mashan, N. I., Shamsudin, M. N., Mohamad, H., Vairappan, C. S., & Sekawi, Z. (2009). Antibacterial activity in marine algae *Eucheuma denticulatum* against *Staphylococcus aureus* and *Streptococcus pyogenes. Research Journal of Biological Sciences*, 4, 519-524.

Arun, N., Gupta, S., & Singh, D. (2012). Antimicrobial and antioxidant property of commonly found microalgae *Spirulina platensis, Nostoc muscorum and Chlorella pyrenoidosa* against some pathogenic bacteria and fungi. *International Journal of Pharmaceutical Sciences and Research*, 3, 4866.

Baleta, F. N., Bolaños, J. M., Ruma, O. C., Baleta, A. N. & Cairel, J. D. (2017). Phytochemicals screening and antimicrobial properties of *Sargassum oligocystum* and *Sargassum crassifolium* Extracts. *Journal of Medicinal Plants Studies*, 5(1), 382-387.

Bonnin, R. A., Rotimi, V. O., Al Hubail, M., Gasiorowski, E., Al Sweih, N., Nordmann, P., et al., (2013). Wide dissemination of GES-type carbapenemases in *Acinetobacter baumannii* isolates in Kuwait. *Antimicrobial Agents and Chemotherapy*, 57(1), 1838.

Bratu, S., Landman, D., Alam, M., Tolentino, E., & Quale, J. (2005). Detection of KPC carbapenem-hydrolyzing enzymes in *Enterobacter* spp. from Brooklyn, New York. *Antimicrobial Agents and Chemotherapy*, 49(2), 7768.

Chowdhury, M. M. H., Kubra, K., Hossain, M. B., Mustafa, M. G., Jainab, T., Karim, M. R. & Mehedy M. E. (2015). Screening of antibacterial and antifungal activity of freshwater and marine algae as a prominent natural antibiotic available in Bangladesh. *International Journal of Pharmacology*, 11 (7), 828-833.

Choudhury, S., Sree, A., Mukherjee, S. C., Pattnaik, P., & Bapuji, M. (2005). *In vitro* antibacterial activity of extracts of selected marine algae and mangroves against fish pathogens. *Asian Fisheries Science*, 18, 285-294.

Chakraborty, K., Lipton, A. P., Raj, R. P., & Vijayan, K. K. (2010). Antibacterial labdane diterpenoids of *Ulva fasciata* Delile from southwestern coast of the Indian Peninsula. *Food Chemistry*, 119:1399-1408.

del Rio, J. M., Reina, G. G., & Peláez, F. (2001). Screening of antimicrobial activities in red, green and brown macroalgae from Gran Canaria (Canary Islands, Spain). *International Microbiology*, 4, 35-40.

Demirel, Z., Yilmaz-Koz, F. F., Karabay-Yavasoglu, U. N., Ozdemir, G., Sukatar, A. (2009). Antimicrobial and antioxidant activity of brown algae from the Aegean Sea. *Journal of Serbian Chemical Society*, 74, 619-628.

Desbois, A. P., Mearns-Spragg, A., & Smith, V. J. (2009). A fatty acid from the diatom Phaeodactylum tricornutum is antibacterial against diverse bacteria including multi-resistant *Staphylococcus aureus* (MRSA). *Marine Biotechnology*, 11(1), 45-52.

Devi, K. S. & Mehta, S. K. (2015). Screening of blue-green algae *Lyngbya* for its antimicrobial activities. *Science Vision*, 15, 98-105.

Diao, J., Song, X., Zhang, X., Chen, L., & Zhang, W. (2018). Genetic engineering of *Crypthecodinium cohnii* to increase growth and lipid accumulation. *Frontiers in Microbiology*, 9, 492.

El-Beih, A. A., Kawabata, T., Koimaru, K., Ohta, T., & Tsukamoto, S. (2007). Monodictyquinone A: a new antimicrobial anthraquinone from a sea urchin-derived fungus *Monodictys* sp. *Chemical and Pharmaceutical Bulletin*, 55(7), 1097-1098.

El Shafay, S. M., Ali, S. S., & El-Sheekh, M. M. (2016). Antimicrobial activity of some seaweeds species from Red sea, against multidrug resistant bacteria. *The Egyptian Journal of Aquatic Research*, 42(1), 65-74.

Fernando, I. S., Kim, M., Son, K. T., Jeong, Y., & Jeon, Y. J. (2016). Antioxidant activity of marine algal polyphenolic compounds: a mechanistic approach. *Journal of Medicinal Food*, 19(7), 615-628.

Fridkin, S., Baggs, J., Fagan, R., Magill, S., Pollack, L. A., Malpiedi, P., & Samore, M. H. (2014). Vital signs: improving antibiotic use among hospitalized patients. *MMWR. Morbidity and Mortality Weekly Report*, 63(9), 194.

Goossens, H. (2003). Susceptibility of multi-drug-resistant *Pseudomonas aeruginosa* in intensive care units: results from the European MYSTIC study group. *Clinical Microbiology and Infection*, 9(9), 980-983.

Gutmann, L., Williamson, R., & Collatz, E. (1984). The possible role of porins in bacterial antibiotic resistance. *Annals of Internal Medicine*, 101(4), 554-557.

Holanda, M. L., Melo, V. M. M., Silva, L. M. C. M., Amorim, R. C. N., Pereira, M. G., & Benevides, N. M. B. (2005). Differential activity of a lectin from *Solieria filiformis* against human pathogenic bacteria. *Brazilian Journal of Medical and Biological Research*, 38(12), 1769-1773.

Hopwood, D. A. (2007). How do antibiotic-producing bacteria ensure their self-resistance before antibiotic biosynthesis incapacitates them? *Molecular Microbiology*, 63(4), 937-940.

Ismail, A., Ktari, L., Ben Redjem Romdhane, Y., Aoun, B., Sadok, S., Boudabous, A., & El Bour, M. (2018). Antimicrobial fatty acids from green alga *Ulva rigida* (Chlorophyta). *BioMed Research International*, 2018.

Kadam, S. U., Tiwari, B. K., & O'Donnell, C. P. (2015). Extraction, structure and biofunctional activities of laminarin from brown algae. *International Journal of Food Science & Technology*, 50(1), 24-31.

Kamei, Y., Sueyoshi, M., Hayashi, K., Terada, R., & Nozaki, H. (2009). The novel anti-*Propionibacterium acnes* compound, sargafuran, found in the marine brown algae *Sargassum macrocarpum*. *Journal of Antibiotics*, 62, 259-263.

Kanekiyo, K., Hayashi, K., Takenaka, H., Lee, J. B., & Hayashi, T. (2007). Antiherpes simplex virus target of an acidic polysaccharide, nostoflan, from the edible blue-green alga *Nostoc flagelliforme*. *Biological and Pharmaceutical Bulletin*, 30, 1573-1575.

Karabay-Yavasoglu, N. U, Sukatar, A., Ozdemir, G., & Horzum, Z. (2007). Antimicrobial activity of volatile components and components and various extracts of the red alga *Jania rubens*. *Phytotherapy Research*, 21,153-156.

Karpiński, T. M., & Adamczak, A. (2019). Fucoxanthin - An antibacterial carotenoid. *Antioxidants*, 8(8), 239.

Kilic, N. K., Erdem, K. & Donmez, G. (2018). Bioactive compounds produced by *Dunaliella* species, antimicrobial effects and optimization of the efficiency. *Turkish Journal of Fishery & Aquatic Science*, 19(11), 923-933.

Kobayashi, N., Nishino, K., & Yamaguchi, A. (2001). Novel macrolide-specific ABC-type efflux transporter in *Escherichia coli*. *Journal of Bacteriology*, 183(19), 5639-5644.

Kojima, S., & Nikaido, H. (2013). Permeation rates of penicillins indicate that *Escherichia coli* porins function principally as nonspecific channels. *Proceedings of the National Academy of Sciences*, 110(28), E2629-E2634.

Kolanjinathan, K. & Stella, D. (2009). Antibacterial activity of marine macro algae against human pathogens. *Recent Research in Science and Technology*, 1, 20-22.

Kotnala, S., Garg, A., & Chatterji, A. (2009). Screening for the presence of antimicrobial activity in few Indian seaweeds. *Pertanika Journal of Tropical Agricultural Science*, 32(1), 69-75.

Kuda, T., Kunii, T., Goto, H., Suzuki, T., & Yano, T. (2007). Varieties of antioxidant and antibacterial properties of *Ecklonia stolonifera* and *Ecklonia kurome* products harvested and processed in the Noto peninsula, Japan. *Food Chemistry*, 103, 900-905.

Lane, A. L., Mulac, L., Drenkarrd, E. J., Shearer, T. L., Engel, S., Fredericq, S., Fairchild, C. R., Prudhomme, J., Le Roch, K., Hay, M. E., Aalbersberg, W., & Kubanek, J. (2010). Ecological leads for natural product discovery: novel sesquiterpene hydroquinones from the red macroalga *Peysonnelia* sp. *Tetrahedron*, 66, 455-461.

Lee, K. Y., Jeong, M. R., Choi, S. M., Na, S. S., & Cha, J. D. (2013). Synergistic effect of fucoidan with antibiotics against oral pathogenic bacteria. *Archives of Oral Biology*, 58(5), 482-492.

Liao, W. R., Lin, J. Y., Shieh, W. Y., Jeng, W. L., & Huang, R. (2003). Antibiotic activity of lectins from marine algae against marine vibrios. *Journal of Industrial Microbiology and Biotechnology*, 30(7), 433-439.

Lima-Filho, J. V. M., Carvalho, A. F. F. U., Freitas, S. M., & Melo, V. M. M. (2002). Antibacterial activity of extracts of six macroalgae from the North Eastern Brazilian coast. *Brazilian Journal of Microbiology*, 33, 311-313.

Livermore, D. M. & Woodford, N. (2006). The beta-lactamase threat in *Enterobacteriaceae*, *Pseudomonas* and *Acinetobacter*, *Trends in Microbiology*, 14(9), 41320.

Lukacik, P., Barnard, T. J., Keller, P. W., Chaturvedi, K. S., Seddiki, M., & Fairman, J. W. (2012). Structural engineering of a phage lysin that targets Gram-negative pathogens. *Proceedings of National Academy of Science USA*, 109, 985762.

Mah, T. F. C., & O'Toole, G. A. (2001). Mechanisms of biofilm resistance to antimicrobial agents. *Trends in Microbiology*, 9(1), 34-39.

Mashhadinejad, A., Zamani, H. & Sarmad, J. (2017). Effect of growth conditions and extraction solvents on enhancement of antimicrobial activity of the microalgae *Chlorella vulgaris*. *Pharmaceutical and Biomedical Research*, 2, 65-73.

Mendiola, J. A., Torres, C. F., Toré, A., Martín-Álvarez, P. J., Santoyo, S., Arredondo, B. O., Señoráns, F. J., Cifuentes, A., Ibáñez, E. (2007). Use of supercritical CO2 to obtain extracts with antimicrobial activity from *Chaetoceros muelleri* microalga. A correlation with their lipidic content. *European Food Research and Technology*, 224, 505-510.

Metzger, P., Rager, M. N., & Largeau, C. (2002). Botryolins A and B, two tetramethylsqualene triethers from the green microalga *Botryococcus braunii*. *Phytochemistry*, 59, 839-843.

Mtolera, M. S. P. & Semesi, A. K. (1996). Antimicrobial activity of extracts from six green algae from Tanzania. *Current Trends in Marine Botanical Research in East African Region*, 1996, 211-217.

Mugilan, V. & Sivakami, R. (2016). Antimicrobial activity of microalgae isolated from fresh water pond, Tamil Nadu, India. *International Journal of Current Microbiology and Applied Science*, 5(6), 588-595

Nainu, F., Asri, R. M., Arsyad, A., Manggau, M. A., & Amir, M. N. (2018). *In vivo* antibacterial activity of green algae *Ulva reticulata* against *Staphylococcus aureus* in *Drosophila* model of infection. *Pharmacognosy Journal*, 10(5), 993-7.

Nair, R., Chabhadiya, R., & Chanda, S. (2007). Marine algae: Screening for a potent antibacterial agent. *Journal of Herbal Pharmacotherapy*, 7, 73-86.

Orhan, I., Sener, B., Atici, T., Brun, R., Perozzo, R., & Tasdemir, D. (2006). Turkish freshwater and marine macrophyte extracts show *in vitro* antiprotozoal activity and inhibit FabI, a key enzyme of *Plasmodium falciparum* fatty acid biosynthesis. *Phytomedicine*, 13(6), 388-393.

Oranday, M. A., Verde, M. J., Martínez-Lozano, S. J., & Waksman, N. H. (2004). Active fractions from four species of marine algae. *International Journal of Experimental Botany*, 2004, 165-170.

Ozdemir, G., Karabay, N. U., Dalay, M. C., & Pazarbasi, B. (2004). Antibacterial activity of volatile component and various extracts of *Spirulina platensis. Phytotherapy Research*, 18, 754-757.

Ozdemir, G., Horzum, Z., Sukatar, A., & Karabay-Yavasoglu, N. U. (2006). Antimicrobial activities of volatile components and various extracts of *Dictyopteris membranaceae* and *Cystoseira barbata* from the cost of Izmir, Turkey. *Pharmaceutical Biology*, 44, 183-188.

Pane, G., Cacciola, G., Giacco, E., Mariottini, G. L., & Coppo, E. (2015). Assessment of the antimicrobial activity of algae extracts on bacteria responsible of external otitis. *Marine Drugs*, 13: 6440-6452.

Patra, J. K., Rath, S. K., Jena, K., Rathod, V. K., & Thatoi, H. (2008). Evaluation of antioxidant and antimicrobial activity of seaweed (*Sargassum* sp.) extract: A study on inhibition of Glutathione-S-Transferase activity. *Turkish Journal of Biology*, 32, 119-125.

Pierre, G., Sopena, V., Juin, C., Mastouri, A., Graber, M., & Maugard, T. (2011). Antibacterial activity of a sulfated galactan extracted from the marine alga *Chaetomorpha aerea* against *Staphylococcus aureus. Biotechnology and Bioprocess Engineering*, 16(5), 937-945.

Plaza, M., Santoyo, S., Jaime, L., Reina, G. G. B., Herrero, M., Señoráns, F. J., & Ibáñez, E. (2010). Screening for bioactive compounds from algae. *Journal of Pharmaceutical and Biomedical Analysis*, 51(2), 450-455.

Pratt, R., Daniels, T. C., Eiler, J. J., Gunnison, J. B., Kumler, W. D., Oneto, J. F., & Smith, J. H. C. (1944). Chlorellin, an antibacterial substance from *Chlorella. Science (Washington)*, 351-2.

Rai, M. K., Deshmukh, S. D., Ingle, A. P., & Gade, A. K. (2012). Silver nanoparticles: the powerful nanoweapon against multidrug-resistant bacteria. *Journal of Applied Microbiology*, 112, 841-852.

Rajauria, G., & Abu-Ghannam, N. (2013). Isolation and partial characterization of bioactive fucoxanthin from *Himanthalia elongata* brown seaweed: a TLC-based approach. *International Journal of Analytical Chemistry*, 2013, 1-6.

Rajasulochana, P., Dhamotharan, R., Krishnamoorthy, P., & Murugesan, S. (2009). Antibacterial activity of the extracts of marine red and brown algae. *Journal of American Science*, 5, 20-25.

Rodolfi, L., Biondi, N., Guccione, A., Bassi, N., D'Ottavio, M., Arganaraz, G., & Tredici, M. R. (2017). Oil and eicosapentaenoic acid production by the diatom *Phaeodactylum tricornutum* cultivated outdoors in Green Wall Panel (GWP®) reactors. *Biotechnology and Bioengineering*, 114(10), 2204-2210.

Sakthivel, K., & Kathiresan, K. (2012). Antimicrobial activities of marine cyanobacteria isolated from mangrove environment of south east coast of India. *Journal of Natural Products*, 5, 147-156.

Sánchez-Saavedra, M., Licea-Navarro, A., & Bernáldez-Sarabia, J. (2010). Evaluation of the antibacterial activity of different species of phytoplankton. *Revista de Biología Marina Oceanografía*, 45(3), 531-536.

Saravanakumar, D. E. M., Folb, P. I., Campbell, B. W., & Smith, P. (2008). Antimycobacterial activity of the red alga *Polysiphonia virgata. Pharmaceutical Biology*, 46, 254-260.

Saritha, S., Kala, K. J., Prashob Peter K. J. Nair, & S. M. (2017). In vitro antibacterial screening of fatty acid fractions from three different microalgae. *International Journal of Pharmacognosy and Phytochemical Research*, 9(11), 1405-1409.

Sasidharan, S., Darah, I., Noordin, M. K. M. J. (2009). Screening antimicrobial activity of various extracts of *Gracilaria changii. Pharmaceutical Biology*, 47, 72-76.

Sasikala, C. & Geetha, R. D. (2017). Comparative study on antimicrobial activity of seaweeds. *Asian Journal of Pharmaceutical and Clinical Research,* 10, 384-386.

Seddek, N. H., Fawzy, M. A., El-Said, W. A. & Ahmed, M. M. R. (2019). Evaluation of antimicrobial, antioxidant and cytotoxic activities and characterization of bioactive substances from freshwater blue-green algae. *Global NEST Journal*, 21 (3), 328-336.

Seenivasan, R., Indu, H. Archana, G. & Geetha, S. (2010). The antibacterial activity of some marine algae from South East coast of India. *American-Eurasian Journal Agriculture & Environmental Sciences*, 9 (5), 480-489.

Selim, R. E., Ahmed, S. M., El-Zemity, S. R., Ramses, S. S., & Moustafa, Y. T. (2015). Antifungal activity and seasonal variation of green alga (*Ulva lactuca*) extracts. *Asian Journal of Agriculture and Food Sciences*, 3(05).

Shanab, S. M. M. (2007). Antioxidant and antibiotic activities of some seaweeds (Egyptian isolates). *International Journal of Agriculture and Biology*, 9, 220-225.

Shannon, E., & Abu-Ghannam, N. (2016). Antibacterial derivatives of marine algae: An overview of pharmacological mechanisms and applications. *Marine Drugs*, 14(4), 81.

Solomon, R. D. J., & Santhi, V. S. (2008). Purification of bioactive natural product against human microbial pathogens from marine sea weed *Dictyota acutiloba. Journal of Microbiology and Biotechnology*, 24, 1747-1752.

Stavri, M., Piddock, L. J. & Gibbons, S. (2007). Bacterial efflux pump inhibitors from natural sources. *Journal of Antimicrobial Chemotherapy*, 59(6), 1247-60.

Sukatar, A., Karabay-Yavasoglu, N. U., Ozdemir, G., & Horzum, Z. (2006). Antimicrobial activity of volatile component and various extracts of *Enteromorpha linza* (Linnaeus) J. Agardh from the cost of Izmir, Turkey. *Annals of Microbiology*, 56, 275-279.

Suzuki, S., Endoh, R., Manabe, R. I., Ohkuma, M., & Hirakawa, Y. (2018). Multiple losses of photosynthesis and convergent reductive genome evolution in the colourless green algae *Prototheca. Scientific Reports*, 8(1), 940.

Taskin, E., Ozturk, M., Taskin, E., Kurt, O. (2007). Antibacterial activities of some marine algae from the Aegean Sea (Turkey). *African Journal of Biotechnology*, 6, 2746-2751.

Thillairajasekar, K., Duraipandiyan, V., Perumal, P., & Ignacimuthu, S. (2009). Antimicrobial activity of Trichodesminum erythraeum (Ehr) (microalga) from South East coast of Tamil Nadu, India. *International Journal of Integrative Biology*, 5, 167-170.

Tuney, I., Cadirci, B. H., Unal, D., & Sukatar, A. (2006). Antimicrobial activities of the extracts of marine algae from the coast of Urla (Izmir, Turkey). *Turkish Journal of Biology*, 30, 171-175.

Vallinayagam, K., Arumugam, R., Kannan, R. R. R., Thirumaran, G., & Anantharaman, P. (2009). Antibacterial activity of some selected seaweeds from Pudumadam coastal regions. *Global Journal of Pharmacology*, 3, 50-52.

Vairappan, C. S., Suzuki, M., Ishii, T., Okino, T., Abe, T., & Masuda, M. (2008). Antibacterial activity of halogenated sesquiterpenes from Malasyan *Laurencia* sp. *Phytochemistry*, 69, 2490-2494.

Vijayabaskar, P., Vaseela, N., & Thirumaran, G. (2012). Potential antibacterial and antioxidant properties of a sulfated polysaccharide from the brown marine algae Sargassum swartzii. *Chinese Journal of Natural Medicines*, 10(6), 421-428.

Vila, J., Marti, S., & Sanchez-Cespedes, J. (2007). Porins, efflux pumps and multidrug resistance in *Acinetobacter baumannii*. *Journal of Antimicrobial Chemotherapy*, 59(6), 1210-1215.

War, W. R., Rizlan, Ross, E. E., Nor Farinna Abdul Rahim., Batrisyia Syazana Faridon. & Khairul Adzfa Radzun. (2018). Antimicrobial activity of marine green algae extract against microbial pathogens. *Malaysian Journal of Biochemistry and Molecular Biology*, 2, 42-46.

Wright, G. D. (2011). Molecular mechanisms of antibiotic resistance. *Chemical Communications*, 47(14), 4055-4061.

Yu, S. H., Wu, S. J., Wu, J. Y., Wen, D. Y., & Mi, F. L. (2015). Preparation of fucoidan-shelled and genipin-crosslinked chitosan beads for antibacterial application. *Carbohydrate Polymers*, 126, 97-107.

Zeid, A. A., Aboutabl, E. A., Sleem, A. A., & El-Rafie, H. M. (2014). Water soluble polysaccharides extracted from *Pterocladia capillacea* and *Dictyopteris membranacea* and their biological activities. *Carbohydrate Polymers*, 113, 62-66.

Zineb, S., Mohamed, L., Mohamed, F., & Khadija, F. Z. (2004). Inhibition of growth and mycotoxins formation in moulds by marine algae *Cystoseira tamariscifolia*. *African Journal of Biotechnology*, 3, 71-75.

Zouaoui, B. & Ghalem, B. R. (2017). The phenolic contents of antimicrobial activities of some marine algae from the Mediterranean Sea (Algeria). *Russian Journal of Marine Biology*, 43, 491-495.

In: Antimicrobial Resistance: Opportunities and Challenges ISBN: 978-1-53617-943-9
Editors: P. Kumar Prabhakar and V. Mishra

Chapter 13

MICROBIAL RESISTANCE: A WORLDWIDE PUBLIC HEALTH PROBLEM

Yachana Mishra[1] and Vijay Mishra[2,*]
[1]Department of Zoology, Shri Shakti Degree College, Sankhahari, Ghatampur, Kanpur Nagar, UP, India
[2]School of Pharmaceutical Sciences, Lovely Professional University, Phagwara, Punjab, India

ABSTRACT

The era of antimicrobial resistance is now upon us. Resistance problems are especially visible with hospitalized patients, as resistant infections readily spread among weakened hosts in close contact. But treatment choices are also diminished for some community infections, such as tuberculosis, pneumonia, and sexually transmitted diseases. Managing this new era in medicine is challenging, in part because we rely so heavily on antimicrobials. The problems are exacerbated by global travel, since resistance is readily disseminated. Calls to reduce consumption have raised general awareness and led to some reduction in the prevalence of resistance, but on a global level antimicrobial consumption remains high, both in agriculture and in human populations. Thus, antibiotic resistance will not disappear soon. Since new antimicrobials are becoming increasingly difficult to find, it is likely that we will need new strategies for suppressing resistance and for incentivizing the discovery and development of new antimicrobial therapies.

Keywords: antimicrobial, bacteria, resistance, tuberculosis, hospital infection

* Corresponding Author's E-mail: vijaymishra2@gmail.com.

1. INTRODUCTION

In recent years, the problem of antimicrobial resistance has been recognized and addressed by international, regional, and national public health agencies, authorities, and professional societies (Carlet et al., 2012; Bell, 2014; Laxminarayan et al., 2014; Shallcross and Davies, 2014). Antibiotics have saved millions of lives and have enabled many other medical advances since their discovery and introduction into clinical practice. The worsening problem of antimicrobial resistance now jeopardizes many of these advances. However, action plans to minimize this threat have been developed by many public health agencies around the globe (CDC, 2013; Kessel and Sharland, 2013; Keown et al., 2014; Paphitou, 2013). This focus is likely to increase among public health agencies in the coming years. Several facets of the problem involve the relationship between human and animal use of antimicrobial agents (Laxminarayan et al., 2014; McIntosh and Dean, 2014). This chapter summarizes the components of the cost of resistance from a public health perspective and contrasts this perspective to other societal perspectives. It then reviews strategies at several different levels of responsibility, ranging from the patient care provider to international agencies. Finally, it considers appropriate public health responses according to the resources available for control.

In an era of globalization, antimicrobial resistance represents an international concern that demands a concerted effort from multiple health and industry sectors. Public health must be at the forefront of these efforts. Antimicrobial resistance is widely recognized as a complex international problem. Antibiotics have saved millions of lives and have enabled many other medical advances since their discovery and introduction into clinical practice. The worsening problem of antimicrobial resistance now jeopardizes many of these advances (Carlet et al., 2011; Carlet et al., 2012; Speck, 2013; Spellberg et al., 2013; Bell, 2014; Keown et al., 2014; Shallcross and Davies, 2014; Paphitou, 2013; Woolhouse and Farrar, 2014). The number of pathogens resistant to multiple classes of antimicrobials has increased worldwide (WHO, 2014).

There have been reports of infections with *Pseudomonas aeruginosa* and *Acinetobacter* species which are resistant to all available antibiotics (Paterson, 2006). Other pathogens, such as *Neisseria gonorrhoeae*, that were previously relatively easy to treat, are now becoming much more difficult to treat due to antimicrobial resistance (Ndowa et al., 2013). There is also evidence that resistant organisms that were previously confined to the acute care hospital setting are now sources of community acquired infections (Dantes et al., 2013). Resistance is not only an issue for bacterial pathogens like Staphylococci, Enterococci, *Escherichia coli*, *Klebsiella*, *Neisseria*, and *Mycobacterium tuberculosis* but also the problem of resistance is increasingly being recognized in nonbacterial pathogens like *Candida*, HIV, malaria, and influenza (WHO, 2014).

The worldwide increase in the prevalence of resistance is a concern because it threatens both the optimal care of patients with infections and the viability of current healthcare systems (DiazGranados et al., 2008). For individual patients, antimicrobial drug resistance has a clear impact on patient morbidity and mortality (Cosgrove, 2006; Neidell et al., 2012; CDC, 2013). Of concern for healthcare systems is the economic impact of resistance (McGowan et al., 2001; Roberts et al., 2009; White, 2011; Neidell et al., 2012). In the United States, resistance is especially costly for the healthcare system and for the third-party payers that support such systems (Roberts et al., 2009; Neidell et al., 2012). Costs are also burdensome for national-based healthcare programs (White, 2011). The incremental cost of caring for patients infected with resistant organisms has several aspects (Cosgrove, 2006; Roberts et al., 2009; Neidell et al., 2012). As the prevalence of resistance increases, physicians often substitute older and less expensive drugs for newer and more expensive agents (Laxminarayan, 2014). Such costs in the United States are absorbed only in part by third-party payers, which often reimburse on the basis of head count, diagnosis related groups, or other formulas unrelated to specific services provided to an individual patient. Thus, most costs associated with resistant infections must be absorbed by the healthcare system itself (Laxminarayan and Malani, 2007). As the prevalence of drug resistant organism increases, these additional costs will mount and become a threat to the financial stability of local, regional, and national healthcare systems, many of which are already struggling to survive.

2. Public Health as a Perspective on Resistance

Public health involves the population as a whole. It focuses on what we as a society do to assure conditions in which we can be healthy (Institute of Medicine, 1988). The goal of public health, the health of a population, can be distinguished from the goals of medical care, the health of a patient (McGowan, 2012). The public health goal, fueled by an aim of social good, is much broader in scope. It can encompass the health of neighborhoods, cities, countries, or even the entire world (McGowan, 2001). Using the public health perspective of health, a long time frame for evaluation is usually appropriate. Since antimicrobials can both prevent and treat infections in society, society considers them a valuable resource (Carlet and Pittet, 2013). As resistance diminishes this resource, a societal goal should be to minimize resistance and therefore to reduce the forces that produce resistance (Chaudhary e al., 2012; Kessel and Sharland, 2013; Carlet et al., 2014). All use of antimicrobials increases the likelihood of resistance developing (Woolhouse and Farrar, 214). From a societal viewpoint, the use of antimicrobials to appropriately treat infections would be an appropriate rate of depletion of this valuable resource (Laxminarayan and Malani, 2007; Bartlett, 2011). Overuse or misuse of antimicrobials would be an inappropriately increased depletion of this resource (Bartlett et al., 2013). The costs of resistance from a public health perspective can be summarized as those resulting from treatment of patients infected with resistant organisms, those resulting from treatment of patients not infected with resistant organisms, and those resulting from antimicrobial use in agriculture, animal breeding, aquaculture, and industry.

The impact of antimicrobial resistance includes an increase in morbidity, mortality, and added costs for patients with resistant organisms. The added costs include those derived from the use of scarce healthcare resources: the care of patients with infections of resistant organisms, preventing transmission, maximizing appropriate empiric therapy, and resistance surveillance. In addition, excess loss of productivity and excess intangible costs such as patient and physician anxiety, pain, suffering, and inconvenience are also increased with antimicrobial resistance. Antimicrobial drug markets may also be affected by increasing resistance. Older, cheaper, and more narrow-spectrum drugs often become less useful, while the marketability of newer, more expensive, and more broad-spectrum drugs is favored (Van Boeckel et al., 2014).

Numerous studies have demonstrated that antimicrobial resistant organisms are associated with a higher morbidity and mortality than susceptible organisms. This has been shown for many different types of resistant organisms (Giske et al., 2008; Lode, 2009; de Kraker et al., 2011a; 2011b; 2011c; Pietersen et al., 2014; Pop-Vicas and Opal, 2014). Invasive infections with methicillin-resistant *Staphylococcus aureus* (MRSA) have frequently been shown to be associated with a higher mortality and an increase in hospital length of stay when compared to methicillin susceptible *Staphylococcus aureus* (MSSA) (Lode, 2009; de Kraker et al., 2011a; 2011b). Similar associations are found when looking at enterococci. Vancomycin-resistant enterococci (VRE) bloodstream infections are associated with a higher mortality than vancomycin susceptible enterococci bloodstream infections (Lode, 2009). Although prior studies had not shown a higher mortality with penicillin-resistant pneumococcal infections (Moroney et al., 2001; Lode, 2009), a more recent meta-analysis has shown penicillin-resistant pneumococcal infections to be associated with a higher mortality compared to penicillin-susceptible infections (Tleyjeh et al., 2006; Lode, 2009). Similar outcomes have been noted for antibiotic-resistant gram-negative infections. *Escherichia coli* and *Klebsiella pneumonia*e strains resistant to extended-spectrum cephalosporins, multidrug-resistant (MDR) *Pseudomonas aeruginosa* strains, and carbapenem-resistant *Acinetobacter* infections have all been associated with an increased length of stay and mortality when compared to their more susceptible counterparts (Giske et al., 2008; de Kraker et al., 2011a; de Kraker et al., 2011c; Lemos et al., 2013). Specifically, this mortality is noted to be higher when study antibiotic-resistant gram negative bacilli in the setting of septic shock (Pop Vicas and Opal, 2014). One plausible reason for this finding is the difficulty in selecting an appropriate empiric antibiotic regimen in the era of highly drug-resistant gram-negative infections. It is well established that the duration of hypotension without appropriate antimicrobial therapy is a crucial factor in the probability of survival in septic shock (Kumar et al., 2006). In addition to gram-positive and gram-negative pathogens, drug-resistant tuberculosis has also been shown to have a higher mortality compared to drug-susceptible tuberculosis (Pietersen et al., 2014).

For all different types of drug-resistant infections, an increase in mortality affects more than just the patient when examined using the public health perspective. Although difficult to quantify, the consequences of premature death include a loss of productivity (number of productive years lost multiplied by the average yearly productivity) and the pain and suffering of family members and friends (Laxminarayan and Malani, 2007).

3. Evolution of Resistance

Antimicrobial resistance is a global dilemma that threatens the health and safety of populations in all countries of the world. Urgent actions are needed to be taken before it reaches a critical stage, when large numbers of people in communities cannot be treated for life-threatening infections due to lack of effective drugs. Although the threat is most imminent from antibacterial resistance to commonly used antibiotics for infections seen regularly in intensive care units and hospitals, it is more prevalent and widespread and involves a wide spectrum of microbes.

To understand the evolution of microbial resistance, it is appropriate to review historical aspects. Development of antimicrobial chemotherapy is usually attributed to Paul Ehrlich ["father of chemotherapy"] based on his quest to find a cure for parasitic infections, toward the latter part of the nineteenth century, with natural dyes and heavy metals [mercury and arsenicals]. Penicillin was subsequently discovered in 1928 and administered clinically in the 1940s; sulphonamides were introduced clinically in 1937 (Greenwood et al., 2007). Thus, the "antibiotic era" was under way by the early 1940s. Penicillin and other antibiotics were initially derived from environmental fungi and bacteria, often with improvements made by chemical synthesis. Hence, the origin of antibiotics is through naturally derived substances produced to antagonize or inhibit the growth of other microorganisms, probably due to an evolutionary process that protects environmental niches of the producing organisms.

Streptomycin and the precursor of cephalosporins were also obtained from soil microbes in the 1940s. Penicillin was first used in 1941, and by 1944 penicillinase producing *Staphylococcus aureus* was described. Streptomycin was introduced clinically in 1944 for treatment of tuberculosis, but resistance soon developed during treatment (Davies and Davies, 2010). By the mid-1950s, most of the major antibiotic families, including aminoglycosides, chloramphenicol, tetracycline, and macrolides, had been developed (Greenwood et al., 2007).

Synthetic chemical agents with antibacterial activity were introduced in the early 1950s with para-aminosalicylic acid and isoniazid as antituberculosis agents; nitrofurantoin was found around the same time, followed by trimethoprim in 1956. Nalidixic acid, discovered in the early 1960s, was the precursor of the fluoroquinolones, with norfloxacin and ciprofloxacin developed in the 1980s. Rifampin was introduced for tuberculosis in 1968. The spectacular success of antimicrobial therapy led to widespread use and emergence of resistance. Pharmaceutical companies saw profit in new, more potent derivatives, which gave rise to broad-spectrum antipseudomonal penicillins and second-generation cephalosporins in the 1970s, with subsequent introduction in the 1980s of third generation cephalosporins. Later, beta-lactamase inhibitors combined with broad spectrum penicillins and carbapenems, new glycopeptides, newer macrolides, later generation quinolones, linezolid [a new class of oxazolidinone], and glycylcyclines [tigecycline] were introduced. Currently, there are more than 100 antimicrobial compounds available.

It is important to understand the evolution of antimicrobial resistance in order to tackle the problem. It was initially thought that antimicrobial resistance was a modern, man-made phenomenon. Although penicillinase-producing bacteria were recognized soon after the discovery of penicillin, it was not recognized as a problem until clinical use of penicillin became widespread. By the end of the 1950s, 80% of *Staphylococcus aureus* isolated from hospital patients were penicillin resistant due to β-lactamase production (Greenwood et al., 2007). This observation led to the development of penicillinase-stable penicillins [methicillin, cloxacillin, and oxacillin] which were introduced in the early 1960s to treat *S. aureus*. But resistance to methicillin was seen within a year or so of its clinical introduction. Thus, this was a sign that microbial resistance would be a problem, but in most cases this was overcome with development of new, more potent compounds. Many agents were developed that initially failed to compete well in the marketplace. The lack of use kept resistance from emerging, and now they are being used more frequently to fill a niche. Vancomycin is one of these agents. It was approved for use in 1958, but it was used sparingly due to concerns of toxicity and efficacy.

When methicillin-resistant *S. aureus* [MRSA] appeared in hospitals during the 1970s and then spread to other health-care facilities, vancomycin became more frequently used, beginning in the 1980s (Levine, 2006). Similarly, polymyxins were developed in the 1950s, but toxicity kept them from widespread use for systemic infections. Now they are gaining clinical use with infections caused by *Pseudomonas aeruginosa*, *Acinetobacter baumannii*, and carbapenemase-producing *Enterobacteria-ceae* that are resistant to all other antibiotics (Landman et al., 2008). Full vancomycin resistance among MRSA strains is rare, probably because mutational changes cause impaired fitness for MRSA as susceptibility decreases. As a result, vancomycin-intermediate resistance [VISA] is slowly becoming a problem. Resistance to polymyxins appears to be emerging readily with increased use.

Over the years, development of new antibiotics to counter resistant bacteria led to appearance of novel resistant strains to these new drugs, which become widespread with increasing use of the antibiotics. This pattern has been seen with every new class of antibiotic developed over the years. There is strong correlation between the frequency and quantity of antibiotics used in humans and animals and the rate of development of antibacterial drug resistance. The logarithmic growth of resistant bacteria since the 1970s is reflected by the number of β-lactamase enzymes identified during the antibiotic era. Before 1970, there were only several β-lactamase enzymes described, and now about 900 β-lactamase enzymes have been identified (Davies and Davies, 2010).

Mobile genetic elements, extra chromosomal self-replicating structures [plasmids] and transposons, found in bacterial cells provide a resistance threat that maybe unconquerable. Our understanding of horizontal transfer of bacterial resistance was heralded by the discovery of antibiotic resistance plasmids that could be disseminated by bacterial conjugation in the mid-1950s (Davies and Davies, 2010). Since 1989 we have gained much greater knowledge of the genetics of bacterial resistance following the discovery of integrons. Integrons are versatile genetic elements, commonly found in bacterial genomes that allow efficient capture and expression of exogenous genes.

Plasmids and transposons are considered mobile integrons. Integrons play a major role in the spread of antibiotic resistance, particularly in Gram-negative pathogens; the majority of these pathogens carry integrons with resistant genes (Gillings, 2014). The process of microbial resistance is complex, and besides plasmids and DNA mutations [acquired or heritable and transferable], other mechanisms include biofilms, which harbor hypermutator bacteria that select for resistance more frequently, and phenotypic tolerance, a situation in which bacteria are not killed by antimicrobials (Nathon, 2012). Antibiotic pressure predisposes to resistance and tolerance. Despite the call for intensive research on mechanisms of microbial resistance and development of novel compounds to counteract the spread in 2012 (Nathon, 2012), no innovative agent is on the horizon.

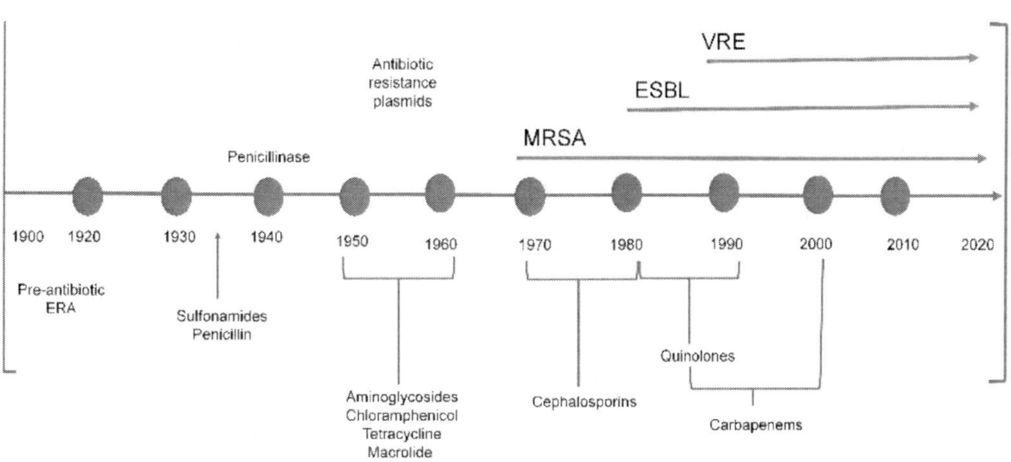

Figure 1. Appearance of antimicrobial resistance. Abbreviations: MRSA methicillin-resistant S. aureus, VRE vancomycin-resistant enterococci, ESBL extended β-lactamase.

The origins, evolution, genetics, and biochemistry of antibiotic resistance have been studied over the last 60 years. Figure 1 outlines the history of antibiotic discovery and subsequent development of antibiotic resistance. The emergence of antibiotic resistance of pathogenic bacteria after antibiotic development and clinical use, plus the absence of resistance in bacteria of the pre-antibiotic era (Hughes and Datta, 1983), suggested that resistance is a modern phenomenon. However, metagenomic analyses of authenticated ancient DNA from 30,000-year-old Beringian permafrost sediments identified genes encoding resistance to β-lactam, tetracycline, and glycopeptides antibiotics (D'Costa et al., 2011). Thus, antibiotic resistance is a natural phenomenon that predates the modern selective pressure associated with clinical and animal use. This explains the rapid emergence of resistance to new antibiotics; resistance will continue to emerge with drugs now in development. Hence, it is predictable that new antibiotics will select for preexistent resistance determinants that have been present and circulating in universal, environmental microbial pangenome for hundreds of years (D'Costa et al., 2011). Thus, the era of antibiotics has shifted naturally to the era of antibiotic resistance. The main challenge is to minimize the problem.

Conclusion

Human migration, animal and vector movement, and food markets may facilitate the spread of antimicrobial resistance across almost any geographic or political boundary. Methicillin-resistant *Staphylococcus aureus* and vancomycin resistant *Enterococcus* were first reported in the United Kingdom. After becoming widespread in North America and Europe, the same resistance patterns were noted in Asia. Today, novel gram-negative carbapenem resistance mechanisms such as OXA-48 and NDM-1 (New Delhi metallo-β-lactamase) were first discovered in Asia and have since spread to Europe and North America. The global spread of antimicrobial resistance demands effort of multiple health and industry sectors (in both low-resource and high-resource countries) to strengthen multinational/international partnerships and regulations. Public health must be at the forefront of these efforts.

References

[1] Bartlett JG, Gilbert DN, Spellberg B. Seven ways to preserve the miracle of antibiotics. *Clin Infect Dis.* 2013; 56:1445–50.

[2] Bartlett JG. A call to arms: the imperative for antimicrobial stewardship. *Clin Infect Dis.* 2011; 53:S4–7.

[3] Bell M. Antibiotic misuse: a global crisis. *JAMA Intern Med.* 2014; 174:1920–1.

[4] Carlet J, Collignon P, Goldmann D, Goossens H, Gyssens IC, Harbarth S, et al. Society's failure to protect a precious resource: antibiotics. *Lancet.* 2011; 378:369–71.

[5] Carlet J, Pittet D. Access to antibiotics: a safety and equity challenge for the next decade. *Antimicrob Resist Infect Control.* 2013; 2:1.

[6] Carlet J, Pulcini C, Piddock LJV. Antibiotic resistance: a geopolitical issue. *Clin Microbiol Infect.* 2014; 20:949–53.

[7] Carlet JM, Artigas A, Niederman MS, Torres A, World Alliance against Antibiotic Resistance. The Barcelona declaration from the world alliance against antibiotic resistance: engagement of intensivists. *Crit Care.* 2012; 16:145.

[8] CDC. *Antibiotic resistance threats in the United States.* (2013). http://www.cdc.gov/drugresistance/threat-report-2013/pdf/arthreats-2013-508.pdf. Accessed 11 Dec 2014.

[9] Choudhury R, Panda S, Singh DV. Emergence and dissemination of antibiotic resistance: a global problem. *Indian J Med Microbiol.* 2012; 30:384–90.

[10] Cosgrove SE. The relationship between antimicrobial resistance and patient outcomes: mortality, length of hospital stay, and health care costs. *Clin Infect Dis.* 2006; 42:S82–9.

[11] D'Costa VM, King CE, Kalan L, et al. Antibiotic resistance is ancient. *Nature.* 2011; 477:457–61.

[12] Dantes R, Mu Y, Belflower R, Aragon D, Dumyati G, Harrison LH, et al. Emerging Infections Program-Active Bacterial Core Surveillance MRSA Surveillance Investigators. National burden of invasive methicillin-resistant *Staphylococcus aureus* infections, United States, 2011. *JAMA Intern Med.* 2013; 173: 1970–8.

[13] Davies J, Davies D. Origins of antibioptic resistance. *Microbiol Mol Biol Rev.* 2010; 74:417–33.

[14] de Kraker MEA, Davey PG, Grundmann H, Burden study group. Mortality and hospital stay associated with resistant *Staphylococcusaureus* and *Escherichia coli* bacteremia: estimating the burden of antibiotic resistance in Europe. *PLoS Med.* 2011a; 8, e1001104.

[15] de Kraker MEA, Wolkewitz M, Davey PG, Grundmann H, Burden Study Group. Clinical impact of antimicrobial resistance in European hospitals: excess mortality and length of hospital stay related to methicillin-resistant *Staphylococcus aureus* bloodstream infections. *Antimicrob Agents Chemother.* 2011b; 55:1598–605.

[16] de Kraker MEA, Wolkewitz M, Davey PG, Koller W, Berger J, Nagler J, et al. Burden of antimicrobial resistance in European hospitals: excess mortality and length of hospital stay associated with bloodstream infections due to *Escherichia coli* resistant to thirdgeneration cephalosporins. *J Antimicrob Chemother.* 2011c; 66: 398–407.

[17] DiazGranados CA, Cardo DM, McGowan Jr JE. Antimicrobial resistance: international control strategies, with a focus on limitedresource settings. *Int J Antimicrob Agents.* 2008; 32:1–9.

[18] Gillings MR. Integrons: past, present and future. *Microbiol Mol Biol Rev.* 2014; 778:257–77.

[19] Giske CG, Monnet DL, Cars O, Carmeli Y, ReAct-Action on Antibiotic Resistance. Clinical and economic impact of common multidrug-resistant gram-negative bacilli. *Antimicrob Agents Chemother.* 2008; 52:813–21.

[20] Historical introduction. In: Greenwood D, Davey P, Wilcox M, editors. *Antimicrobial chemotherapy.* 5th ed. Oxford, UK: Oxford University Press; 2007. p. 1–10.

[21] Hughes VM, Datta N. Conjugative plasmids in bacteria of the 'pre-antibiotic' era. *Nature.* 1983; 302:725–6.

[22] Institute of Medicine. *The future of public health.* Washington, D.C.: National Academy Press; 1988.

[23] Keown OP, Warburton W, Davies SC, Darzi A. Antimicrobial resistance: addressing the global threat through greater awareness and transformative action. *Health Aff (Millwood).* 2014; 33:1620–6.

[24] Kessel AS, Sharland M. The new UK antimicrobial resistance strategy and action plan. *BMJ.* 2013; 346:f1601.

[25] Kumar A, Roberts D, Wood KE, Light B, Parrillo JE, Sharma S, et al. Duration of hypotension before initiation of effective antimicrobial therapy is the critical determinant of survival in human septic shock. *Crit Care Med.* 2006; 34:1589–96.

[26] Landman D, Gergescu C, Martin DA, Quale J. Polymyxins revisited. *Clin Microbiol Rev.* 2008; 21:449–65.

[27] Laxminarayan R, Duse A, Wattal C, Zaidi AKM, Wertheim HFL, Sumpradit N, et al. Antibiotic resistance-the need for global solutions. *Lancet Infect Dis.* 2014; 13:1057–98.

[28] Laxminarayan R, Malani A. *Extending the cure: policy responses to the growing threat of antibiotic resistance.* (2007). http://www.cddep.org/publications/extending_cure_policy_responses_growing_threth_antibiotic_resistance.

[29] Laxminarayan R. Antibiotic effectiveness: balancing conservation against innovation. *Science.* 2014; 345:1299–301.

[30] Lemos EV, de la Hoz FP, Alvis N, Einarson TR, Quevedo E, Castañeda C, et al. Impact of carbapenem resistance on clinical and economic outcomes among patients with *Acinetobacter baumannii* infection in Columbia. *Clin Microbiol Infect.* 2013; 20:174–80.

[31] Levine DP. Vancomycin: a history. *Clin Infect Dis.* 2006;42 (Suppl. 1):S5–12.

[32] Lode HM. Clinical impact of antibiotic-resistant gram-positive pathogens. *Clin Microbiol Infect.* 2009; 1 5:212–17.

[33] McGowan Jr JE. Antimicrobial stewardship—the state of the art in 2011: focus on outcome and methods. *Infect Control Hosp Epidemiol.* 2012; 33:331–7.

[34] McGowan Jr JE. Economic impact of antimicrobial resistance. *Emerg Infect Dis.* 2001; 7:286–92.

[35] McIntosh D, Dean W. Factors associated with the inappropriate use of antimicrobials. *Zoonoses Public Health.* 2014; 61:1–7.

[36] Moroney JF, Fiore AE, Harrison LH, Patterson JE, Farley MM, Jorgensen JH, Phelan M, Facklam RR, Cetron MS, Breiman RF, Kolczak M, Schuchat A. Clinical outcomes of bacteremic pneumococcal pneumonia in the era of antibiotic resistance. *Clin Infect Dis.* 2001; 33:797–805.

[37] Nathon C. Fresh approaches to anti-infective therapies. *Sci Transl Med.* 2012; 4:140-2.

[38] Ndowa FJ, Ison CA, Cole MJ, Lusti-Narasimhan M. Gonococcal antimicrobial resistance: challenges for public health control. *Sex Transm Infect.* 2013; 89:iv3–4.

[39] Neidell MJ, Cohen B, Furuya Y, Hill J, Jeon CY, Glied S, Larson EL. Costs of healthcare- and community-associated infections with antimicrobial-resistant versus antimicrobial-susceptible organisms. *Clin Infect Dis.* 2012; 55:807–15.

[40] Paphitou NI. Antimicrobial resistance: action to combat the rising microbial challenges. *Int J Antimicrob Agents.* 2013; 42:S25–8.

[41] Paterson DL. The epidemiological profile of infections with multidrug-resistant *Pseudomonas aeruginosa* and *Acinetobacter* species. *Clin Infect Dis.* 2006; 43:S43–8.

[42] Pietersen E, Ignatius E, Streicher EM, Mastrapa B, Padanilam X, Pooran A, et al. Long-term outcomes of patients with extensively drugresistant tuberculosis in South Africa: a cohort study. *Lancet.* 2014; 383:1230–9.

[43] Pop-Vicas A, Opal SM. The clinical impact of multidrug-resistant gram-negative bacilli in the management of septic shock. *Virulence.* 2014; 5:189–95.

[44] Roberts RR, Hota B, Ahmad I, Scott 2nd RD, Foster SD, Abbasi F, et al. Hospital and societal costs of antimicrobial-resistant infections in a Chicago teaching hospital: implications for antibiotic stewardship. *Clin Infect Dis.* 2009; 49:1175–84.

[45] Shallcross LJ, Davies SC. The World Health Assembly resolution on antimicrobial resistance. *J Antimicrob Chemother.* 2014; 69:2883–5.

[46] Speck P. Antibiotics: avert an impending crisis. *Nature.* 2013; 496:169.

[47] Spellberg B, Bartlett JG, Gilbert DN. The future of antibiotics and resistance. *N Engl J Med.* 2013; 368:299–302.

[48] Tleyjeh IM, Tlaygeh HM, Hejal R, Montori VM, Baddour LM. The impact of penicillin resistance on short-term mortality in hospitalized adults with

pneumococcal pneumonia: a systematic review and meta-analysis. *Clin Infect Dis.* 2006; 42:788–97.

[49] Van Boeckel TP, Gandra S, Ashok A, Caudron Q, Grenfell BT, Levin SA, Laxminarayan R. Global antibiotic consumption 2000 to 2010: an analysis of national pharmaceutical sales data. *Lancet Infect Dis.* 2014; 14:742–50.

[50] White AR. BSAC Working Party on The Urgent Need: Regenerating Antibacterial Drug Discovery and Development. Effective antibacterials: at what cost? The economics of antibacterial resistance and its control. *J Antimicrob Chemother.* 2011; 66:1948–53.

[51] WHO. *Antimicrobial resistance: global report on surveillance.* (2014). http://apps.who.int/iris/bitstream/10665/112642/1/9789241564748_eng.pdf.

[52] Woolhouse M, Farrar J. An intergovernmental panel on antimicrobial resistance. *Nature.* 2014; 509:555–7.

ABOUT THE EDITORS

Pranav Kumar Prabhakar has received his PhD in Biotechnology from Indian Institute of Technology (IIT) Madras, India during the period of 2011 and he has completed his master in biochemistry from Patna University Bihar, India in 2003. Currently, he is working as Associate Professor at Lovely Professional University Punjab, India. Dr. Prabhakar has qualifies CSIR-NET and Graduate Aptitude Test in Engineering (GATE) and also received travel grants from Department of Science and Technology (DST), Council of Scientific and Industrial Research (CSIR) & World Diabetes Federations to participate and present in International conferences. He is serving as an editorial member of several reputed journals like *Current Diabetes Review, Endocrine, Metabolic & Immune Disorders - Drug Targets, Journal of Applied Pharmaceutics, Bio info-Drug Targets, Journal of Pharmacy and Phytotherapeutics*, & *Asian journal of Phytomedicine and clinical research*, expert Reviewers for journals like *Biofouling* (Taylor & Francis), *Phytomedicine* (Elsevier), *Colloids and Surfaces B: Biointerfaces* (Elsevier), *Journal of Applied Polymer Science* (Wiley), *Advances in Pharmacological Sciences* (Hindawi), *Journal of Microbial & Biochemical Technology* (Omics). He has authored more than 45 research articles/books/ book chapters. He is a member of Indian Science Congress Association, the Royal Society of Chemistry and Asia-Pacific Chemical, Biological &

Environmental Engineering Society (APCBEES). His research has included synergy, Phytomedicine, metabolic disorders.

Vijay Mishra is an Associate Professor at Department of Pharmaceutics, School of Pharmaceutical Sciences, Lovely Professional University, Phagwara (Punjab), India. Dr. Mishra earned his doctoral degree in Pharmaceutical Sciences from Dr. H.S. Gour Central University, (MP), India. He has been a recipient of several internationally acclaimed fellowships and awards including Graduate Research Fellowship (AICTE, New Delhi), UGC-BSR Senior Research Fellowship (UGC, New Delhi) and International Travel Award/Grant (DBT, New Delhi). He has authored more than 52 international publications in reputed peer reviewed journals, 6 book chapters and 3 books on pharmaceutical dosage forms. He is serving as a reviewer and Editorial board member of various journals of high repute including Drug Delivery Letters (Bentham Science) and Anti-Cancer Agents in Medicinal Chemistry (Bentham Science), Current Drug Therapy (Bentham Science). He is a life member of Indian Science Congress Association, Kolkata, India. His current research interests encompass nanomedicine, cancer, toxicology, surface-engineered dendrimers, carbon nanotubes, quantum dots, gold nanoparticles, siRNA delivery as well as controlled and novel drug delivery systems.

INDEX

A

B

C

D

E

F

G

K

L

M

N

O

P

T

U

V